ASSESSMENT OF THE **SCIENTIFIC INFORMATION** FOR THE **RADIATION EXPOSURE SCREENING** AND **EDUCATION PROGRAM**

Committee to Assess the Scientific Information for the
Radiation Exposure Screening and Education Program

Board on Radiation Effects Research
Division on Earth and Life Studies

NATIONAL RESEARCH COUNCIL
OF THE NATIONAL ACADEMIES

D1298677

THE NATIONAL ACADEMIES PRESS
Washington, D.C.
www.nap.edu

THE NATIONAL ACADEMIES PRESS • **500 Fifth Street, N.W.** • **Washington, DC 20001**

NOTICE: The project that is the subject of this report was approved by the Governing Board of the National Research Council, whose members are drawn from the councils of the National Academy of Sciences, the National Academy of Engineering, and the Institute of Medicine. The members of the committee responsible for the report were chosen for their special competences and with regard for appropriate balance.

This study was supported by contract DHHS 232-02-0004 between the National Academy of Sciences and the Health Resources and Services Administration. Any opinions, findings, conclusions, or recommendations expressed in this publication are those of the author(s) and do not necessarily reflect the views of the organizations or agencies that provided support for the project.

International Standard Book Number: 0-309-09610-3 (Book)
International Standard Book Number: 0-309-54931-0 (PDF)
Library of Congress Control Number: 2005929472

Additional copies of this report are available from the National Academies Press, 500 Fifth Street, NW, Lockbox 285, Washington, DC 20055; (800) 624-6242 or (202) 334-3313 (in the Washington metropolitan area); Internet, http://www.nap.edu.

THE NATIONAL ACADEMIES
Advisers to the Nation on Science, Engineering, and Medicine

The **National Academy of Sciences** is a private, nonprofit, self-perpetuating society of distinguished scholars engaged in scientific and engineering research, dedicated to the furtherance of science and technology and to their use for the general welfare. Upon the authority of the charter granted to it by the Congress in 1863, the Academy has a mandate that requires it to advise the federal government on scientific and technical matters. Dr. Ralph J. Cicerone is president of the National Academy of Sciences.

The **National Academy of Engineering** was established in 1964, under the charter of the National Academy of Sciences, as a parallel organization of outstanding engineers. It is autonomous in its administration and in the selection of its members, sharing with the National Academy of Sciences the responsibility for advising the federal government. The National Academy of Engineering also sponsors engineering programs aimed at meeting national needs, encourages education and research, and recognizes the superior achievements of engineers. Dr. Wm. A. Wulf is president of the National Academy of Engineering.

The **Institute of Medicine** was established in 1970 by the National Academy of Sciences to secure the services of eminent members of appropriate professions in the examination of policy matters pertaining to the health of the public. The Institute acts under the responsibility given to the National Academy of Sciences by its congressional charter to be an adviser to the federal government and, upon its own initiative, to identify issues of medical care, research, and education. Dr. Harvey V. Fineberg is president of the Institute of Medicine.

The **National Research Council** was organized by the National Academy of Sciences in 1916 to associate the broad community of science and technology with the Academy's purposes of furthering knowledge and advising the federal government. Functioning in accordance with general policies determined by the Academy, the Council has become the principal operating agency of both the National Academy of Sciences and the National Academy of Engineering in providing services to the government, the public, and the scientific and engineering communities. The Council is administered jointly by both Academies and the Institute of Medicine. Dr. Ralph J. Cicerone and Dr. Wm. A. Wulf are chair and vice chair, respectively, of the National Research Council.

www.national-academies.org

Acknowledgments

During the committee's deliberations, several people provided information to the committee. Their contributions invigorated committee deliberations and enhanced the quality of this report. The committee expresses its appreciation to the Health Resources and Services Administration (HRSA) for sponsoring the study.

The committee and the staff of the Board on Radiation Effects Research (BRER) are grateful for the information provided by invited speakers, who generously contributed their time and participated in the committee's information-gathering meetings: Rebecca Barlow, Alfred Berg, Evelyn Bromet, Douglas M. Brugge, Teresa Coons, Regan Crump, Gerard W. Fischer, David S. James, Richard Kerber, Kiyo Mabuchi, Parthiv Mahadevia, Kimberly Mohs, Karen Mulley, Linda Nelson, Lynne Pinkerton, Regina Ponder, Neil R. Powe, Nettie Prack, Kandace Romero, Steve Simon, Stephanie Singer, Sylvia Echave Stock, Bruce Struminger, Kathleen Taimi, Robert Ursano, and Steven H. Woolf.

The committee is especially grateful for the information provided by downwinders and uranium miners throughout its work. They provided records, explained their concerns, and assisted us in understanding the conditions surrounding the nuclear-weapon tests and the mines. The committee and the BRER staff are appreciative of the information, feedback, and background materials for review provided by invited speakers, the public, the Department of Justice, and HRSA and its grantees. We hope that our work will help to generate changes in the Radiation Exposure Screening and Education Program and Radiation Exposure Compensation Act programs that will make them both more effective.

Finally, the committee thanks the National Research Council staff who worked directly with us, especially Study Director Dr. Isaf Al-Nabulsi for her expertise, dedication, and hard work; for keeping the committee focused and assisting in the writing and preparation of our report; and for her enormous effort in producing a clearly written, well-organized report that reflects the thought of the committee. Dr. Al-Nabulsi was well assisted in the administration of the committee's work by Courtney Gibbs, Doris E. Taylor, and Danielle B. Greene, Banneker-Anderson intern.

Reviewers

This report has been reviewed in draft form by persons chosen for their diverse perspectives and technical expertise in accordance with procedures approved by the National Research Council's Report Review Committee. The purposes of this review are to provide candid and critical comments that will assist the institution in making the published report as sound as possible and to ensure that the report meets institutional standards of objectivity, evidence, and responsiveness to the study charge. The review comments and draft manuscript remain confidential to protect the integrity of the deliberative process. We wish to thank the following for their participation in the review of this report:

John C. Bailar III, The University of Chicago, Washington, DC
Harold L. Beck, Environmental Sciences Division, New York, NY
Joel S. Bedford, Colorado State University, Ft. Collins, CO
Alfred O. Berg, University of Washington School of Medicine, Seattle, WA
Bernard L. Cohen, University of Pittsburgh, Pittsburgh, PA
Kenneth J. Kopecky, Fred Hutchinson Cancer Research Center, Seattle, WA
Jonathan D. Moreno, University of Virginia, Charlottesville, VA
Robert S. Lawrence, Johns Hopkins University, Baltimore, MD
Fred A. Mettler, The University of New Mexico, Albuquerque, NM
Jonathan M. Samet, Johns Hopkins University, Baltimore, MD
Daniel O. Stram, University of Southern California, Los Angeles, CA
John E. Till, Risk Assessment Corporation, Neeses, SC

Although the reviewers listed above have provided many constructive comments and suggestions, they were not asked to endorse the conclusions or recommendations, nor did they see the final draft of the report before its release. The review of this report was overseen by Robert A. Frosch, John F. Kennedy School of Government, Harvard University (Senior Research Fellow) and William J. Schull, University of Texas, School of Public Health (Ashbel Smith Professor Emeritus). Appointed by the National Research Council, they were responsible for making certain that an independent examination of this report was carried out in accordance with institutional procedures and that all review comments were carefully considered. Responsibility for the final content of this report rests entirely with the authoring committee and the National Research Council.

A Note on the Units of Measurement Used in this Report

It has been the custom of the Board on Radiation Effects Research to use the International System of Units (SI) in its reports. The relationships between the units used in this report and the corresponding traditional units and special names are shown below. Decimal multiples and submultiples of the units also are used, for example, *kilo* (K = 1,000 or 10^3), *mega* (M = 1 million or 10^6), *milli* (m = 1/1,000 or 10^{-3}), *micro* (μ = one millionth or 10^{-6}), *nano* (n = one billionth or 10^{-9}), and *pico* (p = one trillionth or 10^{-12}).

Concept	Symbol	Dimensions	Units SI		Traditional	Conversion
Radioactivity	A	Decays/time	(Bq)	Becquerel	(Ci) Curie	1 Ci = 3.7 × 10^{10} Bq
Absorbed dose	D	Energy/mass	(Gy)	Gray	Rad	1 rad = 10^{-2} Gy
Equivalent dose	H = Dw$_r$	Energy/mass	(Sv)	Sievert	Rem	1 rem = 10^{-2} Sv
Effective dose	E = Hw$_T$	Energy/mass	(Sv)	Sievert	Rem	1 rem = 10^{-2} Sv
Working level	WL	Energy/volume	Jm^{-3}		WL	1 WL = 2.08 × 10^{-5} Jm^{-3}
Working level month	WLM	Energy time/ volume	Jsm^{-3}		WLM	1 WLM = 12.7 Jsm^{-3}

Contents

Executive Summary

For more than 20 years during and after World War II, the United States carried out numerous aboveground nuclear-weapons tests. Many of the tests injected substantial amounts of radioactive material into the atmosphere, and some of it reached ground as nuclear fallout. Many people potentially exposed to radiation from the nuclear-weapons testing program later became concerned that radiation exposure had adversely affected their health. In addition, people employed in uranium mining and milling enterprises in support of the US weapons program were at risk for exposure to radiation from inhaled radon and to other airborne hazards in the mines. Experts concluded that those agents increased the incidence of lung cancer and respiratory diseases in miners above that in the general population.

In part to recognize the potential harm of those exposures, Congress issued an apology and passed the Radiation Exposure Compensation Act (RECA), 42 USC 2210 note, on October 5, 1990. RECA provides compensation to people (or their surviving beneficiaries) who have been diagnosed with specified cancers that scientists consider to be radiogenic or other specified chronic diseases that could have resulted from exposure to other agents, such as silica or uranium dust, associated with weapons-program activities. Eligible claimants include civilian *onsite participants* who were involved in aboveground nuclear-weapons tests at various US test sites in the United States and overseas, *downwinders* who lived in areas currently designated by RECA, and *miners* who were exposed to radiation during employment in underground uranium mines and who meet specified residence or exposure criteria. The act provides compensation payments of $100,000 for uranium miners, $75,000 for onsite participants, and $50,000 for downwinders in whom compensable cancer or one of a defined set of other diseases is diagnosed.

On July 10, 2000, Congress passed the Radiation Exposure Compensation Act Amendments of 2000 (PL 106-245), which revised the original act in several important respects. First, two new claimant categories were added—*uranium millers* involved in the crushing, grinding, and leaching of the ore during the uranium extraction process and *ore transporters*, who typically trucked uranium ore from the mine or mill. The 2000 Amendments also specified additional compensable diseases for all claimant categories, reduced the radiation exposure threshold for uranium miners, modified medical documentation requirements, removed some lifestyle restrictions that had limited eligibility for compensation, and expanded the geographic area for the downwinder claimant category.

Further expansion of the program followed with enactment of the Department of Justice Appropriations Authorization Act (PL 107-273), signed into law on November 2, 2002. That legislation included both technical and substantive changes in RECA. In particular, it provided uranium miners with an additional method of establishing exposure to radiation based solely on their duration of employment in a uranium mine.

The RECA amendments of 2000 also amended Subpart I of Part C of Title IV of the Public Health Service Act to add section 417C, on grants for education, prevention, and early detection of radiogenic cancers and other diseases. Section 417C provides the authority for competitive grants to states, local governments, and appropriate health-care organizations to initiate and support programs for health screening, education, medical referral, and appropriate followup services for persons eligible under RECA. People eligible for this program are categorized by the nature of their exposure to radiation as defined by 42 USC 2210 note and sections 4(a)(1)(A)(i) and 5(a)(1)(A) of PL 106-245 and in 28 CFR Part 79. Those categories comprise uranium miners, uranium millers, ore transporters, downwinders, and onsite civilian nuclear-weapons test participants. The Health Resources and Services Administration (HRSA) oversee the grants, which make up the Radiation Exposure Screening and Education Program (RESEP).

In September 2002, in response to a congressional mandate (PL 107-206), HRSA asked the National Research Council's Board on Radiation Effects Research to convene a committee to assess recent biologic, epidemiologic, and related scientific evidence associating radiation exposure with cancers or other human health effects and to determine how such information might affect estimates of the magnitude of the associated health risks. The present committee was formed in response to that request. Under the congressional mandate, HRSA charged the committee to consider the issues and make recommendations, on the basis of scientific knowledge and principles, regarding

> A. technical assistance to HRSA and its grantees on improving accessibility and quality of medical screening, education, and referral services;

B. the most recent scientific information related to radiation exposure and associated cancers or other diseases, with recommendations for improving services for exposed persons; and

C. whether other groups of people or additional geographic areas should be covered under the Radiation Exposure Compensation Act (RECA) program.

HRSA also requested that the committee provide an interim report to the agency and its grantees. The interim report was organized around items A and B, and was to assist RESEP staff to develop an action plan that is consistent with best medical and educational practices and the current state of science. The emphasis in the interim report was preliminary guidance on the ongoing and proposed activities and not on final recommendations.

To address items A, B, and C above, the committee needed to review the history of RECA and the laws, regulations, and objectives that guide it. In addition, important advances in the science and tools available in radiation dosimetry, radiation biology, and radiation epidemiology needed to be considered for their potential effects on determination of whether the exposed populations covered by RECA are likely to be at greater or smaller risk for cancer as the result of radiation exposure than now estimated. Those issues are described and discussed in this report and are reflected in the committee's findings and recommendations.

Much of the committee's effort was directed at the second and third parts of the statement of task—namely, the most recent scientific information related to radiation exposure and associated cancers or other diseases, with recommendations for improving services for exposed persons; and whether other groups of people or additional geographic areas should be covered under RECA. The committee considered a range of possible expansions of the downwinder geographic areas.

CONCLUSIONS

One concern about the RECA program expressed by many downwinders and other involved populations was that their counties or their cancers were not eligible for compensation. The committee discussed such equity issues extensively and concluded that, to be equitable, any compensation program has to be based to a large extent on scientific criteria and has to make the criteria for inclusion and exclusion explicit. Eligibility for compensation needs to be assessed on the basis of criteria that support and are supported by the principle that "like cases are treated alike." The use of scientific criteria is of particular importance because ionizing radiation is not a potent cancer-causing agent, and the risks for radiation-induced disease are generally low at the exposure levels of concerns in RECA populations. For example, the number of cancers observed in the Japanese atomic-bomb survivors that are attributable to radiation is relatively small, even though many in this population received doses much higher than doses received by most of downwinders. Thus, eligibility for compensation needs to be scientifically assessed.

Accordingly, the committee was particularly attentive to the downwinders' complaints about their ineligibility with respect to RECA. It examined the epidemiologic, radiobiologic, and dosimetric information relevant to downwinders' concerns. The scientific evidence indicates that in most cases it is unlikely that exposure to radiation from fallout was a substantial contributing cause to developing cancer. Moreover, scientifically based changes that Congress may make in the eligibility criteria for compensation in response to this report are likely to result in few successful claims. The committee is aware that such conclusions will be disappointing, but they have been reached in accordance with the committee's charge to base its conclusions on the results of best available scientific information.

RECOMMENDATIONS

The committee offers a large number of recommendations that address the main elements of its charge. If implemented, they will improve the compensation program in both a general way and some specific ways. They will also help to reduce screening that does not provide sufficient health benefit to outweigh the risks it poses. And, they will enhance education about programs and services available to affected populations. The recommendations are presented below; extensive discussion of their scientific justification is given in the chapters noted in parentheses.

1. Congress should establish a process using probability of causation/ assigned share (PC/AS) to determine the eligibility of any new claim for compensation for a specified RECA-compensable disease in people who may have been exposed to radiation from fallout from US nuclear-weapons testing. Further, Congress should establish criteria for awarding compensation on the basis of computed distributions of PC/AS for any person making such a claim. (See Chapters 5 and 6.)

   • Prior to implementation of the revised compensation program, the National Cancer Institute (NCI) or other appropriate agencies should perform a population-based preassessment of all radiogenic diseases using PC/AS to provide guidance to individuals who might apply for compensation by determining the likelihood any individuals in a given population of being compensated. This analysis would be determined by disease identified, places of residence at the time of exposure, ages at the time of exposure and at diagnosis, and other demographic factors using the PC/AS criteria (including consideration of the upper credibility intervals) established by Congress. The calculation would use data for the maximal doses that such individuals may have received from fallout. In settings where variability is important in evaluating risk, there may be several such defined populations, and each would be evaluated on its own merits. The criteria for evaluating such population-based preassessments should be the same as those established by Congress for compensation of claims under RECA (Note: two committee members provided their own interpretations of issues related to the preassessment criteria; for detail see Chapter 6.) The preassessments should be made for the following two purposes:

A. To provide guidance to potential claimants and the implementing agency as to which diseases may satisfy the compensation criteria established by Congress.

B. To provide guidance to potential claimants and the implementing agency as to which population groups or geographic areas may satisfy the compensation criteria established by Congress.

• The recommendation applies to residents of the continental US, Alaska, Hawaii, and overseas US territories who have been diagnosed with one of the specified RECA-compensable diseases and who may have been exposed, including exposure in utero, to radiation from US nuclear-weapons testing fallout. Both Nevada Test Site (NTS) fallout and the US fraction of global fallout should be considered.

• PC/AS for any individual should be obtained from an estimate of the radiation dose resulting from US nuclear-weapons testing and the risk estimate associated with such dose.

• Uncertainties in PC/AS cannot be avoided and may be part of the compensation decision process. Because of substantial gaps in the existing data, the uncertainties in estimated doses[1] incurred by people exposed to radiation from fallout, and consequently the uncertainties in the associated PC/AS estimate, are large. This emphasizes the need to choose compensation criteria carefully. For example, a PC/AS value associated with a high percentile of uncertainty could exceed the criteria for compensation even for some very small median doses. The challenge Congress faces will be to decide if it is best to define criteria that avoid rewarding compensation in cases in which there is very low risk, but the uncertainty associated with its PC/AS is very large, because the connection of these cancers with radiation is not well established or the estimated doses are not well known.

---

[1] The dose estimates depend on the measured deposition of radionuclides taken at the time of the nuclear weapons tests. Given the very small number of monitoring stations, most estimates represent interpolations over very large areas. Among the 3000 plus counties in the continental United States, fallout monitoring in areas other than a limited region in Nevada and its neighboring states occurred at never more than 95 stations through the years of aboveground US nuclear weapons testing. (See Chapters 5 and 6.)

To support the use of the PC/AS process for compensation,

• The Centers for Disease Control and Prevention (CDC) and the NCI or other appropriate agencies should complete dose estimates for all significant radionuclides in fallout from US nuclear weapons testing to the population groups identified above. This should include all the major sources of dose related to US nuclear weapons tests considered to have potential health consequences that the CDC-NCI 2001 draft feasibility study described.

• An updated dose calculator, similar to the existing NCI dose calculator for ^{131}I, should be developed for determining dose to the thyroid and other important organs from fallout. Such an updated dose calculator should be directly coupled to a risk calculator similar to IREP Version 5.3 that can compute PC/AS and propagate uncertainties for establishing credibility intervals.

• NCI or other appropriate agencies should maintain and revise the parameters in the models or calculators for estimating PC/AS based on risk estimates recommended by the National Research Council Committee on Biological Effects of Ionizing Radiation, report number 7 (BEIR VII). Over time, the agency should update the PC/AS calculators with the latest risk parameters.

2. The provision which allows individual states not currently covered under Section 5 of the Radiation Exposure Compensation Act to apply for inclusion under RECA if uranium mining occurred in the state during the January 1, 1942 to December 31, 1971 period should be expanded to include not only uranium mining but also uranium milling and ore transportation occurring during that period in support of the US nuclear-weapons program. (See Chapter 6.)

3. On the basis of currently available scientific evidence, no additional diseases should be added to the list of diseases that should be considered for compensation under RECA. (See Chapter 7.)

4. The appropriate agency should review the data on radiation exposure levels obtained inside dwellings constructed from mill and mine tailings. The committee also recommends that its findings regarding potential health consequences of such exposures be evaluated to determine whether the PC/AS values based on these exposures rise to or exceed the levels used in RECA compensation. (See Chapter 7.)

5. The appropriate agency should review historical data on radon concentrations in off-site areas near tailings piles of uranium mills used to produce uranium for the US nuclear-weapons program. The agency should determine whether exposures to those concentrations in off-site areas could result in PC/AS values that meet or exceed the RECA compensation criteria. If so, the agency should take the necessary steps to have these populations included in RECA. (See Chapter 7.)

6. The radiation doses and estimates of risks from the radioactive releases from all NTS nuclear weapons tests, including underground tests that resulted in atmospheric releases, should be included in determining the PC/AS. (See Chapter 7.)

HRSA also asked the committee to assess the agency's screening program and to consider recommendations that could improve access to the program and improve the quality of its educational and referral services for RECA populations. The intent of this report is to ensure that HRSA's action plan is consistent with best medical and educational practices and the current state of science for identifying people who have cancers and other diseases that are compensable under RECA.

On the basis of its review of the RESEP program data and presentations by HRSA officials and RESEP grantees, the committee offers another set of recommendations about medical screening, compensational screening, and education and outreach; they are

7. HRSA should base RESEP medical screening efforts in asymptomatic individuals on robust scientific evidence that such screening improves health outcomes and that its benefits outweigh its risks. (See Chapter 9.)

8. HRSA should not extend its medical screening beyond the generally accepted screening protocols that apply to the US population at large. However, the committee further recommends that uranium miners, millers and ore transporters also be screened for diseases generally recommended for screening in other mining populations and that uranium millers and ore transporters be screened for chronic renal disease. (See Chapter 9.)

9. Once an individual has been shown to be administratively eligible for compensation under RECA (including employment, residence, or a calculated PC/AS at or above some established cutoff criterion), the individual should be offered medical screening recommended in generally accepted protocols that apply to the population at large. The committee notes that HRSA may want to consider screening for depression in its grantees' medical screening protocols (Note: three committee members dissented from this recommendation; for detail see Chapter 9.) (See Chapter 9.)

10. HRSA should regularly monitor and follow screening guidelines developed by the US Preventive Services Task Force and published by the Agency for Healthcare Research and Quality. (See Chapter 9.)

11. HRSA should base decisions about screening primarily for compensation on recommendations drawn from credible scientific evidence that the proposed test provides reliable information about the presence or absence of specified RECA-compensable diseases. (See Chapter 10.)

12. Any screening carried out under RESEP auspices should be preceded by detailed counseling and informed consent that reflects an understanding of and sensitivity to the culture of the potential screenee. The committee also recommends that counselors, when dealing with screening for compensation, ascertain that individuals proposed to be screened fully understand the associated risks, benefits, and likelihood of potential outcomes of screening. (See Chapter 10.)

13. RESEP screening should be undertaken only if individuals satisfy administrative criteria for compensation before screening. (See Chapter 10.)

14. The Department of Health and Human Services should support development of explicit decision models and approaches to shared decision-making and related tools that enhance the ability of patients to participate in decisions that affect their care and prognosis. In particular, the committee recommends that HRSA take responsibility for similar activities in the domain of compensational screening. (See Chapter 10.)

15. If an individual has established eligibility for compensation, RECA should cover the costs of screening, complications of screening, referrals (followup), diagnosis (workup), and treatment for the RECA-compensable diseases for which such eligibility has been established. (See Chapter 10.)

16. HRSA should change its RESEP funding mechanism from grants to contracts. (See Chapter 11.)

17. The Department of Health and Human Services should ensure that the content of public and professional educational programs be consistent across all entities that HRSA supports through its RESEP program. (See Chapter 11.)

18. HRSA should provide information to RECA populations about other radiation exposure compensation programs for which they might be eligible. The committee also recommends that an advisory organization should review all federal compensation programs related to radiation exposure to determine similarities and differences and that HRSA periodically convene representatives of all programs to address inconsistencies among programs and determine the effects of developments over time in radiation biology, risk estimates, legislation, and regulations. (See Chapter 11.)

19. HRSA should ensure that all public informational materials are written so that members of target populations can understand their contents. (See Chapter 11.)

20. HRSA should undertake an enhanced program of education and communication about the risks posed by radiation exposure for people who may have been exposed to radiation from fallout from US nuclear-weapons testing. (See Chapter 11.)

21. HRSA should undertake an appropriately focused educational program explicating the limitations, the benefits, and the risks of medical screening for many RECA diseases. (See Chapter 11.)

22. HRSA should (See Chapter 11):

    A. Use a standardized method to develop outcomes-based goals and objectives for appropriate planning and assessment;

    B. Identify and evaluate the cost and effectiveness of removal of barriers to program implementation; and

    C. Train staff to identify specific barriers to implementation and develop strategies to overcome them.

The committee recognizes that some of its recommendations will be difficult to implement in a short time. Additional information and improved approaches for addressing radiation risk and fallout doses may change compensation programs, medical screening, screening for compensation, and related education and outreach programs. The task of addressing those issues has been difficult, but the committee accepted the challenge because of the critical need for decisions regarding the future of RECA and RESEP.

1

Introduction

Beginning in 1945 and continuing through 1962, the United States conducted a series of aboveground nuclear-weapons tests. Many people potentially exposed to radiation from the testing program later became concerned that their health had been adversely affected by those events. The concerned populations included workers and civilian employees who participated onsite in tests involving the atmospheric detonation of nuclear devices within the boundaries of the Nevada, Pacific, Trinity, or South Atlantic Test Sites and others living in the surrounding areas during the testing period. Uranium miners had also been at risk of exposure to radiation from inhaled radon decay products and other airborne hazards in the mine environment that together were presumed to have caused an increase in the incidence of lung cancer and respiratory diseases among the miners relative to the general population.

On October 15, 1990, the Radiation Exposure Compensation Act (RECA), PL 101-426, was enacted to provide payments to people who developed particular cancers or other diseases as a result of either exposure to radiation released during aboveground nuclear-weapons tests or employment associated with the uranium-mining industry. The cancers specified as compensable under RECA were those that had been determined to be causally associated with radiation exposure on the basis of epidemiologic studies of populations exposed to low to moderate doses at mainly high dose-rates (NRC, 1980).

RECA[1] was amended on July 10, 2000. The amendment broadened the scope of eligibility for benefits to include additional categories of people and

---

[1] In this report we refer to RECA with all amendments attached unless otherwise stated.

modified the criteria for determining eligibility for compensation. The Department of Justice (DOJ) administers RECA as codified by 28 CFR 79.

The RECA of 2000 also amended Subpart I of Part C of Title IV of the Public Health Service Act to add Section 417C—grants for education, prevention of, and early detection of radiogenic cancers and nonradiogenic diseases. Section 417C provides the authority for competitive grants to states, local governments, and appropriate health-care organizations to initiate and support programs for health screening, education, medical referral, and appropriate followup services for persons eligible under RECA. Persons eligible for those programs are categorized by the nature of their exposure to radiation as defined by 42 USC 2210 note and Public Law 106-245, the Radiation Exposure Compensation Act Amendment of 2000, Sections 4(a)(1)(A)(i) and 5(a)(1)(A), and in 28 CFR Part 79. The categories are uranium miners, uranium millers, ore transporters, certain downwinders, and onsite nuclear-test participants. The Health Resources and Services Administration (HRSA) of the Department of Health and Human Services administers the grants as part of the Radiation Exposure Screening and Education Program (RESEP).

In response to a congressional mandate (Public Law 107-206), HRSA asked the National Research Council's Board on Radiation Effects Research to convene a committee to assess the recent biologic, epidemiologic, and related scientific evidence associating radiation exposure with cancers or other human health effects. The committee was asked to consider the issues surrounding the implementation of RECA and to make recommendations on the basis of scientific knowledge and principles. The study began in September 2002.

The following statement of work describes the task set before the committee (HR 107-593):

> On the basis of its information, the committee will make recommendations to HRSA regarding:
>
> A. technical assistance to HRSA and its grantees on improving accessibility and quality of medical screening, education, and referral services;
>
> B. the most recent scientific information related to radiation exposure and associated cancers or other diseases, with recommendations for improving services for exposed persons; and
>
> C. whether other groups of people or additional geographic areas should be covered under the Radiation Exposure Compensation Act (RECA) program.

The committee considered whether additional geographic areas should be added to the previously defined areas[2] on the basis that residents had been simi-

---

[2]The areas previously designated for compensation were: the Utah counties of Beaver, Garfield, Iron, Kane, Millard, Piute, San Juan, Sevier, Washington, and Wayne; the Nevada counties of Eureka, Lander, Lincoln, Nye, and White Pine and the portion of Clark County that consists of townships 13-16 at ranges 63-71; and the Arizona counties of Apache, Coconino, Gila, Navajo, and Yavapai and the part of Arizona that is north of the Grand Canyon.

larly at risk of exposure to fallout from US nuclear-weapons tests. In its report, the committee considers a range of possible expansions of the current downwinder geographic areas to include other areas exposed to high levels of fallout.

In considering task B, the committee focused its review on issues relevant to RECA and RECA populations and did not duplicate nor have access to the detailed final conclusions of the exhaustive efforts of the Committee on the Health Risks of Exposure to Low Levels of Ionizing Radiation (the BEIR VII committee's report will be published in 2005). Reports on task-related topics published since 1990 in the national and international peer-reviewed literature were identified for review. Reanalysis of existing epidemiologic data with alternative methods or models and the collection and analysis of new epidemiologic data were beyond the committee's task.

HRSA also requested that the National Research Council committee provide an interim report to the agency and its grantees. The six grantees funded by HRSA (see Table 11.1) under RESEP in 2002 and in 2003 were at the Dixie Regional Medical Center, the Miners' Colfax Medical Center, the Mountain Park Health Center, the Northern Navajo Medical Center, the St. Mary Hospital and Medical Center, the University of New Mexico Health Sciences Center, and the Utah Navajo Health System, Inc. The interim report was organized around items A and B above, and was to assist RESEP staff to develop an action plan that is consistent with best medical and educational practices and the current state of science (NRC, 2003a).

The committee provides background information in Chapter 3 on recent developments in radiation dosimetry, radiation biology, and radiation epidemiology that influence the risk-assessment process. That approach was taken to enable the committee to determine whether any of the RECA populations are likely to be at more or less risk of cancer as the result of exposure to radiation than previously estimated and whether additional geographic areas should be recommended for inclusion in RECA.

This report constitutes the results of the committee's assessment and its recommendations. It consists of 11 chapters. Chapter 2 describes the Radiation Exposure Compensation Act, including a brief history which led to its creation and recent revisions (addresses item C in the committee's statement of task). Chapter 3 reviews the scientific principles of the physics and dosimetry of ionizing radiation, radiation biology, and epidemiologic methods, and provides basic technical background information in support of Chapters 4, 6, and 7. Chapter 4 reviews and summarizes recent data on radiation epidemiology, dosimetry, and biology (addresses item B). Chapter 5 describes probability of causation (PC) and its use in compensation (addresses item C). Chapter 6 provides information about additional geographic areas that might be included under RECA (addresses item C). Chapter 7 provides information about additional diseases and classes of people that might be included as compensable under RECA (addresses item C).

Chapter 8 discusses an ethical framework as it applies to RECA and RESEP (addresses item C). Chapters 9 and 10 discuss the indications for and implications of screening, both to improve health and to identify persons eligible for compensation (addresses item A). And Chapter 11 addresses education and outreach (addresses item A).

To be consistent with the policies of the National Academies and to fulfill its charge, the committee conducted fact-finding activities involving outside parties in public information-gathering meetings. The committee met in closed sessions including conference calls only to develop committee procedures, review documents, and consider findings and recommendations. It met once in 2002, five times in 2003, seven times in 2004, and once in 2005. Eleven of the fourteen meetings included public information-gathering sessions, and the committee also received and considered other public comments and communications. The information-gathering meetings were structured to solicit information from technical experts and the study sponsor on topics related to the study. At those meetings, the committee heard from representatives of HRSA and its RESEP grantees, the Department of Energy, DOJ, the Department of Labor, scientific, medical and other experts from academic institutions, and other interested parties, and it benefited from the information they provided. The committee appreciated and was impressed by the efforts of the speakers to work with it during the project; their cooperation has been important in the committee's efforts.

The committee also held three public meetings dedicated to information-gathering in St. George, Utah; Window Rock, Arizona; and Salt Lake City, Utah. In addition, the Research Council staff held a public meeting in Boise, Idaho, to gather information for the committee. Notices inviting the public to attend those meetings went to the offices of Senators Orrin Hatch, John McCain, and Larry Craig; to the press in Utah and Arizona; to the Navajo Education Center in Arizona; and to the Salt Lake City Library. Each of the four meetings was a full-day open session at which members of the public and technical experts were invited to express their views and concerns, ask questions, and provide information orally or in writing on issues related to the committee's task. The committee invited those unable to attend the meetings to submit written statements for its information and inclusion in the Research Council's public-access file, which is available on request. Almost 200 people, including members of the mass media, attended each of the public meetings. We found the public eager to assist us. In particular, they seemed to be supportive of our study, interested in learning more about it, and curious about the answers they hoped the committee would be able to provide to a number of questions. Such interaction with the public provided a formal, yet open, exchange of ideas, questions, and responses and proved useful to us.

The oral and written comments, views, concerns, and questions submitted to the committee by members of the public and technical experts were in several broad categories. The committee considered each of those in preparing its report:

1. Personal testimony on the adverse health effects attributed by individuals and families to the activities defined by RECA and on the hardships experienced by them in their treatment and disease-management efforts.

2. Argument in favor of expansion of RECA coverage to include northern Utah counties, all Utah counties, and other areas, such as Idaho, Montana, New York state, and all the United States, based on maps from the National Cancer Institute (NCI), indicative of counties in which the total dose to the thyroid was estimated to be at least as great as that estimated in areas currently covered by RECA.

3. Argument in favor of expansion of RECA coverage to include additional diseases for one or more of the eligible populations.

4. Argument from the Navajo Nation that miners can use affidavits under some circumstances to establish employment history but that millers and ore transporters cannot.

5. Argument from the Navajo Nation that an affidavit should be allowed as proof of presence or residence for downwinder claimants.

6. Argument to expand RECA coverage to residents of Guam who were potentially exposed to fallout and to radiation from the decontamination of naval vessels associated with nuclear tests conducted in the Pacific.

7. Recommendation to support an update of the study of leukemia in the downwinder populations.

8. Reports of immediate-family members (same or different generation) or close relatives with multiple similar or diverse types of cancer (such as breast cancer and thyroid cancer), often among individuals or families with no known history of cancer.

9. Interest in the PC approach to compensation that uses individual doses estimated with the NCI algorithm (difficulty with on-line access to the NCI-PC algorithm was described).

10. Inequity within RECA and among the various federal radiation compensation programs.

11. National or universal health care as a solution (posed as the only equitable solution) to the problem.

12. Concerns about the possibility of the resumption of nuclear testing at the Nevada Test Site (NTS) and elsewhere.

13. Support of full and comprehensive epidemiologic studies among the Navajo Nation.

14. A proposal that the federal government put more money into a trust fund for compensation.

15. The idea that the federal government should conduct more research into the health effects of fallout from nuclear testing.

16. Reports (by medical and other professionals and other citizens of Idaho) of apparent excesses of cases of some radiogenic diseases and others perceived to be attributable to radiation exposure.

Appendix A lists the names of invited speakers, representatives, and other persons who have interacted with the committee through correspondence, by providing information at meetings, or by providing their statements for our use. More than 1,400 people have interacted with the committee. The type of interaction—for example, e-mail, fax, letter, phone call, and attendance at meetings—is also noted.

SCIENCE, VALUES, AND DECISION-MAKING

The review of scientific research, evidence-based medical knowledge, and studies of effective educational strategies presented in this report provides new information, some of which was developed since RECA was enacted and later amended. Applying this new scientific knowledge may require additional societal value-based decisions. In addition, not all the issues presented to the committee fell strictly within its charge. Some issues, such as citizens' concerns about the possibility of the resumption of nuclear testing at the Nevada Test Site (NTS) and elsewhere, clearly fell outside the charges that HRSA put to this committee.

Other issues, although intertwined with the scientific recommendations the committee provides in this report, require debate and deliberation among citizens and their congressional representatives. The committee offers relevant scientific recommendations, but the attendant policy decisions must come from the larger body of citizenry. *Whom* to compensate is one such decision. The method that the committee thoroughly discussed by which a person is eligible, on scientific grounds, to qualify for RECA compensation, will require that Congress make some additional decisions. For example, it calls for further determinations as to whether compensation is to be based primarily on estimated dose or on a composite measure taking both estimated dose and uncertainty into account (see Chapters 5 and 6). These difficult policy decisions lie outside of science *per se*, but science has valuable information to offer in support of them. The decisions are societal judgments based on the acceptance of some scientific consequences over others (Rudner, 1953; Beauchamp and Childress, 2001). Other decisions about compensation include *how* to compensate; an example is whether to focus on communities or individuals. In the latter case, a further decision is *whether* to establish a sliding scale of compensation or to apply a flat rate of compensation to all. These are all societal decisions.

At best, the committee will make transparent, in Chapters 5 and 6, the consequences, values, and assumptions embedded in the various criteria that Congress could adopt to establish eligibility for compensation. The decision rests with Congress.

Citizens' concern to achieve equity occupied much of the committee's deliberations. In providing a scientific basis for establishing justice, the committee still is limited to making recommendations within or directly related to its charge. The committee was not asked to comment on the use of any new criteria for eligibility for compensation for persons in existing RECA-eligible geographic areas.

The committee recognizes that reforms never rest simply on the scientific recommendations of experts. Committee members understand that many other considerations play a role in crafting policies and laws; these include pragmatic consequences, budgetary realities, and competing political goals. The committee intends that its scientific recommendations remain within the parameters of its initial charge and that they are consistent with principles of ethics.

2

Legislation and Compensation

The legislation on radiation exposure compensation, screening, education and outreach is complex and extensive. In this chapter, we describe the Radiation Exposure Compensation Act (RECA), including a brief history which led to its creation and recent revisions. We focus on the topic of compensation as provided by RECA. Chapter 9 includes a description of screening as required by the Radiation Exposure Screening and Education Program (RESEP). Chapter 11 examines RESEP's legislative requirements with respect to education and outreach to RECA stakeholders.

THE RADIATION EXPOSURE COMPENSATION ACT

Events Leading to Legislation

RECA (1990, 2000, 2002) is one of four existing radiation exposure compensation programs[1] that emerged from a variety of legal, political, and social actions in mid-1970s (Walchuk, 2002). During that period, the organized efforts of citizens presumably affected by the government's activities in uranium-mining areas and in areas downwind of the Nevada Test Site (NTS)—100 miles north-

---

[1]These include, in order of enactment, the Veterans Dioxin and Radiation Exposure Compensation Act (1984), the Radiation-Exposed Veterans Compensation Act of 1988 (REVCA), RECA (origin 1990; amended 2000, 2002), and the Energy Employees Occupational Illness Compensation Program Act (EEOICPA) of 2000. The Department of Veteran Affairs administers the first two; RECA, the one at issue in this report, is administered by the Department of Justice. EEOIPCA is administered by the Department of Labor.

west of Las Vegas, Nevada—encouraged members of Congress to advance compensation legislation. The actions and testimony of labor unions, Native American uranium miners, interest groups, and downwinders constitute the background of the RECA legislation. Among these groups are the Oil Chemical and Atomic Workers International, the Office of Navajo Uranium Miners, the Eastern Navajo Agency Uranium Workers, the Northern Arizona Navajo Downwinders, and the Utah Navajo Downwinders. Some organizations, such as Dine Citizens Against Ruining Our Environment (Dine Care) as late as 1998 claimed that "we have been involved in bringing relief to victims of radiation exposure on the Navajo Nation, and in the fight to prevent future mining. Our biggest victory so far has been the reform of the Radiation Exposure Compensation Act" (http://dinecare.indigenousnative.org/about_us.html, accessed December 23, 2004).

Those organizations "approached radiation exposure as a social justice issue, righting government wrongs to the constituent group" (Walchuk, 2002). They have hired Washington, DC, lobbyists to support their efforts in Congress. Active citizen organizations continue to inform their constituents about RECA and its amendments. Many of the groups, such as the Western States RECA Reform Coalition and the Mohave Downwinders, continue to seek further legislative remedies.

While miners and downwinders were organizing around compensation legislation in the late 1970s, lawsuits on behalf of workers were filed against mining companies and the federal government. They met with little success because, in the case of mining companies, worker's compensation precluded suits against employers for on-the-job injury or illness. Most of the worker's compensation claims were denied or never filed. With few exceptions, mining companies have not been held liable (Brugge and Goble, 2002).

A suit was brought against the Atomic Energy Commission (AEC) by John Begay to seek redress for alleged harm from uranium mining activities in *Begay v. United States*. Begay's petitioners argued that the government's special trust responsibilities toward Native Americans should not allow the normal exceptions to apply to negligence toward Navajo uranium miners. Nevertheless, the US District Court in Arizona ruled that there was no subject-matter jurisdiction to proceed because the federal government was shielded from prosecution and any later tort liability by the discretionary-function exception to the Federal Tort Claims Act (FTCA, 28 U.S.C. § 2674 ; 28 USC § 2680). The court also ruled that national-security interests overrode any claim to restitution. "The court concludes that all the actions of various governmental agencies complained of by plaintiffs were the result of conscious policy decisions made at high government levels based on considerations of political and national security feasibility factors. All such decisions, including the PHS epidemiological study, were carried out as directed. Therefore this court lacks . . . subject matter jurisdiction to proceed in these cases. Because the discretionary function exception is dispositive, there is no need to discuss the other legal theories of plaintiffs or defendant" (John N. Begay v. United States, 591 F. Supp. 991, 1984).

In its decision, the court mentions compensatory legislation in concluding (John N. Begay v. United States, 591 F. Supp. 991, 1984) on the basis of information presented to the court that "this tragedy of the nuclear age, however, cries for redress. Such relief should be addressed by the Congress as it was in the case of the Texas City explosion following the decision of the Supreme Court in *Dalehite, supra*; 69 Stat. 707."

The case that opened the door for a legislative remedy for downwinders was *Allen v. United States*. The US district court for the District of Utah, 588 F. Supp 247 (D. Utah 1984), entered judgment against the government on nine of the claims, and the United States appealed. However, downwinder claims against the government for damages under the federal tort act were definitively denied by the US Court of Appeals for the 10th District. On January 11, 1988, the US Supreme Court refused to hear the writ of *certiorari* filed by Irene Allen and others to overturn the federal appeals court decision. Earlier, on April 20, 1987, the US Court of Appeals for the 10th Circuit argued that "while we have great sympathy for the individual cancer victims who have borne alone the costs of the AEC's choices, their plight is a matter for Congress. Only Congress has the constitutional power to decide whether all costs of government activity will be borne by all the beneficiaries or will continue to be unfairly apportioned, as in this case. Until Congress amends the discretionary function exception to the Federal Tort Claim Act (FTCA) or passes a specific relief bill for individual victims, we have no choice but to leave them uncompensated" (*Allen v. United States*, 816 F. 2d 1417 [10th Cir. 1987] cert. denied, No. 87-316 [Jan. 11, 1988], p. 9). By that time, attempts were already under way to seek redress through Congress rather than through tort law and the court system.

Legislative History of the Radiation Exposure Compensation Act

The legislative history of RECA spans several decades. It dates back at least to April 19, 1979, in joint hearings on "Health Effects of Low-Level Radiation" before the Subcommittee on Oversight and Investigations of the House Interstate and Foreign Commerce Committee, the Subcommittee on Health and Scientific Research of the Senate Labor and Human Resources Committee, and the Senate Judiciary Committee. The joint hearings, held in Salt Lake City, Utah, included Governor Scott Matteson's criticism of federal agencies, particularly AEC, for suppressing information and failing to investigate health dangers posed by nuclear-test fallout; the hearings also involved a description of nuclear fallout deposited materials and health effects in Utah and advocacy of independent research on the fallout problem. The general counsel for what was then called the Department of Health, Education and Welfare and later the Department of Health and Human Services (DHHS), the chairman of the Interagency Task Force on Health Effects of Ionizing Radiation, and the director of the National Institutes of Health explained the Carter administration's efforts to address Utah residents' concerns over radiation health effects, focusing on task-force establishment and expansion of research programs. Sheep ranchers and an agricultural agent from Iron City, Utah, described the death and disease incidence among sheep exposed to radiation during nuclear tests, criticized the government's failure to warn the public of nuclear dangers, and described the AEC's explanation of sheep losses. Testimony was heard from residents of St. George, Utah, who were potentially exposed to radiation and were allegedly experiencing radiation-related health problems.

Again in 1979, congressional hearings were held in Grants, New Mexico, against the backdrop of attempts to create legislation modeled on black-lung benefits, that is, a small monthly stipend. On June 10, 1980, a joint hearing before the Subcommittee on Health and Scientific Research and the Senate Judiciary Committee was held to consider S. 1865, the Radiation Exposure Compensation Act of 1979, to provide federal compensation to cancer victims and their survivors for damages attributable to radiation from nuclear tests in Nevada and to provide compensation to uranium miners. Senator Pete V. Domenici (R-NMex) testified on the difficulties of uranium miners and survivors in obtaining just compensation for their losses. Survivors of cancer victims gave personal accounts of cancer cases and deaths that allegedly resulted from uranium mining and nuclear-test fallout, and they described their problems with delayed or denied compensation benefits under worker compensation and other programs. Department of Justice (DOJ) representatives and former Secretary of the Interior Stewart Udall testified to the need to compensate those harmed by exposure to radiation (United States Congress, 1980).

For more than a decade, during the period 1978-1990, hearings focused on the federal government's discretionary-function immunity. Government representatives cited the merits of Federal Tort Claims Act limitations on government liability for personal injury or death claims. During that period, the National Research Council Committees on the Biological Effects of Ionizing Radiations (BEIR) published reports on the risks related to exposure to radiation (NRC, 1980; 1988; 1990). The reports identified cancers that are found at higher rates among uranium workers and downwinders. *BEIR IV* and *BEIR V* also helped established standards of proof to be used in legislation in determining eligibility for compensation (NRC, 1988, 1990). As late as 1989, Stuart Gerson, assistant attorney general in the Civil Division of DOJ, in testimony before the Subcommittee on Administrative Law and Governmental Relations, denied any causal relationship between radiation exposure from US nuclear testing and cancer disorders (United States Congress, 1989).

During the same period, most of the hearings occurred in response to various versions of a radiation exposure compensation act introduced in the Senate or House of Representatives in the following congressional sessions:

- 1979-1981: 96 S. 1827, 96 S. 1865.
- 1981-1983: 97 S. 1483.
- 1985-1987: 99 H.R. 1338.
- 1985-1987: 99 S. 2454.
- 1987-1989: 100 H.R. 1341 100 H.R. 3872, 100 H.R. 5022,
- 1987-1989: 100 S. 612.
- 1989-1991: 101 S. 982, 101 S. 2466,
- 1988-1991: H.R. 2372.

Representative Wayne Owens (D-Utah) with 17[2] cosponsors in the House of Representatives and Sen. Orrin Hatch (R-Utah) with 13[3] cosponsors in the Senate were among those supporting or sponsoring such an act.

On House acceptance of several amendments made by the Senate, H.R. 2372, the bill introduced by Representative Owens, became Public Law 101-426, the Radiation Exposure Compensation Act on October 15, 1990.

The Original Radiation Exposure Compensation Act of 1990

RECA provided for compassionate payments to people in specific classes who had contracted the following diseases:

- lung cancer (with eligibility modified by smoking behaviors);
- leukemia (other than chronic lymphocytic leukemia), provided that initial exposure occurred after the age of 20 years and the onset of disease was between 2 - 30 years of first exposure; and, provided that onset was at least 5 years after first exposure,
  - multiple myeloma,
  - lymphomas (other than Hodgkin's disease),
- primary cancer of:
  - the thyroid (provided that initial exposure occurred by the age of 20 years),
  - the female breast (provided that initial exposure occurred before the age of 30 years),
  - the esophagus (provided that the person had low alcohol consumption and was not a heavy smoker),
  - the stomach (provided that initial exposure occurred before the age of 20 years),
  - the pharynx (provided that the person was not a heavy smoker),

---

[2]Hansen R-UT, Schroeder D-CO, Skaggs D-CO, Richardson D-NM, Bilbray D-NV, Vucanovich R-NV, Kaptur D-OH, Walsh R-NY, Robinson R-AR, Fazio D-CA, Foglietta D-PA, Frost D-TX, Fauntroy D-DC, Condit D-CA, Hatcher D-GA, Gilman R-NY,Slaughter D-NY, and Fuster D-PR.

[3]Hatfield R-OR, Kennedy D-MA, Garn R-UT, DeConcini D-AZ, Reid D-NV, Bryan D-NV, Bingaman D-NM, Domenici R-NM, Pell D-RI, Wirth D-CO, McCain R-AZ, Inouye R-HI, and Gore D-TN.

- the small intestine,
- the pancreas (provided that the person was not a heavy smoker and had low coffee consumption),
- the bile ducts,
- the gall bladder, and
- the liver (except if cirrhosis or hepatitis B was indicated) (PL 101-426).
- nonmalignant respiratory disease, including
  - fibrosis of the lung,
  - pulmonary fibrosis,
  - cor pulmonale related to fibrosis of the lung, and "if the claimant, whether Indian or non-Indian, worked in a uranium mine located on or within an Indian reservation, the term shall also include moderate or severe silicosis or pneumoconiosis" (PL 101-426).

Those diseases were already covered by the Radiation-Exposed Veterans Compensation Act of 1988 (REVCA). They were identified originally on the basis of the findings in the BEIR 1980 report (*BEIR III*, NRC, 1980) and later modified based on the findings of *BEIR IV* (NRC, 1988) and BEIR*V* (NRC, 1990). The eligible classes designated by RECA included occupations in the mining of uranium ore, specific groups onsite during an aboveground nuclear detonation, and downwinders residing in specific counties of specific states (See Map 2.2). RECA required persons in those classes to provide proof that they met eligibility requirements regarding exposure to radiation. Among the requirements were exposure levels among uranium workers, presence during specific periods when aboveground detonations took place, and residence in downwind counties.

After the enactment of the original version of RECA, DOJ published regulations governing its implementation in the *Federal Register* on April 10, 1992 and again on March 22, 1999, "establishing procedures to resolve claims in a reliable, objective, and non-adversarial manner, with little administrative cost to the United States or the person filing the claim." (DOJ Web site http://www.usdoj.gov/civil/torts/const/reca/about.htm, accessed December 23, 2004). Those regulations were the subject of considerable debate among various constituencies.

Period Between the Original and Amended Versions of the Radiation Exposure Compensation Act

From October 15, 1990, to July 10, 2000, complaints were registered with congressional offices regarding the scope and implementation of RECA. The complaints were in three categories: the limited number of diseases and their narrow definitions, narrow or overtechnical DOJ regulations, and exclusion of other mining-industry workers (House Report 106-697). Many highly organized citizens' groups worked behind the scene with lobbyists to change the 1990 RECA. Among the changes sought were expanding coverage of uranium workers to include ore transporters and additional mining and milling occupations;

lowering the standard of proof of exposure of uranium miners from 200 working level months (WLMs) to 40 working level months (WLMs); removing the distinction between smokers and nonsmokers; offering some (albeit lower) compensation payments to injured, deceased, or research-subject uranium workers whose employment-related diseases were not compensable; doubling compensation for eligible uranium workers; expanding the diseases for downwinders; and expanding the geographic areas covered (Brugge and Goble, 2003).

The House Judiciary Committee, which examined the amendments to RECA on June 27, 2000, was convinced by Chapter 18 of the President's Advisory Committee on Human Radiation Experiments report (ACHRE, 1995).[4] The importance of ACHRE's report is that it led the House Judiciary Committee to accept the advisory committee's recommendation to lower the stringent requirements for compensation of exposed miners, eliminating the distinction between smokers and nonsmokers and using length of employment instead of exposure levels to verify eligibility. Length of employment gained support because of the lack of exposure measurements and the uncertainty associated with extrapolations needed to calculate reconstructed exposure times. The expert testimony of Arthur C. Upton, formerly director of the National Cancer Institute (chairman, Department of Environmental Medicine, New York University Medical Center, and chair of the BEIR V Committee), Jonathan M. Samet (then professor and chairman of the Department of Epidemiology, Johns Hopkins University and chair of the BEIR VI Committee), and Duncan C. Thomas (then professor in the Department of Preventive Medicine, University of Southern California, Los Angeles and member of the BEIR V Committee) was another factor that influenced the proposed amendment to expand the list of diseases and eliminate or change exposure levels, age limits, length of latency periods, and alcohol, caffeine, and smoking restrictions among downwinders.

DOJ responded to many of the proposed amendments (Robert Raben, assistant attorney general, DOJ letter to Congressman Henry J. Hyde, R-Ill, January 24, 2000). Although DOJ supported some amendments—including the addition of some diseases, expansion of proof of disease or employment, and the ability to use American Indian law, tradition, and custom in processing claims—the department also expressed concern on three areas.

First, DOJ argued that expanding compensation to millers and ore transporters was premature and should await the results of the National Institute for Occupational Safety and Health (NIOSH) study commissioned by Congress in 1993. The

---

[4]The final report of the Advisory Committee on Human Radiation Experiments (stock number 061-000-00-848-9), the supplemental volumes to the final report (stock numbers 061-000-00850-1, 061-000-00851-9, and 061-000-00852-7), and copies of the Executive Summary (stock number 061-000-00849-7) may be purchased from the Superintendent of Documents, US Government Printing Office. Available at http://www.eh.doe.gov/ohre/roadmap/achre/index.html, accessed December 28, 2004.

House Judiciary Committee claimed in response that the passage of amended legislation should not be delayed and that "furthermore, given the extremely small group of millers being studied, it is projected that the data will have limited statistical significance and will therefore be merely anecdotal in nature" (House Rpt.106-697–Radiation Exposure Compensation Act Amendments of 1999).

Second, DOJ disagreed with the expansion of downwinder regions not defined by the National Cancer Institute (NCI). It stated that "Section 3(b) of S. 1515 would also add several new 'Downwinder' and 'Onsite Participant' diseases. Similarly, S. 1515 would increase the Downwinder 'affected area' to include Wayne and San Juan counties in Utah and the counties of Coconino, Yavapai, Navajo, Apache, and Gila in Arizona. The National Cancer Institute (NCI), the experts in the field, advises us that, at this time, NCI cannot offer any scientific support for the expansion of the RECA program to include these additional diseases, nor are there radiodosimetric studies or other scientific findings to support the inclusion of the proposed areas" (Robert Raben, assistant attorney general, DOJ letter to Congressman Henry J. Hyde, R-Ill., January 24, 2000). The House Judiciary Committee dismissed that objection, stating that NCI had no current studies monitoring downwinder cancer epidemiology. It asserted that "to ignore the written and personal testimonies of the hundreds of victims themselves or survivors concerning their illnesses is unwarranted. The strong evidence they have supplied is sufficient to provide relief" (House Rpt.106-697–Radiation Exposure Compensation Act Amendments of 1999).

Third, DOJ argued that lowering the radiation exposure of uranium millers from 200 WLM to 40 WLM was not based on a sound, scientific approach. DOJ suggested instead, according to the House report, "implementation of a [sic] multi-scale criteria using either the exposure-based or duration of employment models. These models involve computing the WLMs, age, time since exposure, smoking habits, and other factors for each individual prior to evaluating the disease status of the claimant" (House Rpt.106-697–Radiation Exposure Compensation Act Amendments of 1999). The House Judiciary Committee rejected that suggestion, claiming that for most of the claimants, such data would be incomplete. Such burdens would be contrary to the intent of the original RECA.

The Radiation Exposure Compensation Act Amendments of 2000

S. 1515 became Public Law 106-246 known as the Radiation Exposure Compensation Act Amendments of 2000, on July 10, 2000. Stated reasons for the amendments to the earlier RECA include:

- "Regulatory burdens have made it too difficult for some deserving individuals to be fairly and efficiently compensated."
- "Reports of the Atomic Energy Commission and the National Institute for Occupational Safety and Health testify to the need to extend eligibility to States

in which the Federal Government sponsored uranium mining and milling from 1941 through 1971."

• "Scientific data resulting from the enactment of the Radiation-Exposed Veterans Compensation Act of 1988 (38 USC 101 note), and obtained from the Committee on the Biological Effects of Ionizing Radiations, and the President's Advisory Committee on Human Radiation Experiments provide medical validation for the extension of compensable radiogenic pathologies."

• "Above-ground uranium miners, millers and individuals who transported ore should be fairly compensated, in a manner similar to that provided for underground uranium miners, in cases in which those individuals suffered disease or resultant death, associated with radiation exposure, due to the failure of the Federal Government to warn and otherwise help protect citizens from the health hazards addressed by the Radiation Exposure Compensation Act of 1990 (42 USC 2210 note)."

• "It should be the responsibility of the Federal Government in partnership with State and local governments and appropriate healthcare organizations, to initiate and support programs designed for the early detection, prevention and education on radiogenic diseases in approved States to aid the thousands of individuals adversely affected by the mining of uranium and the testing of nuclear weapons for the Nation's weapons arsenal" (PL 106-246 [S. 1515] Jul 10, 2000 Radiation Exposure Compensation Act Amendments of 2000 106 PL 246; 114 Stat. 501).

On August 7, 2002, DOJ issued regulations to implement the RECA amendments (Department of Justice, Part IV 28 CFR Part 79 "Claims Under the Radiation Exposure Compensation Amendments of 2000; Final Rule and Proposed Rule" 79.3). The Health Resources and Services Administration, an agency of DHHS, followed suit on April 30, 2002, with its announcement of a grant program to fund projects designed to carry out Congress's intent to partner with state and local governments in providing screening, referrals for service, and education (PL 107-206).

Continuing Legislative Reforms of the Radiation Exposure Compensation Act

Since the enactment of the RECA 2000 amendments, additional amendments have been introduced into the House and the Senate. These include

• H.R. 1131, the *Paul Hicks Memorial Act.*
• H.R. 1132, *Ensuring Timely Payments Under the Radiation Exposure Compensation Act.*
• S. 898 (no title), to make technical amendments to RECA (42 USC 2210 note), to provide compensation to some claimants under the act, and for other purposes.

- S. 1438, *the Defense Reauthorization Act.*
- H.R. 2215, the *21st Century Department of Justice Appropriations Authorization.*

The Defense Reauthorization Act was signed into law on December 13, 2001 (Public Law 107-107). It authorized appropriations for FY 2002 for military activities. Section 1063 of this broad bill also appropriates to the Radiation Exposure Compensation Trust Fund such sums as may be necessary, not to exceed specified maximums, in FY 2002-2011.

H.R. 2215 was enacted on November 2, 2002, as PL 107-273, the 21st Century Department of Justice Appropriations Authorization Act. It provides technical amendments to RECA, including the reinsertion of a downwinder area in Mohave County, Arizona (which had been inadvertently removed from the 2000 amendments), and a change in eligibility requirements for uranium miners from 40 working levels months (WLMs) of radiation to *either* the 40 WLMs exposure standard *or* the 1-year duration-of-employment standard already applicable to uranium millers and ore transporters. It also removed a requirement (presumably a drafting error) that uranium workers with lung cancer submit evidence of a nonmalignant respiratory disease (a requirement that excluded most lung cancer claimants who did not also suffer from a nonmalignant respiratory disease).

DOJ issued a final rule—28 CFR Part 79 Part II—on March 23, 2004. It constitutes a revision of its existing regulations governing RECA. It is based on both the RECA amendments of 2000 and the 21st Century Department of Justice Appropriations Authorization Act. In it, DOJ responds to many comments, including nearly 50 letters it received on its proposed rule, issued August 7, 2002 (DOJ, 2002), regarding implementation of the 2000 amendments. DOJ also incorporates some technical revisions stemming from the Defense Reauthorization Act. Some public comments suggested clarification, and others requested substantive changes. DOJ makes the important point that it does not have the authority to change a statute in its rule-making. Hence, DOJ could not act on many suggestions, for example, to include additional diseases, because it had (ostensibly) no discretion in this matter.

In the final rule, DOJ examined a question that arose for this committee in its hearings in St. George, Utah: Could exploratory drillers and core drillers be included among uranium miners for compensation for diseases presumably resulting from exposure to radiation? In introducing such distinctions as *in the mine* and *at the mine*, in defining *employment at an aboveground mine*, and in examining *extraction* as a function of core drilling as well as mining, DOJ concluded that to include core drillers would be to effectively change legislation through regulation. DOJ states that "Extracting uranium ore from within a mine is one of the strict definitional limits that cannot be expanded by regulation." At the same time, DOJ clarified that because the 2000 amendments expanded the definition of *uranium mine* to include aboveground mines, miners who worked aboveground stockpiling ore and operating dump trucks would be eligible for compensation, assuming that they satisfy other eligibility criteria (DOJ, 2004, p. 13629).

In its hearing in Window Rock, Arizona, this committee heard persons testifying about their failed attempts at using affidavits in establishing proofs of eligibility. DOJ also clarified that matter in 79.4 of the final rule. Miners can substantiate uranium-mining employment history under some circumstances by affidavit, but millers or ore transporters are not permitted to use affidavits for these purposes (DOJ, 2004, pp. 13630-1). Nor is an affidavit allowed as proof of presence for downwinder claimants (DOJ responses to the committee's questions, March 16, 2004).

The final rule went into effect on April 22, 2004.

In sum, the rule (DOJ, 2004, p. 13628):

    1. "Describes the documentation required to establish proof of employment in a uranium mine or mill or as an ore transporter."

    2. "Describes the medical documentation necessary to establish the existence of renal cancer and chronic renal disease."

    3. "Revises the provision concerning attorney representation of claimant before the Department of Justice with respect to claims brought under the Act."

    4. "Incorporates the following revisions to the regulations:

        a. inserts a portion of Mohave County, Arizona, previously covered under RECA and erroneously stricken from the 2000 Amendments, as a radiation-effected (sic) area for downwinder claimants;

        b. clarifies the requirement that lung cancer be primary for all claimant categories;[5]

        c. adds a duration of employment standard as an alternative to a minimum radiation exposure standard for uranium miners;

        d. amends the documentation required to establish lung cancer for uranium miner, miller, and ore transporter claimants; and

        e. makes other minor revisions consistent with the Appropriations Authorization Act."

COMPENSATION IN THE RADIATION EXPOSURE COMPENSATION ACT

Under RECA, a person is to be compensated if he or she meets two criteria: the person is in a specific class defined by RECA, and the person has developed one of the specific cancers or other diseases specified by RECA. As noted above, the criteria for both the classes of persons and the specific compensable diseases have been modified since the original enactment of RECA in 1990.

---

[5]All cancers covered by RECA are primary cancers.

Populations Covered by the Radiation Exposure Compensation Act

As amended, RECA defines five radiation-exposed populations whose members are eligible for monetary compensation on documentation of specified exposure and disease requirements.

Map 2.1 identifies areas in which activities took place for four of the five populations covered by RECA (onsite participants include those in regions outside the United States).

The following subsections describe the five populations and their requirements for compensation. The descriptions and maps included in this chapter are taken from the DOJ program summary for RECA sent to the committee in May 2004.

Uranium Miners. RECA 2000 specifies a payment of $100,000 to eligible individuals employed in an above-ground or underground uranium mine located in Colorado, New Mexico, Arizona, Wyoming, South Dakota, Washington, Utah, Idaho, North Dakota, Oregon, and Texas at any time during the period beginning on January 1, 1942, and ending on December 31, 1971.

A. Exposure. The claimant must have been exposed to 40 WLMs of radiation while employed in a uranium mine *or* worked for at least one year during the relevant time period.

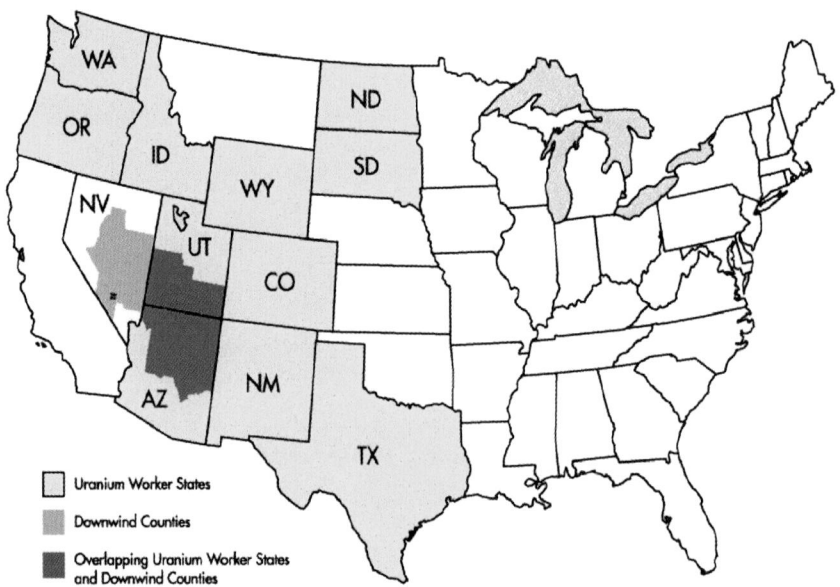

MAP 2.1   Areas covered by Radiation Exposure Compensation Act.

B. Disease. Compensable diseases include primary lung cancer and certain nonmalignant respiratory diseases. (See Table 2.1).

Uranium Millers. RECA 2000 specifies a payment of $100,000 to eligible individuals employed in a uranium mill located in Colorado, New Mexico, Arizona, Wyoming, South Dakota, Washington, Utah, Idaho, North Dakota, Oregon, and

TABLE 2.1   Populations and Diseases Eligible for Compensation under RECA

Diseases and Conditions	Uranium Miners	Uranium Millers	Ore Transporters	Down-winders	Onsite Participants
Malignant Neoplasms					
Bile ducts				√	√
Brain				√	√
Breast				√	√
Colon				√	√
Esophagus				√	√
Gall bladder				√	√
Leukemia[a]				√	√
Liver[b]				√	√
Lung cancer[c]	√	√	√	√	√
Multiple myeloma				√	√
Non-Hodgkins lymphomas				√	√
Ovary				√	√
Pancreas				√	√
Pharynx				√	√
Renal cancer		√	√		
Salivary gland				√	√
Small intestine				√	√
Stomach				√	√
Thyroid				√	√
Urinary bladder				√	√
Nonmalignant Conditions					
Chronic renal disease[d]		√	√		
Cor pulmonale[e]	√	√	√		
Pneumoconiosis	√	√	√		
Pulmonary fibrosis, fibrosis of lung	√	√	√		
Silicosis	√	√	√		

[a]Excluding chronic lymphocytic leukemia.

[b]Except when cirrhosis or hepatitis B is known to be included.

[c]Includes any physiologic condition of the lung, trachea, or bronchus that is recognized as lung cancer by the National Cancer Institute.

[d]Including nephritis and kidney tubal tissue injury.

[e]Relating to fibrosis of the lung.

Texas at any time during the period beginning on January 1, 1942, and ending on December 31, 1971.

A. Exposure. The claimant must have worked for at least one year during the relevant time period.

B. Disease. Compensable diseases include primary lung and renal cancer, certain nonmalignant respiratory diseases, and chronic renal disease including nephritis and kidney tubal tissue injury. (See Table 2.1).

Ore Transporters. RECA 2000 specifies a payment of $100,000 to eligible individuals employed in the transport of uranium ore or vanadium-uranium ore from mines or mills located in Colorado, New Mexico, Arizona, Wyoming, South Dakota, Washington, Utah, Idaho, North Dakota, Oregon, and Texas at any time during the period beginning on January 1, 1942, and ending on December 31, 1971.

A. Exposure. The claimant must have worked for at least one year during the relevant time period.

B. Disease. Compensable diseases include primary lung cancer, certain non-malignant respiratory diseases, renal cancer, and other chronic renal disease including nephritis and kidney tubal tissue injury. (See Table 2.1).

Downwinders. The Act specifies a payment of $50,000 to a civilian individual who was physically present in one of the affected areas downwind of the Nevada Test Site during a period of atmospheric nuclear testing, and who later contracted a specified compensable disease.

A. Exposure. The claimant must have lived or worked downwind of atmospheric nuclear tests in certain counties in Utah, Nevada and Arizona for a period of at least two years during the period beginning on January 21, 1951, and ending on October 31, 1958, or, for the period beginning on June 30, 1962, and ending on July 31, 1962. The designated affected areas are: in the State of Utah, the counties of Beaver, Garfield, Iron, Kane, Millard, Piute, San Juan, Sevier, Washington, and Wayne; in the State of Nevada, the counties of Eureka, Lander, Lincoln, Nye, White Pine, and that portion of Clark County that consists of townships 13 through 16 at ranges 63 through 71; and in the State of Arizona, the counties of Apache, Coconino, Gila, Navajo, Yavapai, and that part of Arizona that is north of the Grand Canyon. Portions of Mohave County, Arizona are reincluded in the 2000 Amendments.

Map 2.2 locates those counties.

B. Disease. After such period of physical presence, the claimant developed one of the following specified diseases: leukemia (other than chronic lymphocytic leukemia), multiple myeloma, lymphomas (other than Hodgkin's disease), and primary cancer of the lung, thyroid, male or female breast, esophagus, stomach, pharynx, small intestine, pancreas, bile ducts, gall bladder, salivary gland, urinary bladder, brain, colon, ovary, or liver (except if cirrhosis or hepatitis B is indicated). (See Table 2.1).

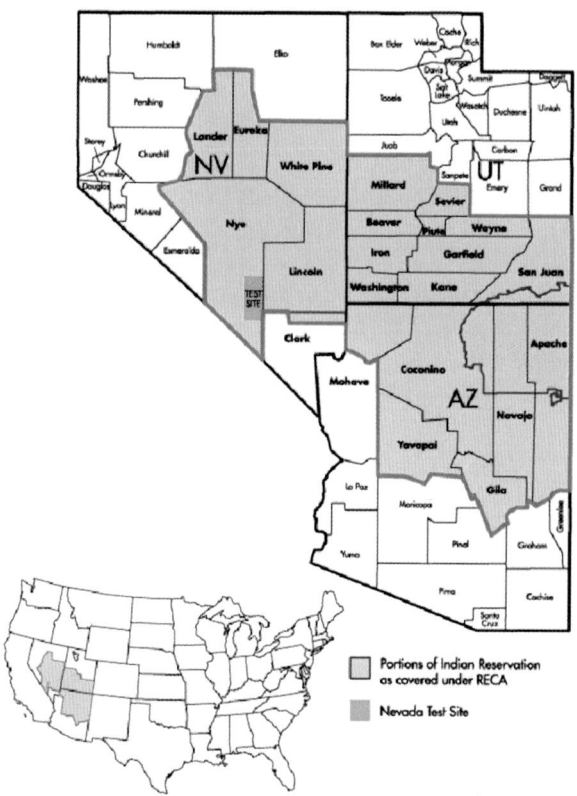

MAP 2.2   Downwinder counties.

Onsite Participants. The Act specifies a payment of $75,000 to individuals who participated onsite in a test involving the atmospheric detonation of a nuclear device, and who later developed a specified compensable disease.

A. Exposure. The claimant must have been present "onsite" above or within the official boundaries of the Nevada, Pacific, Trinity, or South Atlantic Test Sites at any time during a period of atmospheric nuclear testing and must have "participated" during that time in the atmospheric detonation of a nuclear device.

B. Disease. After the onsite participation, the claimant developed one of the following specified diseases: leukemia (other than chronic lymphocytic leukemia), lung cancer, multiple myeloma, lymphomas (other than Hodgkin's disease), and primary cancer of the thyroid, male or female breast, esophagus, stomach, pharynx, small intestine, pancreas, bile ducts, gall bladder, salivary gland, urinary bladder, brain, colon, ovary, or liver (except if cirrhosis or hepatitis B is indicated). (See Table 2.1).

Classes of Diseases Covered

As noted above, the compensable-claim categories of RECA are further defined by whether members of the five populations contracted specific diseases related to their exposures, which are defined by the pathologic conditions that are recognized by NCI. The age of a claimant at the time of initial exposure is relevant in some diseases. The number of years after which onset of disease occurred is also relevant.[6] All the specific diseases for each claimant category covered by RECA are represented in Table 2.1.

Nature and Amount of Compensation

The compensation provided by Congress that is nontaxable is described as partial. "Partial" is undefined in the legislation. Congress may have intended to state that the compensation offered constitutes to only partial restitution. Restitution or rectification is part of the ethical framework that undergirds RECA and, although judged only partial, may function as part of a large package of full restitution. We discuss that in Chapter 8.

The partial compensation is provided in the form of "compassionate payments" or monetary awards. RECA does not cover direct delivery of medical services or payment for medical services, medical insurance premiums, or insurance deductibles. However, certain people who are eligible for compensation under RECA may also be eligible for medical care under other auspices, including other compensation legislation that we discuss below.

Table 2.2 lists the compensatory amounts that an eligible person receives under RECA. A person may receive compensation under only *one subpart* of the regulations for the illness that he or she contracted. A person who contracted more than one of the eligible diseases or contracted a separate disease under different circumstances (for example, was a miner and also worked in a uranium mill) may be compensated only once.

If a person eligible for compensation under a compensable claim category is deceased, his or her surviving beneficiaries may apply to receive compensation on his or her behalf. Those beneficiaries include, in the following order (if a prior beneficiary is deceased), the spouse, a child, a parent, a grandchild, and a grandparent. The benefit is shared equally among the members of each class of benefi-

---

[6]See 28 CFR Part 79 for the details regarding age of claimant and years to onset for each disease.

TABLE 2.2 Partial Compensation under RECA by Population

Population	Partial Compensation
Uranium miners	$100,000
Uranium miller	$100,000
Uranium ore transporter	$100,000
Onsite participant	$75,000
Downwinder	$50,000

ciaries. If the spouse is deceased, compensation is shared among the surviving children; if all the children are deceased, compensation is split between the parents; if both parents are deceased, the grandchildren divide the compensation among themselves; finally, if grandparents outlive all the other beneficiaries, they share the compensation among themselves. If any beneficiary refuses the relevant portion, it is returned to the trust fund rather than being distributed among the remaining beneficiaries in the class (28 CFR Part 79.71).

If a person is both a claimant because he or she contracted one of the diseases in a compensable claim category and a claimant as an eligible survivor, the person may receive more than one payment—one payment for his or her own illness and one payment for each instance of qualifying as an eligible surviving beneficiary (28 CFR Part 79.75).

Claims to Date

The absolute sizes of the five compensable populations are difficult to know. NIOSH and DOJ estimate that 20,000 underground uranium miners may have worked during the period covered by RECA. The current committee is unaware of any published estimates of the sizes of the other four populations or the number of aboveground uranium miners.

The relative sizes of the populations likely to file for compensation can be gauged by using the claims processed. Table 2.3 shows the numbers of claims filed to November 21, 2004 for the four occupational categories of eligible persons and the downwinder exposure category; the numbers of claims approved, denied, and pending for each category; and the amounts of approved compensation.

The table shows that uranium miners and especially onsite participants have higher rates of denial than the entire group of claimants and lower rates of approval; ore transporters have a notably high percentage of pending claims. Downwinders have a substantially higher rate of approved claims and lower rate of denied claims than the entire group.

The 2000 amendments require a report to Congress from the Government Accountability Office (GAO), formerly the General Accounting Office, every 18 months that contains a detailed accounting of DOJ's administration of RECA. The

TABLE 2.3   Total Numbers and Percentages of RECA Claims Approved, Denied, and Pending as of November 21, 2004[a]

Population	Total Claims[b]	Claims Approved (% of total)	Claims Denied (% of total)	Claims Pending (% of total)	Total Amount of Approved Compensation[a]
Uranium miner	5,824	3,130	2,089	605	$312,391,500
($100,000)	(28%)	(54%)	(36%)	(10%)	(37%)
Uranium miller	670	409	124	137	$40,900,000
($100,000)	(3%)	(61%)	(19%)	(20%)	(5%)
Ore transporter	155	96	33	26	$9,600,000
($100,000)	(1%)	(62%)	(21%)	(17%)	(1%)
Onsite participant	2,130	758	1,095	277	$54,437,350
($75,000)	(10%)	(36%)	(51%)	(13%)	(7%)
Downwinder	12,047	8,361	2,546)	1,140)	$418,020,000
($50,000)	(58%)	(69%)	(21%)	(9%)	(50%)
Total	20,826	12,754	5,887	2,185	$835,348,850
		(61%)	(28%)	(10%)	

[a]Source: http://www.usdoj.gov/civil/omp/omi/Tre_SysClaimsToDateSum.pdf, accessed November 21, 2004.
[b]Percentages in first and last columns refer to column totals; percentages in middle columns refer to claims by claim type.

report must contain an analysis of claims, awards, and administrative costs and a budget for DOJ RECA activities or the Radiation Exposure Compensation Program (RECP). GAO submitted its first report on September 17, 2001 (GAO, 2001). GAO then conducted a review of DOJ's administration of RECA from August 2002 through February 2003 and produced a report in April 2003 (GAO, 2003).

In the April 2003 report, GAO noted that pending claims, processing time, and payments of awards had all increased dramatically, in some cases by 300%. Both the Congressional Budget Office (CBO) and DOJ estimated that the trust fund used to cover these costs was underfunded. GAO recommended that the attorney general develop a strategy to address the underfunding for current and projected claims over the period 2003-2011 (GAO, 2003).

In 2004, Congress passed the Consolidated Appropriations Act, 2005, which provided an additional $27.8 million to the Radiation Exposure Compensation Trust Fund in FY 2005, and beginning in FY 2006, makes funding for the Trust Fund a mandatory and indefinite appropriation (PL 108-447).[7]

---

[7]This sentence was added and a quote from the 2003 GAO report was deleted from the report after the prepublication draft was released, to acknowledge the provisions of the Consolidated Appropropriations Act, 2005.

OTHER COMPENSATION PROGRAMS

At least four pieces of federal legislation concerning radiation exposure and the development of nuclear weapons have been enacted in the last two decades that provide some form of compensation. It is helpful to present information about those programs together. Those programs may apply to people who fall into the RECA's compensable-claim categories. Hence, *additional* compensation in the form of payments or medical services is available to some of the same classes of people through other programs. And in some instances RECA compensation is offset by other claims, awards, and payments, including compensation program payments.[8] Offset payments do *not* include claims for worker's compensation.

Two compensation programs apply to military veterans. They sprang from the Veterans Dioxin and Radiation Exposure Compensation Act of 1984 (PL 98-542) and Radiation Exposure Veterans Compensation Act (REVCA) of 1988 (PL 100-321). They are administered by the Department of Veterans Affairs and the Defense Threat Reduction Agency of the Department of Defense. The populations that these programs cover include the onsite participants described by RECA. The magnitude of the awards is described as based on a "complex award formula, not amenable to simple monetary quantification" (Walchuk, 2002; NRC, 2003b). Payments to a military veteran may include disability payments or compensation benefits and any dependency and indemnity compensation payments to survivors. Those awards, settlements, or payments[9] will be subtracted, according to their actuarial present value, from payment under RECA. Additionally, "under certain conditions, veterans who are retroactively awarded service connection may qualify for reimbursement of certain medical expenses (a family member could file on behalf of the deceased veteran) back to the date of the original claim filing. These cases would be referred to as "not previously authorized" claims and must meet generally three conditions: 1) treatment was for a service connected condition or for a condition held to be aggravating an adjudicated service connected disability; 2) a medical emergency; and 3) VA or other

---

[8]These claims must be based on adverse health effects incurred by the claimant on account of the radiation exposures defined above. A claim "includes but is not limited to any request or demand for money made or sought in a civil action or made or sought in anticipation of the filing of a civil action, but shall not include requests or demands made pursuant to life insurance or health insurance contract." (28 CFR Part 79.75). Any such award or settlement payment is subtracted from the payment under RECA.

[9]Those payments do not include active-duty pay; retirement pay; retainer pay; survivor-benefits plan payments, such as death gratuities, or mortgage, life, or health insurance payments, burial benefits or reimbursement for burial expenses, loans or loan guarantees, education benefits and payments, vocational rehabilitation benefits and payments, medical, hospital and dental benefits, or commissary and post exchange privileges.

federal facilities not feasibility available" (correspondence with Department of Veterans Affairs, March 31, 2005).

A third compensation program, based on the Energy Employees Occupational Illness Compensation Program Act of 2000 (EEOICPA) applies to Department of Energy workers and contractor employees. The Department of Labor (DOL) administers the program. DOJ provides information about the program to those classes of people who are also covered by RECA who might be eligible. If a person or his or her survivor had been approved under RECA and was awarded $100,000 in compensation, the recipient must separately file an EEOICPA claim form with DOL. RECA-eligible people do not have to meet the dose requirements (having already met the WLM or 1 year employment criteria), based on the probability of causation and an assigned share, as do other energy employees filing directly with EEOICPA. However, under EEOICPA (but not RECA) a uranium miner, miller, or ore transporter may be eligible for medical benefits related to the condition for which he or she had been approved under RECA. Those benefits begin on the date of filing with DOL and, unlike REVCA, "medical payments are made only to a living energy employee if the claim is approved; no reimbursement is made for prior expenses, and no medical payments are made to survivors if the energy employee is deceased" (correspondence with Department of Labor, April 6, 2005). Downwinders and onsite participants who have been awarded compensation through RECA are not eligible for compensation under EEOICPA.

Table 2.4 illustrates the relationship between RECA compensation payments and those provided by other compensation programs that are also available to some classes of people covered by RECA

When presented together as in the table below, one can easily see the similarities and differences between four radiation exposure compensation programs as well as the relations among the programs. For example, in the area of medical benefits, a uranium worker may elect to claim medical benefits under EEOICPA. Some uranium workers and the class of downwinders do not receive medical benefits under any of the exposure compensation plans. They may receive them directly from the Indian Health Service or through other medical-services delivery mechanisms that may be covered by private medical insurance (for which they would have to pay premiums, copays, deductibles or coinsurance). Clearly, some people who are eligible for RECA's compassionate payments are likely to lack coverage or be underinsured for RECA disease-related medical care.

TABLE 2.4 Comparison of United States Radiation Exposure Compensation Programs

Properties	Radiation Exposure Compensation Act (RECA 1990)	Veterans Dioxin and Radiation Exposure Veterans Compensation Act (1984) and Radiation Exposure Compensation Act (REVCA 1988)	Energy Employees Occupational Illness Compensation Program (EEOICPA 2000)
Legislation/Dates	Public Law 101-426, October 1990; PL 101-510, PL 106-246, July 2000; PL107-107, December, 2001; PL 107-273, November, 2002	Public Law 98-542, January 1984; PL 100-321, May 1988; and PL 102-578, January 1992	PL106-398, October 30, 2000; PL 108-375, October 28, 2004
Administration	Department of Justice	Department of Veterans Affairs	Department of Labor, Energy and Health and Human Services
Specified Diseases	Depending on populations, diseases of leukemia (other than chronic lymphocytic leukemia), cancer of the brain, bile duct, breast, colon, esophagus, gall bladder, liver, lung, multiple myeloma, non-Hodgkin's lymphomas, ovary, pancreas, pharynx, renal cancer, salivary gland, small intestine, stomach, thyroid, urinary bladder, and chronic renal disease, cor pulmonale, pneumoconiosis, pulmonary fibrosis, fibrosis of the lung, silicosis	Presumptive Diseases: leukemia (other than chronic lymphocytic leukemia), cancer of the bile ducts, bone, brain, breast, bronchioloalveolar, colon, esophagus, gall bladder, small intestine, liver, lung, lymphomas (except Hodgkin's disease), multiple myeloma, ovary, pancreas, pharynx, thyroid, salivary glands, stomach, cancer of the urinary track (kidneys, renal pelvis, ureter, urinary bladder and urethra. VA regulations cover a few additional conditions for medical-care access Non-Presumptive Disease: "VA regulations define all cancers as possibly caused by radiation.[a] Other non-malignant conditions might be caused by radiation. These conditions include, posterior subcapsular cataracts; non-malignant thyroid nodular disease; parathyroid adenoma; and tumors of the brain and central nervous system."[b]	A specified disease, as that term is defined in section 4(b)(2) of the Radiation Exposure Compensation Act, including diseases of leukemia (other than chronic lymphocytic leukemia), cancer of the brain, bile duct, breast, colon, esophagus, gall bladder, liver, lung, multiple myeloma, non-Hodgkin's lymphomas, ovary, pancreas, pharynx, renal cancer, salivary gland, small intestine, stomach, thyroid, urinary bladder, and chronic renal disease, cor pulmonale, pneumoconiosis, pulmonary fibrosis, fibrosis of the lung, silicosis; also chronic beryllium disease

TABLE 2.4 Continued

Properties	Radiation Exposure Compensation Act (RECA 1990)	Veterans Dioxin and Radiation Exposure Veterans Compensation Act (1984) and Radiation Exposure Compensation Act (REVCA 1988)	Energy Employees Occupational Illness Compensation Program (EEOICPA 2000)
Exposure Eligibility Criteria	Section 4: Based on residency in a designated county if down-winder; based on employment during specified nuclear detonation periods if onsite participant Section 5: Working Level Month or length of employment	For presumptive diseases, based on exposure in certain situations (see below); For non-presumptive diseases, the upper 99 percent interval of the probability of causation at 0.5 plus other considerations; an assessment as to the size and nature of the radiation dose or doses to be considered in determining exposure to ionizing radiation include: the probable dose, taking into account any known limitations in the dosimetry devices used, the relative sensitivity of the involved tissue to induction, by ionizing radiation, of the specific pathology, the veteran's gender and pertinent family history, age at time of exposure; the time-lapse between exposure and onset of the disease; and the extent to which exposure to radiation, or other carcinogens, outside of service may have contributed to development of the disease	Based on the radiation dose received by the employee(s) at facility and the upper 99 percent interval of the probability of causation at 0.5 in the radioepidemiological tables published under section 7(b) of the Orphan Drug Act (42 U.S.C. 241 note), as such tables may be updated under section 7(b)(3) of such Act from time to time; If, and only if, the cancer specified was at least as likely as not related to employment take into consideration health-related factors (for example, smoking)
Eligible Populations	Uranium miners Uranium millers Uranium ore transporter Onsite participant Downwinder	For presumptive Diseases: Veterans who participated in atmospheric nuclear tests by the U.S., as well as certain underground tests at Amchitka Island, Alaska, prior to January 1, 1974, who served with the U.S. occupation forces in Hiroshima or Nagasaki, Japan, between August 1945 and July 1946, who were prisoners of war in Japan, or some who served at the gaseous diffusion plants in Paducah, KY, Portsmouth, OH, and area K25 at Oak Ridge, TN. Non-Presumptive Diseases: Radiation-exposed veterans with exposure situations or conditions other than mentioned above may be eligible for compensation.	Employees of the Department of Energy or Department of Energy contractors or subcontractors; RECA eligible populations under Section 5, including uranium miners, millers and ore-transporters and certain individuals under Section 4, including onsite participants and downwinders Special Exposure Cohort

Also, some veterans who received nasopharyngeal (NP) radium therapy have enhanced eligibility for medical care for cancers of head and neck but not for presumptive compensation

Amount of Compensation	$50,000–$100,000 depending on population Sec. 5: Miner — 100K Sec. 5: Miller — 100K Sec. 5: Ore Transporter — 100K Sec. 4: Onsite participant — 75K Sec. 4: Downwinder — 50K	The amount of money provided in the monthly payments depends on the degree of disability (and loss of the ability to work). The degrees range from 0 to 100 percent. They are expressed in 10 percent rating increments. A veteran with an 80 percent rating would receive considerably more than one with a 50 percent rating and much less than a veteran with a 100 percent rating. Compensation rates change frequently, and are set by Congress. The monthly rate in 2004 for a 50 percent disability rating was $646; for 80 percent, $1,195, and for 100 percent, $2,239."[c] For deaths in 1993 and after, survivors are paid a flat rate; additional amounts are based on other factors[f]	$150,000 lump sum payment Federal payments instead of worker's compensation (see Part E of EEOICPA) $50,000 added to RECA eligible populations under Section 5 and certain individuals under Section 4 of RECA
Multiple Programs	May also receive under EEOICPA	Veterans who were onsite participants may also receive compensation under RECA	May not receive full compensation under both RECA and EEOICPA[e]
Diagnostics Costs	Not covered	Covered	Not covered

TABLE 2.4 Continued

Properties	Radiation Exposure Compensation Act (RECA 1990)	Veterans Dioxin and Radiation Exposure Veterans Compensation Act (1984) and Radiation Exposure Compensation Act (REVCA 1988)	Energy Employees Occupational Illness Compensation Program (EEOICPA 2000)
Medical Benefits for Approved Condition	No	Eligible for VA treatment, including hospital care, medical services, and nursing home care under Priority Category 6, not subject to copayments. Veterans who received nasopharyngeal (NP) radium therapy are also eligible for VA treatment as related to cancers of the head and neck but cannot enroll under Priority Category 6. For other veterans with occupational exposures, there is no special eligibility for VA treatment but they may receive VA treatment under the VA health care system.	Yes—the services, appliances, and supplies prescribed or recommended by a qualified Physician, including transportation and other incidental expenses, for specified occupational illnesses; Beryllium sensitivity monitoring A covered uranium employee shall receive medical benefits for the illness for which that employee received $100,000 under section 5 of RECA
Medical Benefit Provider	Not applicable	Department of Veteran Affairs	United States medical officers and hospitals, or, at the individual's option, by or on the order of physicians and hospitals designated or approved by the President.
Offset of Compensation by:	Final awards or settlement on a claim against any person, not including worker's compensation.	Final Awards or settlement on a claim against any person; "Payment to any individual under the provisions of the Radiation Exposure Compensation Act (RECA) based upon disability or death resulting from a specific disease shall bar payment, or further payment, of compensation or dependency and indemnity compensation to or on behalf of that individual based upon disability or death resulting from the same disease."	Final awards or settlement on a claim against any person, not including worker's compensation and insurance payments

Survivor Benefits	Yes; no age or dependency restriction	Yes. Age and dependency restrictions apply	Yes; age and dependency restrictions apply
Total $ Awarded by 12/31/2004	$853,081,387[f]	Not available	$975,640,617[g]

[a]Radiogenic disease shall not include polycythemia vera.

[b]Ionizing Radiation Review, Vol 1, No. 1, February, 2004 at http://www1.va.gov/irad/docs/IRADnewsletterFeb04.pdf, accessed January 5, 2005.

[c]VA RADIATION PROGRAMS INFORMATION for Veterans Health Administration (VHA) Environmental Health Clinicians/ Coordinators, April 15, 2003 at http://www1.va.gov/irad/docs/VARADPROGRAMSINFORMATION.doc, accessed January 5, 2005.

[d]38 CFR 3.715, page 298 (Authority: 42 U.S.C. 2210 note) [58 FR 25564, Apr. 27, 1993]

[e]Election to accept payments under section 4 of RECA (either $50,000 for "downwinders" or $75,000 for "onsite participants") will disqualify one from receiving any benefits under EEOICPA for which one may be eligible. If a "downwinder" or "onsite participant" claimant accepts payment under section 4 of RECA, he or she will not be eligible to receive any payments under the EEOICPA.

[f]DOJ http://www.usdoj.gov/civil/omp/omi/Tre_SysClaimsToDateSum.pdf, accessed January 5, 2005.

[g]DOL http://www.dol.gov/esa/regs/compliance/owcp/eeoicp/WeeklyStats.htm, accessed January 5, 2005.

CONCLUSION

This concludes our description of the background and historical develop-ment of RECA and our description of the various aspects of compensation that it includes. In an upcoming chapter (Chapter 8) we explore the ethical framework that undergirds the compensation legislation. Such issues as equity in compensa-tion and the ethics of screening are treated, along with other ethical concerns that surfaced in our public meetings with stakeholders. We insert this discussion after our recommendations regarding the most recent scientific information related to radiation exposure and associated cancers or other diseases (Chapters 4 and 7) and whether other groups of people or additional geographic areas should be covered under the Radiation Exposure Compensation Act (RECA) program (Chapters 5 and 6).

Those recommendations are based in sound science. Those in Chapters 9, 10 and 11 are grounded in evidence-based medical practice and studies regarding effective educational strategies. We turn now to the scientific matters about which the committee has been charged.

3

Basic Concepts in Radiation Physics, Biology, and Epidemiology

The scientific issues related to radiation and associated health effects are complex and may be confusing for persons not professionally involved with them. The topics are even more complicated in the context of the Radiation Exposure Compensation Act (RECA) and the Radiation Exposure Screening and Education Program (RESEP). This chapter will give concerned readers an opportunity to become familiar with the terminology and concepts used in the radiological sciences. It is limited to scientific topics directly related to the basic charge presented to the committee. The chapter is divided into three sections. The first presents the principles of physics related to ionizing radiation. The second presents the biology necessary for understanding how radiation affects cells and the mechanisms of radiation injury and repair. The third section describes the methods used to identify and measure the risks to persons who are exposed to radiation.

RADIATION PHYSICS

Definition of Radiation

Observable matter is made up of discrete components known as atoms and molecules. Atoms are divisible into particles, such as electrons, protons, and neutrons. Other elementary particles are part of the fabric of nature, but they are more elusive and do not directly form stable atoms or molecules. When a particle or group of particles is accelerated, it can reach high energies and travel a large distance in a very short time. Radiation can be defined as any collection of elementary particles that have sufficient energy to interact with and transfer some of their energy to objects or materials that intercept their path.

Ionizing Radiation

Many different types of interactions can take place when radiation strikes an object. For instance, atoms in an irradiated object are neutral; they each consist of a positively charged nucleus (made up of protons and neutrons) surrounded by negatively charged electrons. The process of removing an orbital electron from an atom is called ionization.

Some types of radiation can transfer energy in a manner that creates ionization in the object. X rays and gamma rays are particles called photons that can create ionization. Microwaves, ultraviolet radiation, visible light, and infrared are also photons, but they do not result in ionization and are referred to as nonionizing radiation.

Ionization created by radiation in living systems can have unique biologic consequences that are different from those caused by nonionizing radiation. RECA is related specifically to diseases found to have an association with exposure to ionizing radiation.

The process that accelerates particles to form radiation can occur naturally. For example, the sun continuously emits particles that reach the atmosphere and result in a continuous shower of elementary particles on the surface of the earth. Some sources of radiation are man-made, such as x-ray machines, particle accelerators used for cancer therapy, and nuclear power reactors used to generate electricity.

Radioactivity

Radioactivity is another important source of ionizing radiation. Every element such as hydrogen, oxygen, or iron are defined by the number of protons in the nucleus. However, atoms of the same element can have a different number of neutrons in the nucleus. These are called isotopes. Isotopes are identified by the name of the element and the total number of protons and neutrons in the nucleus. For example, the element hydrogen has one proton, ^{1}H. There is another isotope of hydrogen with one proton and one neutron, ^{2}H, called deuterium and also one proton and two neutrons, ^{3}H, called tritium. Some nuclei are unstable, and these can transform (decay) into more stable nuclei by emitting particles—a process called radioactive decay. The emitted particles are a form of radiation originating from radioactivity.

Every element in the periodic chart has at least one isotope that is radioactive. For instance, sodium-23 (^{23}Na) is stable, but sodium-22 (^{22}Na) and sodium-24 (^{24}Na) are radioactive; similarly, iodine-127 (^{127}I) is stable, and iodine-131 (^{131}I) is radioactive. A salt containing natural potassium will always contain some radioactive potassium-40 (^{40}K). Potassium is an essential mineral in our diet. Some of the ingested potassium is absorbed in tissue. That process is not limited to potassium, but can occur with iodine, sodium, radium, and so on. Therefore, all persons contain some radioactivity.

Each radioactive isotope has unique properties. One property is the type of particles emitted, and another is the energy of the particles emitted. No two radioactive isotopes emit the same combination of particles and energies. Therefore, one can identify the presence of a specific isotope at a given location by measuring the types and energies of the emitted particles.

Radioactive decay is a random process: it is impossible to determine when a given nucleus will decay. However, it is possible to estimate how many nuclei in a group will decay during a given period. The half-life of an isotope is the time it takes for half the nuclei in a group or sample to decay. Thus, isotopes with short half-lives decay rapidly and those with long half-lives decay more slowly. No two isotopes have the same half-life. For example, the half-life of nitrogen-16 (^{16}N) is 7.3 seconds; that of radon 222 (^{222}Rn), 3.8 days; that of ^{131}I, 8 days; and that of uranium-238 (^{238}U), 4.5 billion years.

Radioactivity specifically refers to the rate at which decays occur. The amount of radioactivity present depends on the number of radioactive atoms and their corresponding half-life. The rate at which atoms are decaying is proportional to the number of atoms divided by the half-life. This decay rate is described in units of either Becquerels (Bq) in the International System, SI, of units or Curies (Ci) in the traditional system of units used in the United States; 1 Bq is equal to 1 decay per second, and 1 Ci is equal to 37 billion decays per second. The amount of radioactivity is often stated in terms of a millicurie (mCi), which is one thousand times smaller than a Curie. One microcurie (μCi) is one million times smaller than a Curie and one picocurie is one trillion times smaller than a Curie. The amount of radioactivity at any time is reduced by one-half in a period of time equal to one half-life.

Radioactivity generates radiation by emitting particles. Radioactive materials outside the body are called external emitters, and radioactive materials located within the body are called internal emitters.

Types of Ionizing Radiations

Radioactive nuclei can emit several kinds of particles, but there are three primary types: alpha particles (α), beta particles (β), and photons that are either x rays or gamma rays (γ). Several properties distinguish those particles from one another. One is electric charge; alpha particles are emitted with a positive charge of 2, beta particles are emitted with either 1 negative charge (electron) or 1 positive charge (positron), and x rays and gamma rays have no charge and are thus neutral.

Another important property is penetration of the particles through matter. Alpha particles lose energy rapidly and stop in a very short distance. Most travel no more than 3-5 centimeters in air and only about 30-50 microns in water or tissue. They cannot penetrate clothes or skin. Alpha particles must be emitted very close to biologic targets to produce an effect. External alpha emitters therefore are generally not considered to pose a health hazard. However, radioactive materials can enter the body through inhalation, ingestion, or transfer through cuts and wounds. Some of this radioactive material passes through the body and is eliminated, and some remains in tissues that might contain radiosensitive cells. The distribution of the radioactive material in the body depends on the chemistry of the radioactive element. For example, radium has chemical properties similar to those of calcium, and the alpha-particle emitter radium-226 (^{226}Ra) will accumulate with calcium in bone.

Beta particles are electrons that lose energy rather slowly when passing through materials. A high-energy beta particle can travel several centimeters through water and tissue. Lower-energy beta particles travel some fraction of that distance. External emission of low-energy beta particles, as in the decay of tritium, which is an isotope of hydrogen (^{3}H), or carbon-14 (^{14}C) is not considered a health hazard, whereas external emission of high-energy beta particles from strontium-90 (^{90}Sr) reach some regions of the body that are sensitive to radiation. As in the case of alpha-emitters, the distribution of internal beta-emitters depends on the chemistry of the radioactive element. Strontium has chemistry similar to that of calcium, and ^{90}Sr will accumulate in bone. Most of the iodine in the body that is not excreted will accumulate in the thyroid. Beta particles from ^{131}I can originate in the thyroid and deposit most of their energy there.

Photons can be very penetrating. High-energy x rays and gamma rays travel many meters in air and through many centimeters of concrete, iron, and tissue. Thus, external gamma rays can penetrate and deposit energy throughout the body. The distribution of internal gamma-emitters depends on the chemistry of the radioactive element. Internally emitted gamma rays can deposit energy in the tissue of residence or neighboring tissues. For example, cesium-137 (^{137}Cs) deposited in soft tissues, and the entire body is exposed uniformly to gamma rays.

Radiation Measurements and Units

Radiation can be described and measured in many ways. For purposes of radiobiology and radiation protection, the concept of absorbed dose, D, is most commonly used. It does not measure each particle but describes the energy deposited in a specified region. Absorbed dose is the energy absorbed in a volume of material divided by the mass of the material. It is the result of the physical interactions of the ionizing radiation within the volume of material. An absorbed dose can be delivered by any type or combination of types of radiation in any type of material.

The units of absorbed dose are the gray (Gy) in the SI and the rad in the traditional system often still used in the United States; 1 Gy is equivalent to 100 rad. The centigray (cGy) is a unit of convenience often used in cancer therapy that is equivalent to 1 rad.

Dose rate refers to the distribution of dose as a function of time. It can be expressed as Gy per second (Gy s^{-1}), per minute (Gy min^{-1}), per hour (Gy h^{-1}), and per year (Gy y^{-1}). A protracted dose is one received over a long period of time. A given dose delivered within 1 h often will have different consequences than the same total dose delivered over a period of one year. In some cases, if the dose rate is constant for long periods, it is referred to as continuous exposure to radiation. A dose rate can change with time; radiation could occur in the form of random pulses or vary periodically.

Dose fractionation describes the case in which a dose is delivered in segments or fractions over a specified period. For example, in radiation therapy for cancer, a total dose of 50 Gy might be delivered at a high dose rate of 2 Gy min^{-1} for only 1 minute per day over a period of 25 days (5 weeks, excluding weekends).

Equivalent Dose

The concept of absorbed dose, D, was created to estimate biologic effects of ionizing radiation. Scientists hoped that absorbed dose could serve as a universal predictor of biologic effects and corresponding risks to humans from exposure to ionizing radiation. However, it was soon discovered that similar doses of radiation from different particles produced different amounts of biologic damage. In some cases, up to 1 Gy of gamma rays is needed to produce the same effect as 0.1 Gy of alpha particles. That was observed for many biologic systems and was ultimately referred to as relative biological effectiveness (RBE).

RBE is related to the density or rate of ionization produced by a particle as it passes through matter. Linear energy transfer, LET, is a measure of the rate of energy loss and therefore ionization along the track of a particle. Alpha particles have short tracks, but create large amounts of ionization along the track and are referred to as high LET radiation. Electrons and beta particles are sparsely ionizing and are referred to as low LET radiation. X rays and gamma rays create electrons when they interact in materials and are also considered to be low LET radiation. To a first approximation, RBE increases with LET.

Rules for and regulation of radiation protection of humans must be related to the risks associated with exposure to ionizing radiation. RBE makes it impossible to base a system of regulations on absorbed dose alone. It was necessary to include the type of radiation in a consistent manner that reflected changes in the biology as well as the physics. For this reason, the concept of equivalent dose was established for purposes of radiation protection. Equivalent dose (H_T) in a tissue or organ, T, is the product of absorbed dose averaged within a tissue (D_T) and a radiation weighting factor (w_R), and thus $H_T = D_T \times w_R$.

The radiation weighting factor is used to adjust the absorbed dose to reflect the RBE for radiation of type R. It is thus related to LET. Alpha particles have a w_R of 20. Beta particles, x rays and gamma rays have a w_R of 1.0. Equivalent dose is described in sievert (Sv) or rem.

Effective Dose

Some tissues and organs are more sensitive to radiation than others. When the entire body is irradiated uniformly, all organs receive a dose and contribute to the total risk of a health effect, such as cancer. In some cases, particularly with internal emitters, only one or two organs receive a dose, and the other organs are not at risk. When one needs to know the combined risk for such a case, it is necessary to include a factor that is related to the risk to each of the exposed organs. The equivalent dose, H_T, in each tissue, T, is multiplied by a tissue-weighting factor, w_T. The effective dose, E, is then the sum of $H_T w_T$ for all exposed tissues. Effective dose is a risk averaged dose that serves as a measure of risk including adjustments for both the type of radiation, w_R, and the tissues exposed, w_T. Effective dose is expressed in sievert (Sv) when the absorbed dose is measured in Gy, or in rem when the dose is measured in rads; 1 Sv = 100 rem.

The International Commission on Radiological Protection (ICRP, 1991) has made recommendations for values of w_T on the basis of the occurrence of cancer and hereditary effects observed in exposed populations. The currently accepted values are shown in Table 3.1.

One way to interpret Table 3.1 is for a large population of persons irradiated uniformly. Some people might develop cancer as a result of the absorbed dose received. The types of cancer associated with radiation would be distributed according to the fraction represented by w_T in Table 3.1. ICRP makes recom-

TABLE 3.1   Currently Recommended Tissue Weighting Factors, $w_T{}^a$

Tissue	w_T
Gonads	0.20
Bone marrow	0.12
Colon	0.12
Lung	0.12
Stomach	0.12
Bladder	0.05
Breast	0.05
Liver	0.05
Esophagus	0.05
Thyroid	0.05
Skin	0.01
Bone surfaces	0.01
Remainder[a]	0.05
Total	**1.00**

[a] w_T for the remainder is divided equally between adrenals, brain, upper large intestine, small intestine, kidney, muscle, pancreas, spleen, thymus, and uterus.

mendations for revising the values as new evidence on cancer incidence and tissue sensitivity becomes available.

Natural Background Radiation

All persons are exposed to ionizing radiation from natural sources. Sources of background radiation can be outside the body (external radiation) or inside the body (internal radiation). The primary contributions to external radiation from natural background are cosmic rays and penetrating gamma rays emitted by radioactive materials in rocks and soil, in particular ^{40}K, ^{232}Th and ^{238}U. The primary contributions to internal radiation from natural background are radioactive materials that enter the body through the diet—^{40}K, carbon-14 (^{14}C), ^{226}Ra— and inhaled radioactivity originating from ^{222}Rn.

Natural background radiation can have large variations. Exposure rates around the world depend on geography, geology, and housing environments. Table 3.2 shows a summary of the average annual effective dose received from natural background radiation by persons in the United States and the average received by persons residing near the mountains in the western part of the country (NCRP, 1987).

There are other exposures to ionizing radiation. The most common sources are medical examinations that prescribe diagnostic x rays and computed tomography (CT) scans. Table 3.3 shows the effective dose received from several types of diagnostic examinations (NCRP, 1987).

In addition to medical examinations, the general population may be exposed to radiation from industrial applications and consumer products. Figure 3.1 shows the relative contribution to effective dose for an average person in the United States from natural background and man-made sources (NCRP, 1987).

In Figure 3.1, *cosmic* refers to the contribution from external radiation from penetrating particles originating in the atmosphere. *Terrestrial* refers to the contribution from external gamma rays originating in radioactivity in soil, rocks, and building materials. *Internal* radiation refers to the contribution from radioactivity deposited throughout the body from diet and inhalation. *Radon* represents

TABLE 3.2   Average Annual Effective Dose Received by People in the United States from Natural Background Radiation

Source		United States (mSv/year)	Mountains (mSv/year)
External	Cosmic rays	0.3	0.6
External	Radioactivity	0.3	0.6
Internal	Radioactivity	0.4	0.4
Radon	Inhalation	2.0	3.4
	Total	3.0	5.0

TABLE 3.3   Effective Dose Received from Diagnostic Examinations of
Specific Organs and Tissues

Examination	mSv
Arms and legs	0.10
Chest	0.08
Pelvis	0.44
Upper gastrointestinal tract	2.40
Mouth (Dental)	0.03
Breast (mammography)	0.40
Head and body (CT)	1.11

the contribution from inhalation and deposition of radioactivity in the lung that
originates from radon gas. *Medical* represents the contribution from diagnostic
medical examinations. *Other* represents the contribution from man-made sources
of ionizing radiation, such as the nuclear-power industry and consumer products
(for example, smoke detectors, CRT monitors, porcelain, and tobacco).

Uranium

The original nuclear-weapons program depended on exploration, mining,
and milling of natural uranium. At that time, most of the known uranium deposits
were deep underground and required extensive mining operations that were labor-
intensive. As mentioned earlier, uranium is radioactive and has a very long half-
life. When uranium decays, it emits an alpha particle. The remaining nucleus,
thorium, is also radioactive. It promptly decays by emitting a beta particle. That
radioactive sequence continues for 13 decays until a stable isotope of lead is

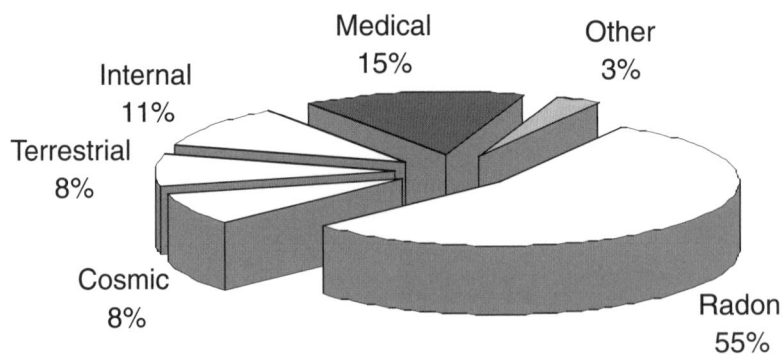

FIGURE 3.1   Relative contribution to average effective dose received by persons living
in the United States. The striped sections are from man-made sources of ionizing radiation. The other sections are from natural background radiation.

formed. Thus, alpha (α), beta (β), and gamma (γ) radiations are present in underground mines. Their presence can result in external exposure to gamma rays and internal exposure to alpha, beta, and gamma radiations from inhalation and unintentional ingestion of ore dust.

Risk to Miners

In general, the most hazardous exposure pathway for underground miners is not related to the ore dust itself or external gamma rays. About halfway through the uranium decay process, ^{226}Ra decays into ^{222}Rn. Radon is an inert gas that escapes from the rocks and begins to accumulate in the mine. It ultimately decays. This initiates a prompt series of decays that occur within a matter of minutes:

$$^{222}\text{Rn} \rightarrow \text{polonium-218 (}^{218}\text{Po)} \rightarrow \text{lead-214 (}^{214}\text{Pb)} \rightarrow$$
$$\text{bismuth-214 (}^{214}\text{Bi)} \rightarrow \text{polonium (}^{214}\text{Po)} \rightarrow$$

Those four short-lived radioactive descendants or decay products of radon have historically been called radon daughters. They can become suspended in air and are respirable. Inhaled radon is rapidly exhaled whereas radon decay products can deposit in the airways. Alpha particles emitted by ^{218}Po and ^{214}Po can deliver a large amount of energy and result in a large dose to cells in the airways. Those processes have been directly associated with the development of lung cancer in uranium miners. They are also the reasons for concern in family dwellings that have high concentrations of indoor radon from natural background.

The concentration of the short-lived decay products of radon is measured in working level (WL). A person's exposure at a given location is based on the concentration of decay products and the amount of time the person spends at the location. That exposure is expressed in working level month (WLM). For the purposes of this definition, 1 month is considered to be 170 h. Thus, 1 WLM is equivalent to 1 WL for 170 h, 2 WL for 85 h, 5 WL for 34 h, and so on. The risk of radiation-induced lung cancer is related to the exposure in WLM. For comparison purposes, 1 WLM delivers an effective dose of about 10 mSv (1 rem) to the trachea bronchial region of the respiratory tract.

Risk to Ore Transporters and Millers

After uranium ore is extracted from the mine, it is shipped to a mill, where it is crushed into fine sand and subjected to a chemical process to remove uranium selectively from the ore. The final product, uranium oxide (U_3O_8), often called yellowcake, is used for the production of weapons or as fuel for nuclear reactors. The remaining sands, called uranium mill tailings, are placed into a tailings pile close to the mill.

Yellowcake has a much higher concentration of uranium than the original ore removed from the mine. However, because uranium has an extremely long half-life, the radioactivity is not the principal hazard. The most serious hazard is heavy-metal chemical toxicity because of ingestion or inhalation.

Risk from Mill Tailings

The fine silt and sands in mill tailings contain all the other radioactive isotopes in the ore except the uranium. In effect, that represents all the radioactivity in the uranium decay series, including radon. Because radon is a noble gas, it can escape from the sands and is a potential route of exposure of persons residing near the mill or of exposure later when the tailings are used for construction or landfill around homes. Exposure to external radiation and radon decreases rapidly with distance from the tailings.

Nuclear Weapons and Fallout

Yellowcake is an oxide of natural uranium. Natural uranium consists of the isotopes ^{238}U (99.3%), ^{235}U (0.7%), and ^{234}U (trace). ^{235}U is necessary for a nuclear weapon. Thus, the yellowcake must undergo another process to increase the proportion of ^{235}U. That is called enrichment, and the desired product is enriched uranium. The remaining byproduct is called depleted uranium and is almost exclusively ^{238}U.

When a nuclear weapon is detonated, energy is released through a process called fission. Fission occurs when a heavy nucleus absorbs an additional neutron and then violently splits into two pieces and a few extra neutrons. If neutrons survive to produce another fission, the process can sustain itself. Weapons are designed to generate enough fissions to initiate an explosion in a fraction of a second following detonation. The first nuclear weapons released energy equivalent to 15,000 tons (15 kiloton) of TNT. Later versions used either ^{235}U or plutonium-239 (^{239}Pu) to produce fission yields over 1,000 kilotons.

The two nuclear fragments that remain after fission are called fission products. Many possible combinations of fragments can occur. One or both of the fission products can be radioactive. Some have very short half-lives and so decay within seconds or minutes. Others have half-lives of days (for example, ^{131}I) or years (for example, ^{90}Sr, and ^{137}Cs).

Fission products are propelled into the atmosphere by the force of the explosion. They can remain suspended and transported by winds. Eventually, the radioactive fission products settle back toward the surface of the earth and are called fallout. Fallout can be increased locally by precipitation wash-out (Beck, 2002; Bennett, 2002). Fallout can be responsible for both external and internal exposures of people in the vicinity. More than 150 fission products have half-lives greater than 1 h. Some of the important radioactive fission products in fallout and their principal exposure pathway are listed in Table 3.4.

Dosimetry

Dosimetry is the process of determining the effective dose received by persons exposed to ionizing radiation. The most accurate way to determine dose to an individual is to make measurements with a dosimeter assigned to each person. That is required today for radiation workers that might be exposed during routine occupational activities. Area monitors measure external radiation or radioactivity suspended in the air at specific locations. No dosimeter can directly measure the dose to the lung from the inhalation of radioactive materials, so area monitors are the principal instruments used for measuring and controlling internal exposure in underground mines.

Personal dosimeters were not available for all persons who might have been affected by fallout from atmospheric weapons testing. The US Atomic Energy Commission collected fallout on gummed film at more than 100 locations in the United States and its territories. The film was collected regularly and analyzed to estimate radioactivity deposited on the ground (Beck et al., 1990). The resulting data and weather patterns were used to create maps of fallout across the country.

Dose reconstruction is a computational process for estimating the dose to persons in situations when direct measurements are incomplete or unavailable. The National Cancer Institute has developed maps that show concentrations of radioactivity deposited in the United States from fallout during the period of atmospheric

TABLE 3.4   Some Important Fission Products in Fallout and Their Exposure Pathways. They are Ordered with Increasing Atomic Mass

Fission Product	Symbol	Half-Life	Emissions	External Exposure	Internal Exposure
Manganese	^{54}Mn	300 days	γ	√	
Strontium	^{89}Sr	52 days	β		√
Strontium	^{90}Sr	28 years	β		√
Zirconium	^{95}Zr	64 days	β,γ	√	
Zirconium	^{97}Zr	17 hours	β,γ	√	
Ruthenium	^{103}Ru	39 days	β,γ	√	
Ruthenium	^{106}Ru	368 days	β		√
Iodine	^{131}I	8 days	β,γ	√	√
Iodine	^{133}I	22 hours	β,γ	√	√
Cesium	^{136}Cs	13 days	β,γ		√
Cesium	^{137}Cs	300 days	β,γ	√	√
Barium	^{140}Ba	13 days	β,γ	√	√
Neptunium	^{239}Np	2.4 days	β,γ	√	
Plutonium	^{241}Pu	14.4 years	α		√

weapons testing. These data can be used to estimate the dose from internal and external radiation to persons living downwind of a test site. A more extensive description of these maps and dose calculators is presented in Chapter 4.

RADIATION BIOLOGY

When people are exposed to ionizing radiation from sources outside or inside the body, the radiation may interact with molecules in cells in their path. As described earlier in this chapter, some ionizing radiation can travel through a few or several layers of cells (beta-particle radiation) or through many cell layers into and through tissues deep within the body (x and gamma radiation), whereas alpha-particle radiation has short paths or tracks. The rate at which radiation loses energy along its tracks is referred to as linear energy transfer (LET) and depends on its track length. Thus, beta-particle radiation and the electrons associated with x and gamma rays, which are sparsely ionizing, are described as low-LET radiation, and alpha-particle radiation, which is densely ionizing, as high-LET radiation.

Biologic Actions of Ionizing Radiations

The main target of importance with respect to radiation damage is the deoxyribonucleic acid (DNA) in the cell's nucleus. The interactions between ionizing radiation and DNA can be direct or indirect.

Direct interactions occur when the radiation is deposited or transfers its energy directly to DNA. However, the probability of direct interactions is low because the volume of DNA is small relative to the total volume of the cell. Direct interactions occur more commonly when the radiation is of the densely ionizing type, such as alpha- or beta-particle radiation, than when it is less densely ionizing, such as gamma and x radiation.

Radiation interacts indirectly with DNA by first interacting with water molecules in the vicinity of the DNA, causing ionizations that result in the formation of free hydroxyl radicals. The free radicals can then diffuse to the vicinity of the DNA and can cause alterations in it. About 60% of the DNA damage caused by radiation is the result of indirect interactions. However, few of the many interactions that occur result in DNA damage, because most of the free radicals disperse and deposit their energy without interacting with DNA.

Biologic Sensitivity to Ionizing Radiations

An important concept in radiation biology is that the most rapidly dividing cells are the least well differentiated and are the most sensitive to radiation and thus are the most vulnerable to radiation-induced death and injury. The concept of radiosensitivity was formulated by Bergonie and Triboneau (1906). Some proliferative cells in the testis, red bone marrow, and intestinal mucosa, are among the most radiosensitive. Cells that divide more slowly, if at all, and cells that are highly differentiated, such as mature red blood cells and muscle and nerve cells, typically are relatively insensitive to radiation. Large lymphocytes (a type of white blood cell) divide more frequently than do small lymphocytes, but they are both highly sensitive to radiation. One of the earliest clinical effects of an acute whole-body dose of radiation—over about 250 mSv (25 rem)—in humans is a rapid fall in the number of large lymphocytes, beginning within 24 h. Because small lymphocytes divide infrequently radiation-induced changes in their DNA are more persistent, so aberrations in them can persist for many years after a large radiation dose (Goans et al., 2001).

A radiation dose delivered all at once or within a short period has a greater biologic effect than the same total dose delivered in small amounts over a period of weeks (fractionation) or in very small amounts continuously over a long period (protraction). In the latter cases, fewer cells are likely to be killed or lethally damaged at one time. DNA repair can proceed in the intervals between the successive exposures of a single cell from a fractionated exposure, or may be sufficient to counteract damage occurring during a protracted exposure, so that low dose rates allow for cell recovery or replacement.

Radiation-Induced Biological Damage

External exposure of the whole body or a substantial part of the body to penetrating radiation, such as gamma and x rays, can damage DNA in the cells of tissues deep within the body. External radiation dose is deposited independent of differential uptake in cells and sub-cellular regions due to ongoing local metabolic processes. Inhomogeneous dose distribution is more characteristic of internal emitters than external radiation sources. When an exposed person leaves the vicinity of an external source of radiation, no further dose is received from that source.

High-energy alpha- or beta-particles deposited on or close to the skin can penetrate the outer layers of dead and aging skin cells to reach the deeper or germinal layer in which cells are actively dividing. Radioactive particles that enter the body are distributed through many organs according to the nature of the metabolism of the particles, and the functions of the different organs. Only rarely are they distributed uniformly throughout the body; most are deposited in target tissues or organs; for example, ^{131}I, like stable iodine, targets the thyroid gland. The dose deposited in different organs is the best measure of radiation to use in correlations of internal dose with observed and expected effects. Doses to different organs from radioactive particles in the body are likely to be quite heterogeneous; large differences between organs are based on metabolic factors. Radioactivity taken into the body persists until it decays away or the radioactive element is eliminated from the body.

Repair of Radiation-Induced Damage

Repair of DNA damage caused by radiation from sources outside or inside the body is an effective, normal biologic process. This highly efficient repair process, which has evolved over many millennia, enables organisms, including humans, to survive and thrive despite constant exposure to background levels of radiation in the environment that in earlier millennia were much higher than they are now. However, ionizing radiation is more likely to damage both strands of DNA simultaneously than are normal metabolic processes. That is because ionizations may occur close together along the tracks of charged particles (electrons, protons, and alpha particles), thereby damaging both DNA strands and producing DNA double-strand breaks or other damage affecting both DNA strands in close proximity.

Repair of radiation-induced damage is usually complete and accurate, restoring damaged DNA to its full function. But if the damage is irreparable and the cells die immediately or are unable to divide to produce new cells of the same type, cell systems become depleted; and if the rate of depletion exceeds the rate at which the body can replace the lost cells, the underlying radiation-induced biologic damage is likely to become clinically evident in the form of adverse health effects. Radiation biologists describe such effects as deterministic effects because their type and severity are determined by the nature and magnitude of the radiation dose received. DNA repair also can be incomplete or inaccurate, in which case cells survive and divide but with some probability of changes, or mutations, in some of their genes. In time, such mutations may result in other adverse health effects, primarily cancer. Radiation biologists describe these late or delayed-onset effects as stochastic effects because their occurrence follows some random probability distribution or pattern; that is, they are effects that occur at random with some degree of probability that is related to a person's radiation dose.

Human Health Effects of Radiation-Induced Biologic Damage

The onset of deterministic health effects may be acute or delayed, depending on the type.

Acute or Early Deterministic Effects

Acute or early deterministic effects become clinically evident within minutes up to about 2 months after an acute radiation exposure of the whole body or partial body of sufficient magnitude to cause a critical number of cells in individual tissue systems, such as the blood-forming tissues, to die prematurely or to lose their ability to divide. The higher the acute radiation dose, the earlier the deterministic effects occur after the exposure and the more severe they are. Clini-

TABLE 3.5   Estimated Threshold Absorbed Doses for Selected Deterministic Effects of Acute Exposure to low LET Radiation[a]

Health Effect	Organ	Dose (mSv)	Reference
Temporary infertility	Testis	150	ICRP, 1984
Depression of blood cell formation process	Bone marrow	500	ICRP, 1984
Reversible skin effects	Skin	1,000-2,000	UNSCEAR, 1982
Permanent sterility	Ovary	2,500-6,000	ICRP, 1984
Temporary hair loss	Skin	3,000-5,000	UNSCEAR, 1982
Permanent sterility	Testis	3,500	ICRP, 1984
Cataract	Lens of eye	5,000	ICRP, 1984

[a]SOURCE: Adapted from IOM, 1999.

cal, epidemiologic, and animal studies have shown that threshold doses of radiation are required to cause specific deterministic effects, that is, dose thresholds below which specific types of deterministic effects are not seen (Mettler and Upton, 1995). The minimum or threshold doses necessary to cause specific deterministic effects depend on the radiation sensitivities of the exposed cell systems. Estimated threshold absorbed doses for selected deterministic effects of acute exposure to low-LET radiation are shown in Table 3.5.

The spectrum of early signs and symptoms observed after a whole- or partial-body dose of 0.5-1.0 Gy or more is known as the acute radiation syndrome (ARS). The clinical features of the ARS have been described in detail by Young (1987). On the basis of the committee's review of information about reconstructed radiation dose estimates of downwinders and onsite participants, it is considered highly unlikely that people in the RECA populations received acute whole- or partial-body doses of gamma radiation of sufficient magnitude to cause deterministic effects, including the ARS (Lloyd et al., 1990; Henderson and Smale, 1990; Till et al., 1995; Caldwell et al., 1983).

Exposure to ionizing radiation at natural background levels normally present in the environment does not result in discernible deterministic health effects in humans.

Late Deterministic Effects

Some types of deterministic effects may appear many months or years after an exposure to a relatively high dose of radiation; these effects result from cell death or injury that occurred at the time of the exposure but which do not become clinically evident until a long period has passed. This category includes radiation-induced cataract, fibrosis, fibrovascular atrophy, thyroid dysfunction, and effects in an exposed embryo or fetus.

Cataract is one of the few health effects of radiation exposure that essentially is pathologically characteristic, at least in its early stages, of radiation injury.

Cataracts of the posterior subcapsular type have been described as being clinically detectable and distinguishable from cataracts due to other causes after doses to the lens of about 2 Gy of low LET radiation and a minimum latent period of about 10-12 months.

The threshold doses of radiation to localized areas of the body sufficient to result in radiation-induced fibrosis, fibrovascular atrophy, and thyroid dysfunction are considerably higher than the threshold dose for cataract induction.

Exposure of a pregnant woman to radiation may cause nonspecific deterministic effects in the embryo or fetus. Such in utero effects may be expressed clinically in the embryo or fetus or after the child's birth. The nature of these effects and their severity are related to the radiation dose to the embryo or fetus and the period of the pregnancy (gestation) in which the exposure occurred (Brent, 1999) (see Chapter 7).

Stochastic Effects

Radiation-induced damage that is incompletely or incorrectly repaired increases the probability of genetic mutations in affected cells. If the affected cells are of the somatic type, that is, the type of cell that is not handed on to a person's offspring, the probability is increased for stochastic (late) effects such as cancer, appearing in irradiated people years or even decades after exposure. If the affected cells are of the reproductive type—that is, they are transmitted to the next generation—there also is a small probability of radiation-induced heritable genetic effects in the progeny of those exposed. Such effects, which are not peculiar to radiation, occur randomly with frequencies and probabilities that increase with increasing dose. Their severity is unrelated to dose. In the absence of definitive biologic or epidemiologic data to the contrary, it is assumed that there is no dose threshold below which the risk of stochastic effects is zero.

Cancer and the Carcinogenic Effect of Radiation   Cancer is a collective term used to describe many types of malignant diseases. Their induction and development follow a multistage process that is not yet fully understood but is known to be influenced by many factors inside and outside the body. Cancer occurs mainly in older people. The American Cancer Society estimates that 40-45% of the US population develop some form of cancer during their lifetime and that cancer accounts for about 25% of deaths in the United States (Jemal et al., 2004). Exposure to radiation has been shown to increase the cancer risk in the exposed population by some amount that is often related to the dose and to the normal or background risk in the nonexposed population.

After exposure to radiation, mutations induced in somatic cells (cells whose genes are not passed on to the next generation) of an exposed person may alter cell proliferation and result in benign or noncancerous tumors. Additional mutations may then cause malignant changes whereby a benign tumor becomes malignant. Theoretically, radiation-induced mutations in a single somatic cell can eventually result in the cell and its progeny becoming malignant or cancerous; this progression is complex and depends on a variety of factors, only some of which have been characterized. On the basis of animal and epidemiologic studies, factors known to influence radiation induction of tumors include age at the time of exposure, sex, genetic background, and immune status; these host factors and other known factors are discussed in more detail in Chapters 4 and 7.

In the absence of definitive data, scientists generally assume that all types of cancers are susceptible to induction by ionizing radiation. However, animal and epidemiologic studies have shown some cancers to be more likely to have been caused by radiation than others. Various types of cancer grouped by the strength of their statistical association with radiation and available risk estimates obtained in analyses of data from epidemiologic studies of populations at risk of exposure are shown in Table 3.6.

The time between the induction of any disease and its clinical detection or diagnosis is known epidemiologically as the latent period. Because we do not know precisely when a tumor is induced after a radiation exposure, the latent period of a radiation-induced tumor in an exposed person generally is taken to be the time between exposure and detection or diagnosis of the tumor. On the basis of epidemiologic data, the minimum latent periods for radiation-induced leukemia and most solid cancers usually are taken to be about 2 years and 10 years, respectively. For thyroid and bone cancers, the minimum latent periods are estimated to be about 5 years. Age-at-exposure and the magnitude of the radiation dose have been shown in epidemiologic studies to influence the latent periods of some specific tumor types that have been causally associated with radiation exposure.

The relative risk (RR) of developing leukemia (all types except CLL) after radiation exposure appears to rise to a plateau about 15 years after exposure and then about 25 years after exposure to begin a gradual decline toward the risk in the general, or nonexposed, population. The RRs for solid cancers appear to increase to a plateau at about 25 years after exposure and to remain at that level for an extended period—possibly for life, depending on the type of cancer.

Radiogenic cancers, cancers that can be attributed to radiation exposure, are histopathologically and clinically indistinguishable from spontaneous, or naturally occurring, cancers in nonexposed populations. As is discussed later in this chapter, attribution of cancer in general or of specific cancer types to radiation therefore must depend on the observation of statistical differences between their frequencies in populations exposed and those not exposed to radiation (other than background exposures).

When a specific type of cancer is described as radiogenic it does not mean that every cancer of that type was caused by radiation; rather, it means that it is a type of cancer that has been statistically associated with radiation exposure in studies of exposed populations. Similar findings for a specific cancer type in

TABLE 3.6   Susceptibility of Cancers to Induction by Radiation, Grouped by Strength of Association with Radiation[a]

High susceptibility
Bone marrow (leukemia other than chronic lymphocytic leukemia, CLL)
Breast (female)
Salivary glands
Thyroid

Moderate susceptibility
Urinary bladder
Colon
Stomach
Liver
Lung
Ovary
Skin

Low susceptibility
Bone
Brain
Connective tissue
Kidney
Larynx
Nasal sinuses

Very low or absent susceptibility
Cervix of the uterus
Body of the uterus (endometrium)
Chronic lymphocytic leukemia
Oral cavity
Esophagus
Melanoma
Prostate
Pancreas
Rectum
Gallbladder
Hodgkin's disease
Lymphatic system and myeloma
Testes
Muscle
Small intestine

[a]SOURCE: Adapted from Mettler and Upton, 1995.

several epidemiologically valid population studies confirm the association as causal.

Most populations for which strong associations have been found between increased risks of specific cancers and radiation, were exposed to moderate to high doses of radiation at high dose rates. The findings of updated studies of several of those populations are discussed in Chapter 4.

Heritable Genetic Effects. Radiation-induced mutations in the reproductive cells of exposed people may lead to increases in the risk of genetic diseases in their children or descendants. That effect was observed originally in Drosophila (fruit flies) by Muller (1928) and later in other animal species (Russell et al., 1960). However, in the absence of measurable increases in the risk of genetic diseases in the offspring of the atomic-bomb survivors (Schull et al., 1981), estimates of the risk of radiation-induced heritable genetic effects in humans are based largely on data from laboratory animal studies (UNSCEAR, 2001).

RADIATION EPIDEMIOLOGY

Much of what is known about the long-term effects of ionizing radiation in humans has been learned from epidemiologic studies of exposed populations at risk of exposure. Epidemiology is the study of determinants and distributions of disease in human populations. Humans cannot be exposed to radiation under the same conditions as are used in experimental studies or clinical trials. Thus, investigators who want to conduct studies of radiation effects on human health have had to take advantage of situations in which groups or populations of humans have already been exposed to radiation under conditions over which the investigators have had no control. Such situations are described as natural experiments, and the studies are described as observational. Radiation-epidemiology studies conducted to date have involved populations previously exposed to radiation accidentally, in military operations, medically, or occupationally.

Like epidemiologic studies in general, radiation-epidemiology studies can be "descriptive" or "analytic." Descriptive studies are conducted to generate hypotheses to evaluate cause-effect relationships as a basis for risk estimation. Analytic studies can then be used to test the hypotheses and to estimate exposure-specific risks of disease or death. Radiation epidemiology's primary objective is to estimate risks as related to radiation exposure or doses, and its risk estimates take several forms.

Epidemiologic Study Designs

The epidemiologic studies discussed in later chapters are of several types. It is important to understand the strengths and limitations of each type of study design in evaluating the relative importance of studies. The four types of epidemiologic studies commonly used in radiation research are cohort, case-control, cross-sectional, and ecologic designs. We briefly describe the designs in decreasing order of importance and reliability.

Cohort Study

A cohort study is an observational study of a defined group of people who are followed for the purpose of comparing outcomes (usually death or disease) between exposed subsets and unexposed or low-exposed subsets. These studies generally involve large samples, compare multiple outcomes (different causes of death or disease), are the least susceptible to bias, and examine changes in outcome patterns over time, so they are considered the most informative studies. When individual exposures or doses are available, cohort studies typically use a dose-response model and report estimates of relative or absolute risk as a function of exposure or dose. When individual estimates of dose are not available, mortalities often compared with that in an external referent population, usually the US or state rates, and the result is a standardized mortality ratio (SMR). When incidence is used, the analogous metric is the standardized incidence ratio (SIR). Although the cohort study is often described as prospective, there are also retrospective or historical cohort studies where the cohort and its exposures are defined from records. True prospective studies are generally preferable because all data regarding cohort characteristics needed for the eventual analysis can be collected at the beginning of the study, and changes in subject characteristics are added as they occur. This design, however, is expensive, and it generally takes a long time for cases to accumulate, particularly in a cancer study. The atomic-bomb survivors study and the Colorado Plateau Uranium Miners study are examples of prospective cohort studies.

The historical cohort has the advantage that a cohort is identified many years in the past and is followed forward from that time with existing records. This type of cohort study can be completed in much less time and for less money, but with much the same research strength as the true prospective study. The studies of workers exposed to radiation while engaged in nuclear-weapons development in Department of Energy facilities are examples of historical, or retrospective, cohort studies. The disadvantage of this study is that exposure records not originally designed for epidemiologic research must be used to estimate exposures or doses for individual cohort members.

Case-Control Study

A case-control study is an observational study that identifies the persons with a disease of interest (cases), such as lung cancer, and a suitable control group without the disease. Comparisons are made between cases and controls with respect to their exposure. The comparisons are calculated as odds ratios (OR) which are discussed in the next section. If the disease is rare, these studies provide an unbiased estimate of the RR with a major gain in efficiency over a cohort study because far fewer subjects who do not develop the disease of interest are needed to obtain an estimate of RR. Another advantage is that case-control studies are retrospective (cases occurred in the past), whereas cohort studies are often prospective so that the investigator must wait for cases to occur. Accordingly, case-control studies can generally be done much more quickly than prospective cohort studies. Case-control studies have two major disadvantages: they can be used to study only one disease at a time, and, when the disease is not rare, the OR overestimates the RR that would have been obtained with a cohort study (unless it is corrected).

Cross-Sectional Study

A cross-sectional study is an observational study in which exposure and disease are determined simultaneously at one time. This design is also referred to as a prevalence study. When the study is designed primarily to estimate the association of the presence or absence of disease and exposure to a hazardous agent, the risk measure most commonly used is the prevalence OR. The principal disadvantage of the cross-sectional design is that temporal order of exposure and disease cannot be determined; therefore, causal relationships cannot be determined, only associations between disease and exposure. Another weakness of this design is that the duration of the disease has a substantial effect on the relationship between the prevalence and the incidence of the disease. The prevalent cases may not be representative of all cases that developed in the population over some period, either because the condition is detectable for only a short time, or because it is lethal and affected persons are removed from the study by premature death.

Ecologic Study

The least reliable study for estimating risk posed by an exposure is the ecologic study. An ecologic study is an investigation of the association of a disease and an exposure in groups. For example, Cohen (Cohen, 1995) has attempted to link average radon exposures in US counties to county lung-cancer rates. Using an ecologic design, he has reported negative exposure-response relationships; that is, the higher the county radon average, the lower the lung-cancer rate. The problem with ecologic studies is that they cannot determine whether the people who have the disease also had higher exposures. This error in ascribing to the members of a group characteristics that they do not have is known as the ecologic fallacy. These studies are often highly biased, so the committee has chosen to avoid citing them in discussions of radiation risk estimates.

Measures of Risk

Relative Risk

The term relative risk (RR) is used in several ways in epidemiologic studies. In general, RR is the ratio of the risk of disease or death among the exposed population to the risk of disease among the unexposed. In practice, RR can be estimated in a number of ways, depending primarily on study design. In a cohort study, which compares the overall or cause-specific mortality with that in an external referent population, such as the US population, the estimate of average RR is often the SMR. The SMR is the ratio of the number of observed deaths due to some disease to the number expected in the referent population after stratification on such factors as age, calendar year, race, and sex. The SMR is usually reported as a summary RR without regard to individual exposure levels. The summary ratio is reported as either the ratio of the observed to the expected number of deaths or as that ratio multiplied by 100. Since this inconsistency may be confusing to the reader of the committee's literature review, we will adopt the convention of reporting the SMR as a ratio, regardless of the metric used by the author(s). When mortality or morbidity is examined in a cohort study with RR models, the RR is estimated on the basis of the ratio of mortality or incidence among those exposed to a given level of exposure to that among the unexposed or low-exposed group. In such analyses, the referent pro‸ ‸s often a subset of the cohort (internal comparison).

In case-control studies or cross-sectional studies, the RR is usually estimated on the basis of the OR. The OR is defined as the ratio of the odds of an event in one group to the odds of the event in another group. The odds is the number of times (or proportion) that the event occurred divided by the number of times (or proportion) that it did not occur. That is in contrast with a probability, which is the number of times the event occurred divided by the total number of times that it could have occurred. The OR is an accurate estimate of the RR when the disease or condition being studied is rare in both the exposed and unexposed populations, for example, when the incidence or prevalence is less than 10%. When the prevalence is high, the OR will overestimate the RR when the RR is over 1 and underestimate the RR when the RR is less than 1. In this report, the cancers and diseases are sufficiently rare that any cited OR is considered to be a good estimate of RR.

Although the term relative risk is sometimes used in a general sense, in this report the committee attempts to specify the type of RR estimate for any study cited.

Absolute Risk

Absolute risk (AR) is the expression of risk in terms of the proportion of persons with disease in some defined population or the number of cases of the disease in a population of some defined size. In epidemiologic studies, AR is often discussed in terms of the difference in risks or rates, in contrast with RR which is a ratio of risks or rates. AR is generally reported as either a difference in proportions (difference in percentages) or a difference in rates (difference in number of cases per population size per unit time). Understanding the nature of those various measures of comparative risk is important for comprehending epidemiologic study results. When a disease is relatively rare, the RR can be large and the AR reasonably small. For example, if the annual baseline rate of leukemia is 1 per 10,000 persons and the rate among a radiation-exposed population is 2 per 10,000, the RR is 2.0, but the AR is only one additional case per 10,000 persons or 0.0001 (10^{-4}). In contrast, common diseases may appear to have a low RR in an epidemiologic study, but the number of additional cases may be large. For example, a study of cardiovascular mortality may report a RR of 1.2, which appears to be rather small, but the AR (often not reported) may indicate hundreds of excess deaths in a large study population.

Excess Relative Risk vs Excess Absolute Risk

In discussing the results of epidemiologic studies, the terms *excess relative risk* (ERR) or *excess absolute risk* (EAR) are often useful. These are estimates of the amount of risk due to the exposure of interest when the effects of all other exposures are removed. When RR estimates are used, ERR is defined as

$$ERR = RR - 1$$

When AR estimates are used, EAR is defined as

$$EAR = AR_1 - AR_0$$

where AR_1 = total number of observed deaths or cases of disease per population at risk in a specified period, and AR_0 = number of deaths or cases of disease in the unexposed population at risk in the same period.

For example, if the estimated RR is 2.5 for lung cancer in a group of uranium miners, their ERR is $2.5 - 1 = 1.5$. If the AR is 200 deaths from leukemia per 10,000 persons up to the age of 70 years in a group exposed to external radiation and AR_0 is 150 per 10,000 persons up to the age of 70 years among those exposed only to background radiation, then

$$EAR = 200 - 150 = 50 \text{ additional leukemia deaths per 10,000 persons up to}$$
the age of 70 years

Most radiation studies have reported their results in ERR, but the importance of EAR in complementing the ERR estimates has been widely recognized (Preston et al., 2003). The latest publications of risk estimates for the atomic-bomb survivors report both types of estimates. In considering the results of radiation studies cited in Chapters 5 and 6, it is useful to examine how different the patterns of ERR and EAR can be in the same population. Figures 3.2 and 3.3 show hypothetical examples of patterns in lung-cancer rates by age. Lung cancer mortality in the US population in 1990-1994 by age at death was used as the referent population. Figure 3.2 illustrates a hypothetical situation where the number of additional lung-cancer deaths due to radiation (EAR) was a constant

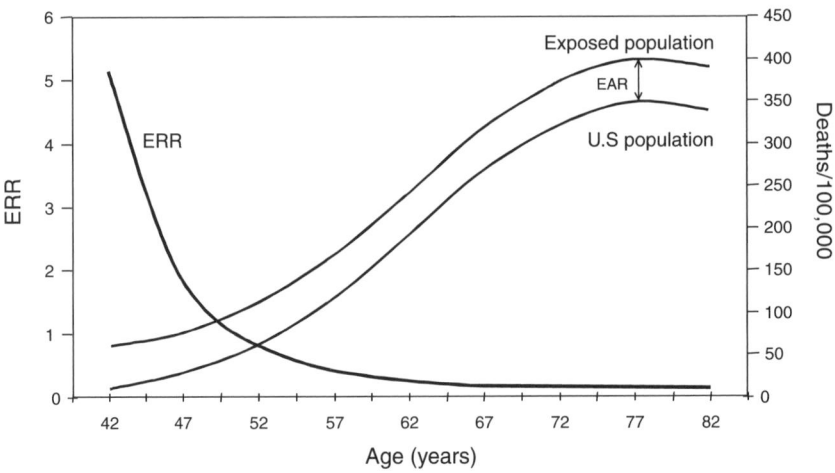

FIGURE 3.2   Lung-cancer decreasing ERR vs constant EAR.

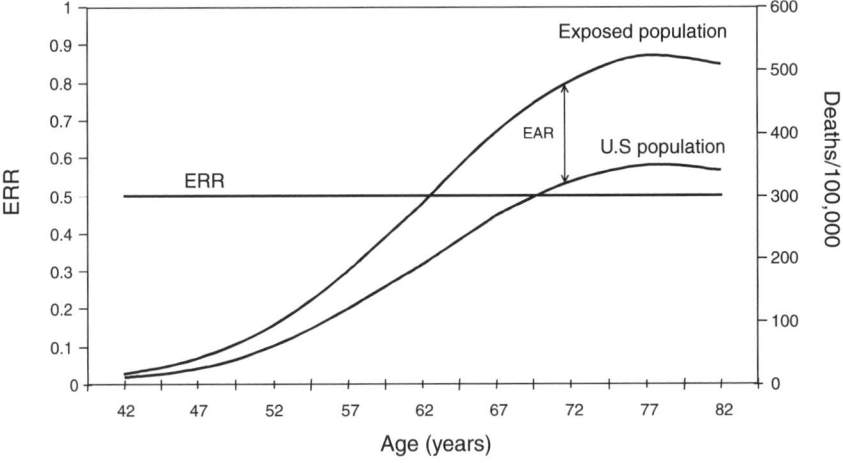

FIGURE 3.3   Lung-cancer constant ERR vs increasing EAR.

50 deaths per 10,000 at each year of age. The background (overall US lung-cancer death rate) rises sharply with advancing age, so the constant-EAR scenario results in ERRs that decrease rapidly with age. In contrast, Figure 3.3 illustrates a hypothetical situation where ERR is a constant 0.50; that is, RR is 1.5 for all ages above 40 y, before which lung cancer is extremely rare. In this scenario, EAR would rise dramatically with age because the excess number of

deaths would have to increase rapidly to maintain a 50% increase above the increasing background rate. That is indicated by the widening gap between the US rate and the exposed-population rate in Figure 3.3.

It is important to understand the relationship between ERR and EAR because many of the studies discussed in later chapters report that ERR decreases with age. Few studies report both ERR and EAR, so it is often impossible to discuss both types of estimates when describing study results in later chapters. The decreasing ERR estimates, however, do not necessarily imply that EAR is also decreasing with age. As can be seen by examining Figures 3.2 and 3.3, a decreasing ERR could be present in a population in which EAR is constant or even increasing with age. In short, when the mortality rate in the unexposed population rises with age, as is the case with most health outcomes, a constant ERR results in an increasing EAR, while a constant EAR implies a decreasing ERR.

Mortality and Incidence Studies

Many of the studies discussed in this report use death from a particular cancer as the end point. Virtually all lung-cancer and pancreatic-cancer studies are mortality studies, because these cancers are often rapidly fatal. Few incident cases of pancreatic and lung cancer are missed in a longitudinal study. In general, mortality studies provide a good estimate of disease incidence when the case-fatality rate is high and the duration of the disease is relatively short. For diseases that are not commonly fatal, the most useful end point is incidence (diagnosis) of the disease. An example of the latter type of disease in radiation research is thyroid cancer. A mortality study of thyroid cancer could be biased and have marginal statistical power because most of the incident cases would be missed.

Cancer-incidence studies measure directly the rate at which cancers occur in a population over time, so they would seem to be the design of choice. There are difficulties, however, in conducting cancer-incidence studies in the United States. To conduct an incidence study, one has to be able to identify each new diagnosis of the disease in the study population. That is possible only in states where cancer registries are available and have been in place throughout the period of study. Because many states do not have registries or the registries are relatively new, cancer-incidence studies are often difficult to implement. Even if a study is done in a state with an existing registry, cases may be missed because of emigration to states without registries. Accordingly, mortality studies are more commonly used in the United States because of the ease of identifying cases nationwide through the National Death Index.

Uncertainty in Risk Estimates

Risk estimates are based on limited information that reflects a lack of perfect knowledge concerning the factors used to calculate the estimate. For that reason, risk estimates are not always precise and always have an element of uncertainty.

Uncertainty in a risk estimate may be the result of the amount of data available, the quality of the data, or both. For example, in an epidemiologic study of the relationship between radiation dose from an identified source and disease frequency, little information might be available on the doses that people received. In addition, the number of exposed persons in a study might be too small to produce precise risk estimates. Each of those situations increases the uncertainty of the estimates of risk per unit dose that investigators can develop.

Uncertainty Intervals and Confidence Intervals

Because of the uncertainty in risk estimates that are made for different radiation dose levels, scientists often include an interval surrounding a risk estimate. The intervals can be confidence intervals or uncertainty intervals. Confidence intervals provide an upper and lower bound on the point estimate of risk that accounts for sampling variability that causes error in the point estimate. For example, a 95% confidence interval (CI) associated with an estimate of RR from a particular study means that if a study of the same size and age and dose distributions were replicated 100 times, we would expect 95 of the intervals to contain the true RR and 5 not to. The width of a confidence interval decreases with increasing sample size.

Uncertainty intervals are much wider because they attempt to account for the uncertainty in all the factors that are used to estimate the risk. These intervals are also referred to as credibility intervals, and Bayesian methods are often used to calculate the credibility bounds. That is done by specifying uncertainty distributions for all risk factors and then drawing random samples from this family of distributions in a Monte Carlo analysis using hundreds or thousands of realizations. The credibility interval is then defined by percentiles of the Monte Carlo sample.

Challenges for Studies on Radiation Health Effects

Several challenges arise in designing, conducting, and evaluating studies of possible links between radiation exposure and specific illnesses in a selected population. They include the following

• Health effects can have causes other than radiation exposure. Although studies have established a strong link between radiation and some types of cancer, radiation is not the only cause of these cancers. Lung cancer is a good example: smoking is a stronger risk factor than all but the highest radon exposures in miners.

• Generally, dose estimates are not available for individuals. Without accurate estimates, assessing reliably whether a detected health problem is connected directly to radiation exposure is difficult.

• People are exposed to a variety of radiation sources. In addition to specifically identified radiation exposures, people are exposed to natural sources of radiation and radiation from medical and dental procedures, consumer goods (such as tobacco products), and fallout from global nuclear-weapons testing.

• People are different, and their circumstances change. People eat different foods, have different lifestyles, and change their habits as they age. All those factors, and many others can directly or indirectly influence the probability of radiogenic disease in individuals and populations.

• Health effects of low-level radiation exposure cannot be detected immediately. The delay between the time of exposure and the time when a health effect occurs can be long. That period of time, the latent period, varies among diseases and among individuals. For leukemia, the latent period can be as short as about 2-5 years. Thyroid cancer generally takes at least 5 years to grow enough to be diagnosed. Most thyroid cancers would be expected to appear within 10-20 years after exposure; in some people, the delay could be much longer.

Derivation of Radiation Risk Estimates

General approaches for estimating health risks (most specifically cancer risks) posed by exposure to ionizing radiation have been developed by ICRP and in the United States by the National Council on Radiation Protection and Measurements (NCRP). Those and other bodies have used the risk estimates as input data to develop dose limits for radiation-protection purposes (ICRP, 1991; NCRP, 1997).

The epidemiologic data that have been used for estimating tumor risks posed by low-dose exposures (a few mSv) have been obtained for much higher doses (10s of mSv) as well. The sources of the data are survivors of the atomic-bombs at Hiroshima and Nagasaki and people exposed occupationally and medically.

To use those data on medium to high doses to estimate responses at much lower doses, scientists have developed extrapolation methods. The generally accepted model for such an extrapolation for solid tumors is the linear non-threshold (LNT) model (NCRP, 2001), which assumes that from very low doses to much higher doses the stochastic cancer response is linear and that there is no radiation dose that poses no risk of inducing cancer, however small it might be. The best fit curve using the data for solid tumors at high and intermediate doses for the atomic-bomb survivors is linear and the extrapolation to low doses is considered to be conservative such that it does not underestimate the low-dose risk (Pierce and Preston, 2000). A similar conclusion holds when the new dosimetry system of the atomic-bomb survivors, DS02, is used (Preston et al., 2004).

Given the relative paucity of human data and the extrapolation approaches used, some degree of uncertainty is inevitable in the risk estimates calculated. For leukemia, the dose-response curve based on the atomic-bomb survivor data is nonlinear over the dose range studied. In general terms, the slope of the curve increases as the dose increases—an upwardly curving response (Preston et al., 1994).

The slope of the line that is the best fit to the total solid-cancer mortality data on the Japanese atomic-bomb survivors in the Life Span Study (LSS) has been used to derive the risk coefficient for solid cancers in an exposed population (ICRP, 1991). The coefficient is presented as fatal cancers per 10,000 persons per Sv of external radiation; we discuss this more fully below. A similar approach has been used for developing risk coefficients for specific tumor types, and these values are presented in Table 3.7. An effort as part of the LSS is under way to develop a similar set of risk coefficients for cancer incidence.

The total risk in a population exposed to external radiation is generally considered in terms of a population of working age (adults) and the whole population (adults and children). For a population of working age composed of both sexes, the lifetime risk of death from cancer is 8×10^{-2} per sievert for high doses and high dose rates of external radiation and 4×10^{-2} per sievert for low doses and low dose rates (ICRP, 1991). The comparable values for the whole population are slightly higher because of the increased sensitivity of young persons: they are 10×10^{-2} per sievert for high doses and dose rates and 5×10^{-2} per sievert for low doses and dose rates. Those values need to be considered in the context of a background lifetime risk of about 15% of dying as a result of one of the cancers listed as being radiogenic in RECA.

TABLE 3.7   Contributions of Organs to Total Cancer Risk for Population[a]

Site of Cancer	Rate of Fatal Cancer (per 10,000 person-Sv)
Urinary bladder	30
Bone marrow	50
Bone surface	5
Breast	20
Colon	85
Liver	15
Lung	85
Esophagus	30
Ovary	10
Skin	2
Stomach	110
Thyroid	8

[a]SOURCE: Adapted from Table B-20 in ICRP Publication 60, 1991.

The need to conduct epidemiologic studies for low exposure levels (in the range of a few mSv) is clear. Such work can yield a more direct measure of the risk posed by low doses of radiation. Such low-dose studies have been conducted with cohorts of workers involved in the development of nuclear weapons in the United States and several other countries.

In addition to the risk estimates described above for external radiation exposure, scientists have developed separate estimates for specific exposure conditions. For example, the risk of lung cancer from radon exposure (for example, from uranium mining) has been calculated (NRC, 1999). Similar approaches have also been used to calculate the thyroid-tumor risk from exposure to ^{131}I (discussed in NCRP, 1999, pages 155-162).

In general, risk assessments for deterministic noncancer effects (effects resulting from tissue injury) are estimated on the basis of a threshold dose-response such that, over the range of several hundred mSv, the radiation-induced response does not differ from the response to background levels for any particular adverse health outcome (ICRP, 1991). Thus, for the range of doses estimated for the great majority of environmental and occupational exposures, no increase in deterministic noncancer effects is expected.

An informative general source that provides an overview of risk-assessment practices is *Radiation Carcinogenesis* (Hall, 1994, chapter 19).

Probability of Causation/Assigned Share

In 2003, the National Cancer Institute and the Centers for Disease Control and Prevention (NCI-CDC) produced an updated set of radioepidemiological tables that associate a given dose of ionizing radiation with the likelihood that exposure caused a given type of cancer (NCI-CDC, 2003). The tables make a broad set of assumptions of susceptibility and exposure that are applied more properly to a subpopulation than to an individual. Such an association has been termed the probability of causation (PC), although it expresses strength of association rather than a mechanism of causation. The tables generate PCs as a function of a person's estimated dose, sex, age at exposure, and age at diagnosis. The tables are available as Web-based calculators for 22 specific cancers at http://irep.nci.nih.gov.

In an exposed population of people who have developed a specific type of cancer, PC is the ratio of their excess risk of that cancer to their overall risk of that cancer. More specifically,

$$PC = ERR/(1 + ERR)$$

The concept of PC has been applied to people (for example, in the Energy Employees Occupational Illness Compensation Program Act) in much the same way that clinical-prediction decision rules are routinely applied to individual patients in the clinical arena. It differs, however, in that most clinical rules are applied prospectively to predict the likelihood of an outcome or event whereas the PC is applied retrospectively to analyze the likelihood that a known cancer was caused by radiation. In that respect it is akin to the statistical concept of attributable risk. In other words, clinical-prediction rules are conditioned on a set of factors and a prior probability to predict a future event whereas the PC is conditioned on the event's having already occurred. In some sense, clinical predictions are a priori chances, whereas PCs are a posteriori chances. Both are probabilities, and both range from 0 to 1. In recent years, when these analytic likelihoods (PCs) have been applied to individuals, they have been called the assigned share (AS) of the risk, and the sum of all such assignments (base plus excess) equals unity. The PC/AS concept is developed in detail in Chapter 5.

4

Review of Recent Data on Radiation Epidemiology, Biology, and Dosimetry

The statement of task from the Health Resources and Services Administration (HRSA) to the committee requests that we assess the most recent scientific information related to radiation exposure and associated cancers to determine whether there is new information that could affect the magnitude of radiation cancer-risk estimates. If there is, it would provide part of the information base that is needed for considering the inclusion of new populations and new geographic areas in the Radiation Exposure Compensation Act (RECA) populations.

The risk estimates for human cancers after exposures to low-LET ionizing radiation are based on human tumor frequencies, which come mainly from cancer mortality data on the survivors of the atomic-bomb detonations at Hiroshima and Nagasaki (NRC, 1990; ICRP, 1991; NCRP, 2001; reviewed in Wakeford, 2004). Risk estimates for high-LET radiation are based on mortality data on uranium and other underground miners exposed to radon (NRC, 1999) and on the radium-dial painters (NRC, 1988; reviewed in Wakeford, 2004). The responses at very low exposures to low-LET radiation are estimated by extrapolation of data on atomic-bomb survivors over the available low- to moderate-dose range (0.005–2 Sv). The extrapolation model used is the linear nonthreshold (LNT) one (NCRP, 2001) that is discussed in Chapter 3. Support for the use of the LNT model for estimates of cancer risks posed by low-LET radiation comes from human epidemiologic studies (medical and occupational), experimental-animal tumor studies, and cellular-radiation studies (NCRP, 2001). The data from similar but fewer studies involving high-LET exposures support the use of the LNT model here also (NCRP, 2001). The same types of studies are used to provide estimates of the effects of dose fractionation and dose protraction (NCRP, 2001). Epidemiologic studies are also used for estimating risks to specific exposed populations, such as underground miners exposed to radon (NRC, 1999) and populations exposed to iodine-131 (^{131}I) (UNSCEAR, 2000).

The International Commission on Radiological Protection (ICRP) and the National Council on Radiation Protection and Measurements (NCRP) are moving to use tumor incidence, rather than mortality, in their revised cancer risk estimates. Using tumor-incidence data for developing risk estimates provides an additional useful measure of risk because morbidity entails health, emotional, and financial costs to the individual and society.

In this chapter, we consider and present the evidence from new or updated epidemiologic studies, radiation-biology advances, or dosimetry approaches that could result in significant changes in the risk estimates for human cancer induced by ionizing-radiation exposure. This chapter brings together information that could influence compensation for diseases currently covered by RECA legislation. In Chapter 7, we discuss additional diseases brought to our attention by members of the public at a series of hearings held in response to community invitations with a view to whether eligibility for coverage should be extended thereto. The following sections discuss what is new in those fields of study.

RECENT DEVELOPMENTS IN RADIATION EPIDEMIOLOGY

Epidemiologic studies of the Japanese survivors of the atomic bombs and of other populations exposed to radiation medically, occupationally, or accidentally have characterized the long-term health effects of radiation (see Chapter 3). Risks estimates for radiogenic cancers and nonmalignant diseases now compensable under RECA come primarily from epidemiologic studies of uranium and other underground miners exposed to radon and from studies of the atomic-bomb survivors. The mining populations were exposed primarily to radon internally while the atomic-bomb survivors were exposed primarily to external gamma rays. Risk estimates for thyroid cancer also come from populations exposed to external x and gamma rays, and internally to radioiodine. Studies of worker populations exposed to low or very low doses of low LET radiations over long periods provide radiogenic-cancer risk estimates with which the more precise estimates obtained from the atomic-bomb survivors can be compared to evaluate their applicability to populations chronically exposed to low radiation levels. Extensive and detailed reviews of those studies have been reported previously (NRC, 1990, 1998; 1999; ICRP, 1991; UNSCEAR, 1993, 2000; IARC, 2000; 2001).

A comprehensive reassessment of risk estimates is included in a companion, forthcoming report from the National Research Council Committee on Biological Effects of Ionizing Radiation (BEIR) specifically, the Committee on Health Risks from Exposure to Low Levels of Ionizing Radiation (BEIR VII).

Risks to the Health of Miners, Millers, and Ore Transporters

Studies of Uranium Miners

Epidemiologic studies of underground miners have identified an increased risk of primary lung cancer associated with exposure to alpha-particle radiation from decay products of inhaled radon (NRC, 1988). Those studies generally need relative-risk (RR) models; estimates are discussed below. Although absolute risk is important from a public-health perspective, we choose to discuss RR and excess relative risk (ERR; ERR = RR − 1) because of their use in the cited literature.

The most recent and widely recognized lung-cancer risk estimates associated with radon exposure were reported in the BEIR VI report (NRC, 1999). An important finding of the BEIR VI committee relevant to some of the RECA populations—identified as uranium miners, uranium millers, and ore transporters—is that the ERR of radiogenic lung cancer decreases with increasing attained age and time since exposure. The eligible people now seeking compensation are generally more than 60 years old and have been out of the mines for 30 years or more. Accordingly, they are at much lower RR for radiogenic lung cancer now than they were in earlier years after retiring from mining uranium. Using data on 11 international cohorts, the BEIR VI committee estimated that uranium miners 65-74 years old have about 25% of the ERR of radon-induced lung cancer that miners in their 50s have. The most recent analysis of the Colorado Plateau uranium-miner data (Hornung et al., 1998) estimated that the ERR for lung cancer in miners in their 70s was less than 10% of that of miners in their 50s. Similarly, the BEIR VI committee estimated that miners of the same age who have been out of the mines for more than 25 years have less than half the lung-cancer ERR of recently retired miners. The analysis of the Colorado Plateau miner lung-cancer data indicated a 65% reduction in ERR for miners who have been out of the mines for more than 25 years.

Those analyses also have shown a synergistic relationship between exposure to radon and cigarette smoking. That is, the ERR of radiogenic lung cancer in smoking miners is greater than the sum of the ERRs of lung cancer associated with smoking alone and radon alone. The most recent analysis of the Colorado Plateau uranium miners cohort (Hornung et al., 1998) and the pooled analysis of 11 miner cohorts in BEIR VI (NRC, 1999) each found that the joint effect was greater than additive but less than multiplicative. The nature of the interaction was that never-smokers had about 3 times the ERR per WLM of ever-smokers in both analyses. These findings were supported by a study of non-smoking uranium miners in the Colorado Plateau (Roscoe, 1997) who had an SMR = 12.7 for lung cancer compared with the overall SMR = 5.8 in the entire cohort.

In the most recent update of all cancer mortality in the Colorado Plateau uranium miners' all-cause mortality study (Roscoe, 1997), the cohort of 3,238 white male miners was followed to determine certified causes of deaths in 1960-1990. Their mortality experience was compared with the combined mortality in neighboring states. Most of the findings of the study were consistent with those of previous studies of this miner population. The standardized mortality ratios (SMRs) for lung cancer and pneumoconiosis continued to show statistically significant increases (371 deaths, SMR = 5.8, 95% CI [confidence interval] = 5.2-6.4 and 41 deaths, SMR = 24.1, 95% CI = 16.0-33.7, respectively). The SMRs for lung cancer and pneumoconiosis increased with increasing level of radon-decay products and with duration of employment in the mines. Roscoe (1997) concluded that lung cancer and pneumoconiosis remain the most important long-term causes of death in this cohort.

The most definitive study of cancer other than lung cancer among miners exposed to radon was the meta-analysis of data on the 11 international miner cohorts reported by Darby et al. (1995). The men in those cohorts (N = 64,209) had been employed in underground mines for an average of 6.4 years; they had an estimated average annual cumulative exposure to radon of 155 working-level months (WLM) and an average followup of almost 17 years. The RR of all cancer causes of death combined other than lung cancer (N = 1,179) was similar to the expected value (RR = 1.01, 95% CI = 0.95-1.07), on the basis of the mortality of the general populations in areas around the mines. Those results should be interpreted cautiously since they are likely to underestimate the true RR in the uranium miner population due to the Healthy Worker Effect. The authors concluded that the study provided strong evidence that high concentrations of radon in air do not cause a substantial risk of mortality from cancer other than lung cancer.

Studies of Uranium Millers and Ore Transporters

Risks to the health of uranium millers and ore transporters from occupational exposure have not been as well characterized as the risks to miners' health because of smaller sample sizes and little or no data on individual exposures. Exposures to millers were primarily from inhalation of dusts containing uranium, silica, and vanadium. Their internal exposure posed potential health hazards from radiation (alpha particles) and from the chemical toxicity of uranium compounds arising during the conversion of uranium ore to yellow cake (see Chapter 3).

A study of mortality among 662 millers from the Colorado Plateau who were followed from 1950 through 1967 (Archer et al., 1973) found four deaths from lymphatic and hematopoietic cancers combined (excluding leukemia), for a small and nonstatistically significant increase over the rate in the US general population. A later larger and more powerful study evaluated mortality among an expanded cohort of millers in the same area (N = 2,002) who were followed through 1971 (Waxweiler et al., 1983). They found no statistically significantly increased RRs of mortality from any malignant (radiogenic or other) neoplasm, including renal cancer. The only statistically significant increase in disease risk in that cohort was for nonmalignant respiratory disease (55 deaths, SMR = 1.63, 95% CI = 1.23-2.12); however, there was no evidence that the risk increased with increasing length of employment. A nonstatistically significant ERR of death from chronic (nonmalignant) renal disease (6 deaths; SMR = 1.67, 95% CI = 0.60-3.5) was also found, but it did not appear to be related to work in the mills.

Pinkerton et al. (2004) updated the Waxweiler et al. study by extending the vital-status followup by 27 years December 31, 1998. The authors completely reviewed and updated all work histories and recoded errors found in previous files. They also limited the study cohort to men who met the original cohort definition, never worked in uranium mines, and worked in one or more of seven mills whose personnel records were originally microfilmed. That redefinition of the study cohort resulted in a reduction in the size of the cohort from 2,002 in the Waxweiler et al. study to 1,485. Because exposure estimates were not available for individual workers, Pinkerton et al. used life-table analyses to compare mortality in the workers with that in the general US population.

Mortality from all causes combined (810 deaths, SMR = 0.92, 95% CI = 0.86-0.99), including all cancers (184 cancer deaths observed, SMR = 0.90, 95% CI = 0.78-1.04), was less than expected on the basis of US rates. A statistically significant increase in nonmalignant respiratory disease mortality was found (100 deaths, SMR = 1.43, 95% CI = 1.16-1.73). No statistically significant increase was found in mortality from lung cancer (78 deaths, SMR = 1.13, 95% CI = 0.89-2.35) or chronic renal disease (6 deaths, SMR = 1.35, 95% CI = 0.58-2.67). No positive trend in excess mortality from these or any other types of cancer with duration of employment was found.

There have been few studies of morbidity among uranium millers. Thun and colleagues examined renal toxicity in a group of 39 uranium millers compared with 36 cement-plant workers (Thun et al., 1985). They found a weak dose-response relationship for excretion of beta-2-microglobulin among millers working in the yellowcake drying and packaging area, the area with the highest exposure to soluble uranium. They concluded that the results suggested reabsorbtion of low-molecular-weight proteins consistent with uranium nephrotoxicity.

More recently there have been two studies of uranium workers that were engaged in production activities using the uranium coming from the mills. A study of uranium enrichment workers (McGeoghegan and Binks, 2000a) in the UK found no overall excess mortality or morbidity due to any cancer when compared to non-radiation workers. They did find, however, a significant dose-response relationship for bladder cancer when external radiation dose was lagged by 20 years. A similar study by the same investigators regarding workers involved in the production of nuclear fuels and uranium hexafluoride (McGeoghegan and Binks, 2000b) found no significant association of radiation exposure and any cancer with the exception of Hodgkin's disease (both mortality and morbidity).

They also reported a significant association with morbidity due to nonHodgkin's lymphoma. They noted that these associations were not likely to be causal.

The committee is unaware of any epidemiologic studies of ore transporters. Like the millers' exposure, their primary potentially hazardous exposure was to ore dusts, probably with a greater risk of chemical toxicity than radiation toxicity. The nature of their work makes it unlikely that their body burdens of soluble uranium compounds exceeded renal thresholds for chemical toxicity or that their exposure to radiation from the ores substantially exceeded normal background levels.

Risks to Downwinders and Onsite Participants at US Nuclear Tests

Several populations have been at risk of exposure to ionizing radiation of types similar to those of downwinders and onsite test participants. Followup studies of the other populations provide information about the long-term health effects of such exposure; some also provide data from which estimates of the risks of radiation-related or radiogenic diseases, primarily malignant diseases, are calculated. We discuss here new information from specific population studies that adds to the knowledge and understanding of the types and magnitude of the health risks for which downwinders and onsite test participants currently are compensated.

Radiogenic Cancers and Other Diseases

Information on radiation risks is summarized in many of cited sources (UNSCEAR, NRC, and NCRP) and chapters in several textbooks dealing with the subject (Mettler and Upton, 1995; Hall, 2000). Updated information is scheduled to appear shortly in a report from the BEIR VII committee. The risk estimates in BEIR VII take into account DS02 data for the atomic-bomb survivors that were not available to this committee, and those data should be used whenever there are significant discrepancies between findings we survey from literature published in the last 20 years, and the BEIR VII update based on reanalysis of current data. Long-term studies of irradiated populations continue to provide new information on effects from internal and external sources of exposure. Effects of high-dose-rate exposure are chronicled in reports of findings in the Japanese atomic-bomb survivors supplemented by data from several large studies of radiation workers exposed to low-dose-rate radiation. The lower dose-rate exposures received by worker populations, along with data from medical-therapy populations add to the current status of knowledge of the risks in humans of the different radiogenic diseases with respect to rate and amount of radiation dose to body organs and the total body.

Dose from internal emitters is protracted because it is delivered over the decay time of the particular radionuclide. Effects of internal emitters of low linear energy transfer (low LET) are less than those of comparable doses delivered in a single high-dose-rate exposure, because there is continuing repair of sublethal damage when a dose is delivered at a low dose rate. The need to expand knowledge of radiation effects of ^{131}I has led to many studies, some of which continue. The dose to the thyroid from ^{131}I per unit intake is about 1,000 times higher than the dose received by other normal organs. The dose to different body organs from other fallout radionuclides is much lower because of low uptake and retention in different organs (CDC-NCI, 2001). Increased incidence of thyroid cancer has been observed in children who received high ^{131}I doses, but no increase in leukemia from the lower bone marrow doses received following ^{131}I doses from fallout has been statistically confirmed. Continuing studies of health effects in persons resident in Southern Utah during the high NTS fallout years reveal marginally significant increases in thyroid neoplasms and leukemia in children.

The studies of disease in the Japanese atomic-bomb survivors provide the most reliable information for risk assessment for several reasons:

- They received a wide range of dose; and, unlike medical subjects, the population is composed of people with a typical range of health conditions prior to their exposure.
- Large numbers of subjects in well-defined cohorts have been studied over many years. Very good followup involving a range of ages and both sexes has resulted in many person-years of followup which is needed for valid statistical analyses.
- Good estimates of dose have been calculated for each member of the cohort as a result of in-depth dosimetry studies (Dosimetry System 02, DS02). The new dosimetry system recently introduced incorporates refinements taking into account shielding histories and new information on neutrons (Preston et al., 2004; DS02 to be published in 2005).

Periodic publications update findings from the joint US-Japanese Radiation Effects Research Foundation studies of the several a priori-defined cohorts and subcohorts of the survivors of the atomic bombs dropped in 1945. The best established information on cancer mortality and cancer incidence comes from the large Life Span Study (LSS) cohort, buttressed by results of special studies of cancer in children born of irradiated parents (Izumi et al., 2003), and of leukemia mortality in children who were in utero at the time of the bombs (Delongchamp et al., 1997). In the absence of statistically meaningful data from fallout-exposed populations themselves, risk estimates from the atomic-bomb survivors are the best data we have to assess the magnitude and kinds of effects expected in downwinders and onsite test participants. Thyroid cancer is discussed in a separate section, which compares the results of studies in different irradiated populations.

Atomic-Bomb Survivor Studies

Important cancer-mortality findings reported since 1990 and the results of new incidence studies are summarized below.

Cancer Mortality   Cancer mortality through 1990 was analyzed on the basis of the DS86 dosimetry system. The major findings include

- Most of the excess deaths from leukemia occurred in the first 15 years after exposure.
- For solid cancers, the excess risk was consistent with a life-long increase in age-specific cancer risk.
- The excess relative lifetime risk per sievert for solid cancers in persons exposed at the age of 30 was about three times greater than for persons exposed at age 50, and the projected lifetime risks for those exposed at age 10 were 1.0 to 1.8 times higher than the estimates for those exposed at age of 30 years.
- Excess risks of solid cancers were linear up to about 3 Sv, but they were nonlinear for leukemia, for an estimated risk at 0.1 Sv of about 1/20 the risk at 1.0 Sv (Pierce et al., 1996).

More recently, the findings were extended through 1997 (Preston et al., 2003). The study included 9,335 deaths from solid cancer and 31,881 deaths from noncancer diseases on the basis of a 47-year followup. About 440 (5%) of the solid-cancer deaths were attributed to the radiation exposure. The excess risks of solid cancer were linearly related to dose down to the lowest dose studied (0-150 mSv). Results demonstrated that ERRs declined with increasing attained age (age at death); another was that the ERR was highest for those exposed as children. There was no direct evidence of radiation effects after doses less than about 0.5 Sv (Preston et al., 2003).

Cancer Incidence   Cancer incidence in the atomic-bomb survivors is based on data in the Hiroshima and Nagasaki tumor registries.

Among 79,972 individuals in the extended Life Span Study (LSS-E85), 8,613 had a first primary solid cancer diagnosed between 1958 and 1987 (Thompson et al., 1994). Cancer cases occurring among members of the LSS-E85 cohort were identified in the Hiroshima and Nagasaki tumor registries and special efforts were made to ensure complete case ascertainment, data quality, and data consistency in the two cities. Dosimetry System 1986 (DS86) organ doses were used for computing risk estimates.

Ron et al. (1994) compared results from an analysis of 9,014 first primary incident cancers diagnosed in 1958-1987 in LSS cohort members and compared incidence with mortality rates based on analysis of 7,308 death certificates that listed cancer as the underlying cause of death. When deaths were limited to those occurring in the same interval in persons living in Hiroshima or Nagasaki, there were 3,155 more incident cancer cases than cancer deaths overall and 1,262 more incident cancers of the digestive system than deaths from cancers of this system. For many cancers, the incidence series was at least twice as large as the comparable mortality series, and both had significant dose-response relationships. For all solid tumors, the estimated ERR at 1 Sv (ERR_{1Sv}) for incidence ($ERR_{1Sv} = 0.63$) is 40% larger than the ERR based on mortality data from 1950-1987 in all of Japan ($ERR_{1Sv} = 0.45$). The corresponding excess absolute risk (EAR) point estimate is 2.7 times greater for incidence than for mortality. For some cancer sites, the difference in the magnitude of risk between incidence and mortality is greater. The differences reflect the greater diagnostic accuracy of the incidence data and the lack of full representation of radiosensitive but relatively nonfatal cancer, such as breast, skin, and thyroid cancers—in the mortality data. Incidence and mortality data provide complementary information for risk assessment (Ron et al., 1994).

The observations made in the tumor-registry studies are summarized in Table 4.1.

A survey of breast-cancer incidence in the LSS population found 1,093 breast cancers diagnosed during 1950-1990. A linear and statistically highly significant radiation dose-response relationship was found. Exposure before the age of 20 years was associated with higher ERR_{1Sv} than exposure at greater ages, with no evidence of consistent variation with age of exposure for ages under 20 years. ERR_{1Sv} was observed to decline with increasing attained age, with the largest drop around the age of 35 years (Land et al., 2003). The EAR was not reported, but it probably changed in the opposite direction, but to a lesser extent.

The incidence of leukemia, lymphoma, and myeloma in the LSS cohort from late 1950 through the end of 1987 was analyzed on the basis of followup of

TABLE 4.1   Tumor Incidence Rates Observed in the Japanese Atomic-Bomb Survivors (1994)

Thompson et al., 1994	ERR_{1Sv}	EAR 10^{-4} PY Sv	Ron et al., 1994	ERR_{1Sv}
All solid cancer	0.63	29.7	Significant increased risk	0.63
Stomach	0.32		Significant increased risk	
Colon	0.72		Significant increased risk	
Lung	0.95		Significant increased risk	
Breast	1.59		Significant increased risk	
Ovary	0.99		Significant increased risk	
Urinary bladder	1.02		Significant increased risk	
Thyroid	1.15		Significant increased risk	
Liver	0.49		Significant increased risk	
Nonmelanoma skin	1.0		Not stated	
Salivary gland			Significant increased risk	

93,696 survivors accounting for 2,778,000 PY (Preston et al., 1994). The analyses added 9 years of followup for leukemia and 12 years for myeloma to previous reports, and included the first analysis of lymphoma incidence in this cohort. The leukemia registry and the Hiroshima and Nagasaki tumor registries, included a total of 290 leukemia, 229 lymphoma, and 73 myeloma. The primary analyses were restricted to first primary tumors diagnosed among residents of the cities or surrounding areas with DS86 dose estimates of 0 - 4 Gy (231 leukemia, 208 lymphoma, and 62 myeloma) and used time-dependent models for the EAR. Separate analyses were reported for acute lymphocytic leukemia (ALL), acute myelogenous leukemia (AML), chronic myelocytic leukemia (CML), and adult T-cell leukemia/lymphoma (ATL). There were few cases of chronic lymphocytic leukemia (CLL) in the Japanese population independent of radiation exposure, so CLL was excluded from later leukemia risk analyses. There was strong evidence of radiation-induced risks for all subtypes except ATL, and there were substantial subtype differences with respect to the effects of sex and age at exposure and in the temporal pattern of risk. The AML dose-response function was nonlinear, whereas there was no evidence against linearity for the other subtypes. When averaged over the followup period, the EAR estimates (in cases per 10^4 PY Sv) were 0.6, 1.1, and 0.9 for ALL, AML, and CML, respectively. The corresponding estimated average ERRs at 1 Sv are 9.1, 3.3, and 6.2 respectively. There was some evidence of an increased risk of lymphoma in males (EAR = 0.6 case per 10^4 PY Sv) but no evidence of any excess in females. There was no evidence of an excess risk of multiple myeloma in these analyses.

Mortality from Leukemia and Solid Cancers in Children Exposed in Utero

Cancer mortality through 1992 was assessed in 807 atomic-bomb survivors exposed in utero and in 5,545 survivors who were less than 6 years old at time of exposure (Delongchamp et al., 1997). Doses in both groups were at least 0.01 Sv. Mortality was compared with that in low-dose group (10,453 persons with little or no exposure). Ten cancer deaths were observed among in utero-exposed persons, with a statistically significant dose-response relationship and an ERR per sievert of 2.1 (90% CI = 0.2-6.0). That estimate did not differ substantially from that for survivors exposed during the first 5 years of life. The cancer deaths among those exposed in utero included leukemia (two), female-specific organs (three), and digestive organs (five). Nine of the deaths occurred in females (ERR/ Sv = 6.7, 90% CI = 1.6-17), and much of the effect was due to female-specific cancers (ERR/Sv = 9.7, 90% CI = 0.7-42). Those risks did not differ significantly from those seen in females exposed as children. No deaths from solid cancer occurred in males exposed in utero. Mortality in males and females differed even when female-specific cancers were excluded from the comparison. There were only two leukemia deaths among those exposed in utero, but the leukemia death rate in this group was still marginally higher than in the comparison group (p = 0.054). The authors expressed caution in the interpretation of the data because of the number of cancer deaths was small, and because of the unexplained difference in mortality from solid cancer between sexes (Delongchamp et al., 1997). Their tentative conclusions were that the study provided support for a somewhat higher risk during the first trimester of pregnancy, that the increased risk persisted through childhood into the adult years, and that the pattern of diseases was similar after in utero and childhood exposure. Because of the wide uncertainty range, they concluded that their data did not exclude the possibility that the cancer risk from in utero exposures could be several times higher than the risk from childhood exposure.

A comprehensive review of the uncertainties contained in the different published studies is provided by Boice and Miller (1999). They discuss the confounding of reasons for referral with the risk of pelvimetry and conclude that "although it is likely that in utero radiation presents a leukemia risk to the fetus, the magnitude of the risk remains uncertain." They found the causal nature of the risk of cancers other than leukemia to be less convincing, and the similar relative risk (RR = 1.5) for virtually all forms of childhood cancer suggested an underlying bias. Chapter 8 in Mettler and Upton (1995) also provides a broad review of current knowledge regarding the effects of radiation exposure in utero.

Conclusion   Continuing investigations in the Japanese atomic-bomb survivors confirm and extend the evidence defining cancer mortality and risk after total-body high-dose-rate exposure. The radiation risk is better defined than previously based on the analysis of the incidence data classified by types of cancer, by age and sex at time of exposure. The high risk from thyroid cancer in children is consistent with the results of other studies (see thyroid cancer section). Data on cancer incidence and mortality from ongoing studies of the youngest survivors, all of whom are now over 60 years old, will be important as they emerge from studies. Although the risk of particular cancers posed by radiation is better described by incidence than by mortality, the number of documented cases in each disease category are still small, so the uncertainty range is wide. Continuing followup will be needed to increase confidence in disease-specific risk coefficients. The newest risk estimates are based on longer followup and better dosimetry.

Thyroid Cancer

Thyroid cancer is a relatively rare disease, with about 1,000 deaths certified and about 13 times as many new thyroid cancers reported each year in the United States (http://seer.cancer.gov/csr/1973_1998/thyroid.pdf, accessed February 17, 2005). A definite trend of increasing thyroid-cancer incidence during most of the last 60 years has been attributed in part to radiation therapy of the head and neck given for benign conditions, a practice that has been discontinued since the 1950s and 1960s. Despite the increased incidence, primarily of the relatively benign papillary form of the disease, thyroid-cancer mortality declined in most of the ensuing years. In recent years, the upward trend has been influenced by increased case-finding brought about by routine use of ultrasonographic imaging of the thyroid. There are excellent reviews of the subject, including NCRP (1985, 1991, and 2001); ICRP (1991); UNSCEAR (1994 and 2000); Chapter 5, *Thyroid Cancer* section, in Mettler and Upton (1995); Shore et al., (1993); and Thomas et al., (1999).

New information since RECA was enacted in 1990 reveals a wider geographic distribution of dose from ^{131}I than was generally recognized when Congress identified selected counties as affected areas for downwinder eligibility. It is now known that persons living in other states and in other counties in Utah could have received as high or higher thyroid doses as did those living in areas specified in RECA (NRC, 2003c). Recognition by the public of disparities in compensation created a need for further consideration of risk to the health of persons in the affected areas posed by fallout. The following text reviews the current state of knowledge concerning the risk of thyroid cancer after exposure to radiation from fallout and other sources of radiation exposure of the head and neck. For information on the distribution of dose from Nevada Test Site (NTS) weapons tests to the US population, see Figures 4.1-4.4.

Studies of Populations Exposed to Fallout from Nuclear-Test Releases

Fallout from Nevada Test Site

Several epidemiologic studies of Utah schoolchildren exposed to fallout from the NTS weapons tests have been conducted. The first showed no increase in thyroid disease (Rallison et al., 1974). A second followup study found a marginally significant increase when thyroid cancer was grouped together with benign thyroid nodules (Rallison et al., 1990). The 1990 cohort study compared the prevalence of thyroid abnormalities in children born between 1947 and 1954 who lived near the NTS in two counties, one in Utah and the other in Nevada, with a group selected from an Arizona county that was presumed to have had little or no fallout from the NTS.

Thyroid nodules were found in 76 of the 4,818 children examined (15.8/1,000). Of the 76 thyroid nodules, 22 were diagnosed as neoplasms. The rate of thyroid neoplasm in the Utah-Nevada cohort (5.6/1,000) was higher than the rate observed in the Arizona subjects (3.3/1,000) (RR = 1.7), but the difference was not statistically significant. In a 1985-1986 re-examination of the original study subjects, thyroid nodules were found in 125 people (44.2/1,000), and 65 were classified as neoplasms (benign and malignant). The rate of thyroid neoplasm in the Utah-Nevada cohort (24.6/1,000) was again slightly higher than in Arizona cohort (20.2/1,000) (RR = 1.2), but the difference was not statistically significant (p = 0.65) (Rallison et al., 1990).

A third study (Kerber et al., 1993), estimated individual radiation doses and current thyroid-disease status in members of the same cohort of 4,818 school-children studied by Rallison et al. The investigators collected questionnaire data on dietary intake during the fallout period and estimated thyroid doses from ^{131}I for 2,473 of the subjects. RR models adjusted for age, sex, and location (state) were used to estimate prevalence data on thyroid carcinomas, benign neoplasms, and nodules. Doses ranged from 0 to 4,600 mGy, and averaged 170 mGy in Utah. There was a statistically significant excess of thyroid neoplasms (benign and malignant; n = 19) and an increase in ERR of 0.7% per mGy. A relative risk of thyroid neoplasia of 3.4 was observed among 169 subjects exposed to doses greater than 400 mGy. Positive but nonsignificant dose-response slopes were found for thyroid carcinomas and nodules (Kerber et al., 1993).

The only other cancer for which an increase attributed to NTS releases has been suggested is leukemia (not including CLL). Early studies by Lyon et al. (1979) and Machado (1987) posited an increase, but statistical evidence failed to support a significant dose-response relationship (Stevens et al., 1990). However, they did find a statistically significant association between leukemia and dose for those who died at age 20 and those dying in the period 1952-1957, which is surprisingly early, given the distribution for latent periods for nonCLL leukemia observed in the atomic-bomb survivors. The median dose for all subjects was 3.2 mGy (Simon et al., 1995). When all subjects were included weak but non-statistically significant dose relationship was found (Gilbert et al., 2002). The most important source of radiation dose to the bone marrow is not ^{131}I taken into the body, but external exposure to sources of radiation deposited on the ground (Beck and Krey, 1983). The maximum dose (internal and external) to the bone marrow from NTS fallout was estimated to be 3 - 10 mGy (see Figures 5.3 and 5.4 and the draft feasibility study of Bouville et al., 2002). The leukemia dou-bling dose was estimated at 1.1 Sv (EPA, 1999). There is no evidence to support a statistically significant increase in leukemia or any other cancer besides thyroid cancer from NTS releases in the heavily exposed southern Utah downwinders. It is unlikely that an increased incidence of leukemia or cancer, other than possibly thyroid cancer, from NTS fallout in residents exposed to comparable or lower doses in more distant locations in the United States would be detectable.

Marshall Island Studies

The Marshall Islanders resident on Rongelap and Utirik atolls received much larger doses than those exposed to NTS fallout, largely from the BRAVO test, one of the series of tests conducted by the United States at the Bikini atoll. The population has been under study since 1954. The Rongelap population received the highest thyroid dose, with estimates as high as 52 Gy in a 1 year old child, and as high as 13 Gy in an adult female. An adult cancer patient on Utirik had an estimated thyroid dose of 6.8 Gy. The thyroid dose was 85% from short-lived radionuclides of iodine, and about 15% of the dose was from [131]I. Thus, the type of exposure differed substantially in amount and kind from that experienced by the population living near the Nevada Test Site. Medical followup of the most heavily irradiated residents of Rongelap and Utirik was conducted by the Department of Energy and its predecessors through 1991 (Howard et al., 1997). These studies are reported in a series from the Brookhaven National Laboratory (BNL) (Howard et al., 1995, and Howard et al., 1997). A special issue of Health Physics (volume 73, 1997) was devoted to a review of the health consequences of nuclear testing in the Marshall Islands. Subsequent followup has been conducted by other international teams.

The small size of the group exposed on Rongelap and Utirik, the low fraction of the thyroid dose from ^{131}I, uncertainties in the dosimetry, the intermittent use of thyroxin suppression after 1965, and the absence of ultrasound screening prior to 1994, taken together diminish the credibility of numerical risk estimates drawn from these studies. Thyroid nodules were first detected by palpation in 1963. Adenomatous nodules in Rongelap (17/67) and Utirik (10/167) were diagnosed and treated surgically. Of the 12 individuals exposed in utero (4 in Rongelap, and 8 in Utirik), 3 (12%) were found to have thyroid nodules, and 1 had a suspect papillary thyroid cancer. Most of the adenomatous nodules occurred in children under age 15 in Rongelap, with a risk estimated at 0.83 per 10^4 persons per mGy per year. There were 6 thyroid cancers diagnosed among the Rongelap exposed (7.0%), and 11 in the Utirik exposed (6.6%); 7 of the 17 cancers were classified as occult (microscopic) papillary cancers. Risk was estimated at 0.15 per 10^4 persons per mGy per year. The ratio of benign to malignant disease in the Utirik population was 3.5:1 in person's age < 10 years at time of exposure (ATE), and 6.5:1 in those > 10 years ATE. In Rongelap exposed persons over age 10 years ATE, the benign to malignant ratio of nodules was 18:1. This suggested a smaller cancer risk when the thyroid dose exceeded 20 Gy, a decline that was presumed to be due to cell killing (Howard et al., 1997, NRC, 2000).

Marshallese living on the many atolls which comprise the Republic of the Marshall Islands (RMI) were exposed to a wide range of levels of fallout from the many tests at Bikini. A 10-year study examined 7,172 Marshallese (1993-1997) Takahashi et al., 2001. The investigators estimated thyroid dose based on recorded data for Utirik exposed individuals, but used ^{137}Cs levels on the soil as a surrogate for estimating dose from ^{131}I to the persons living on the other atolls. Exposed individuals in Rongelap were excluded in order to avoid non-linearity noted due to cell killing at the higher dose levels. They used ultrasound in their investigations and found 38 new thyroid cancers adding to the 30 reported previously. Summing over all the histological types, papillary variants comprised 77% of the 68 thyroid cancers with an additional 13% not classified as to cell type. Thyroid cancer was approximately twice as frequent in females as males.

TABLE 4.2   Risk Factors of Thyroid Cancer Among 3,378 People Alive at the BRAVO Test for Whom Dose Estimates Could Be Derived[a]

Weighted Median Dose (cGy)	Number of cases of thyroid cancer (%)			Adjusted Odds Ratio (for Total) (95% CI)
	Male	Female	Total	
0 – 3.4 (2.33)	3 (0.7)	8 (1.9)	11 (1.3)	1.0
3.4 – 7.5 (5.6)	6 (1.3)	4 (1.1)	10 (1.2)	0.99 (0.41-2.42)
7.5 – 18.7 (10.2)	2 (0.6)	12 (2.4)	14 (1.7)	1.37 (0.59-3.14)
18.7 – 677.7 (77)	4 (1.0)	11 (2.5)	15 (1.8)	1.67 (0.73-3.83)

[a]SOURCE: Adapted from Takahashi et al., 2001.

The absolute risk of thyroid cancer was not higher in persons exposed as children than as adults, but they were unable to correct analytically for temporal differences in ascertainment. Thyroid cancer risk was not significantly correlated with dose in 3,378 people for whom dose estimates could be made. Odds ratios were greater than 1.0 for the two highest dose quartiles, but the trend was not statistically significant ($p = 0.15$). The odds ratio for sex (female/male) was strongly positive, 2.11 (1.14-3.89) (See Table 4.2)

Clinical study findings approximately 40 years after the BRAVO test indicated that

• Disorders of thyroid function, such as hypothyroidism and Graves disease, were infrequent and had rates lower than or comparable with those in most other countries.

• Autoimmune thyroiditis was rare in the Marshall Islands.

• There was a high prevalence of thyroid nodules (size > 4mm diameter), in the Marshall Islanders (in about 50% of women over 60 years old).

• The frequency of thyroid nodules did not decrease with distance (a surrogate for dose) from Bikini, as had been suggested by Hamilton et al. (1987).

Conclusion   The risk of thyroid nodules and thyroid cancer (papillary) on the islands of Rongelap and Utirik was increased but uncertainties in dose, and the fact that only 15% of the dose is believed to come from ^{131}I, limits the quantitative inferences that can be drawn for numerical comparisons of risk with other ^{131}I exposed groups. The evidence is strong for an increase in thyroid cancer and thyroid nodules in the high-dose Rongelap group. RECA compensation is not relevant since residents of these islands are already covered for compensation by separate legislation.

Semipalatinsk Test Site (STS) in Kazakhstan (former Russian Nuclear Test Site)

Persons who lived downwind of the nuclear testing at the Semipalatinsk Test Site (STS) in Kazakhstan (nuclear test period, 1949-1962) are being surveyed for radiation-related thyroid disease. An initial small screening study was done in 1998 (1,990 subjects) and dosimetry has not been completed (Land et al., 2004).

Studies of Persons Exposed to Environmental Releases of Radioactivity from Nuclear Plants and Accidents

Chornobyl Studies

The largest increase in absolute thyroid cancer risk after the Chornobyl accident has been in children who were under 5 years old at the time of the accident, with a progressive decrease in observed risk to the age 18 years (Thomas et al., 1999). Ecologic studies have reported significant correlations between thyroid-cancer incidence and radiation exposure, but only two small published case-control studies (Astakhova et al., 1998; Davis et al., 2004a; Stepanenko et al., 2004) have shown higher estimated doses in the cases than in the controls. The Astakhova et al. study, in Belarus, found a strong trend for increased thyroid cancer with increasing dose (preliminary dose estimates). The Davis et al. study, in Russia, found a highly significant regression between thyroid cancer and dose (p < 0.009). The number of cases was small (26), and the doses imprecise, and some bias may have been due to unblinded interviewers about disease status; but the results were internally consistent and in agreement with other observations. Larger cohort and case-control studies include children in Ukraine and Belarus and are in progress based on measured thyroid doses. The method used to estimate individual thyroid doses in Belarus has been published (Gavrilin et al., 2004). Dosimetry methods and findings in Ukraine have been published (Likhtarov et al., 2005) with a forthcoming publication containing thyroid-cancer risk estimates from a cohort study in Ukraine (Tronko et al., Submitted). When published, those studies should provide well-grounded risk estimates of ^{131}I that can be compared with those derived from external exposures (atomic-bomb survivors, Table 4.3; and medically treated subjects, Table 4.4).

The prevalence of noncancer thyroid disease in Chornobyl exposed children was surveyed among children in the Bryansk and Kaluga regions who were 10 years old or under at the time of the accident. Dose was estimated in about 2,500 of the children who were examined and had ultrasonography and thyroid-function biochemical tests. The diseases considered were thyroid nodules, cysts, and chronic thyroiditis. Diffuse goiter in young men (25 years old at the time of examination) was the only positive finding (the odds ratio [OR] at 1 Gy was 1.36 (95% CI = 1.05-1.99) (Ivanov et al., 2005). In contrast, a similar study of Nagasaki atomic-bomb survivors exposed to external radiation did not show a significant correlation with diffuse goiter but did have a significant dose-response relationship for nodule prevalence (Nagataki et al., 1994).

Hanford Releases: Hanford Thyroid Disease Study

The Hanford Thyroid Disease Study (HTDS) was mandated by Congress in 1988. The epidemiologic study was designed to examine whether rates of thyroid disease were higher than normal among people exposed to releases of radioactive iodine from the Hanford site during the period of highest releases, 1944-1957. The study was conducted by a team of investigators at the Fred Hutchinson Cancer Research Center. It covered 5,199 people identified from records of births during 1940-1946 to mothers whose place of residence was in one of seven higher-dose counties in Washington state. The study was a screening study consisting of a cohort selected on the basis of presumed past exposures to various levels of ^{131}I released to the atmosphere from Hanford operations.

The major end points were thyroid cancer, benign thyroid nodules, hypothyroidism, and autoimmune thyroiditis. For each of those four categories, the study found that people with high doses had about the same amount of disease as people with low doses. There was no evidence of a statistically significant increase in any of the four diseases with increased radiation dose to the thyroid (Davis et al., 2004b). Problems associated with the dose correlations are inherent in environmental epidemiology studies and are discussed below.

Retrospective dosimetry: It is difficult to accurately estimate, and validate absorbed dose in environmental epidemiology studies in general and in each of the studies we have reviewed. The major uncertainties include the amount of radioactivity taken into the body and interindividual variation in metabolism and anatomy. An estimate of the average dose to people in a region is better defined than the dose to particular individuals. Validated person-specific dose estimates await the development of accurate tissue specific biomarkers. In addition, when different radiation sources are involved, other problems arise, as noted below.

Relative Biological Effectiveness: The effects from short-lived radionuclides of iodine, ^{131}I, result from x and gamma rays known to differ in amount, but not in kind. The absorbed dose distribution differs significantly between alpha-, beta-, and gamma-emitting radionuclides, and short-range electrons emitted convey intense dose to nearby structures. The cancer risk coefficients for external x and gamma rays to the thyroid are based on better thyroid absorbed-dose estimates than those derived from internal-emitter studies, so the relative biological effectiveness of ^{131}I vs x rays is still an open question; but current estimates place it at values near 1 for cancer induction from low LET radiation (UNSCEAR, 1993, and 2000). The dose from short-lived radionuclides of iodine from the Chornobyl accident is believed to be relatively minor (Gavrilin et al., 2004), but it is presumed to have had a substantial influence in the Marshall Islanders, in whom it constituted about 85% of the dose (Lessard et al., 1984; Lessard et al., 1985).

Conclusion The large increase in thyroid cancer observed in young children exposed to ^{131}I intake from the Chornobyl accident is the first reliable evidence in humans of an increased thyroid-cancer rate after relatively large exposure to [131]I. It adds to the information derived from observations of the Marshall Islanders, who received even larger doses, mostly from radioactive isotopes of iodine other than [131]I.

Studies of Thyroid Cancer after External Irradiation

The first data on radiation-induced thyroid cancer came from x-ray therapy of the head and neck in children. Followup of many medically irradiated populations has contributed much of the information on the magnitude of the risk, especially to children. Data on the risk to both children and adults also come from studies of the Japanese atomic-bomb survivors. Table 4.3 contains average ERR (AERR) at 1 Sv, and average excess absolute risk (AEAR) (10^4 PY per Sv) and summarizes the observed thyroid cancer risks estimated since 1990 in the major studies.

A 1994 incidence study of persons in the Life Span Study (LSS) includes 817,600 person-years of followup (Thompson et al., 1994). There were 132 observed cases of thyroid cancer, with higher risk coefficients in males than in females and a stepwise decrease with age in both.

The LSS cohort sample included 375,600 person years at risk for persons who were 0-19 years old at the time of the bomb (ATB). There were 59 thyroid-cancer deaths vs 22.2 expected, with a mean dose of 0.26 Sv. The $AERR_{1Sv}$ was 6.3 (95% CI = 5.1-10.1) and the AEAR was 3.8 (95% CI = 3.8 (2.7-5.4), values within the range of those observed in the incidence study; see Table 4.3.

The various medical studies summarized in Table 4.4 all found a significant increase in thyroid cancer after doses of 0.1-12.5 Sv. Variations in the risk coefficients may reflect differences in radiation sensitivity of children to different diseases for which radiation was administered and dosimetry uncertainties. Other

TABLE 4.3   Thyroid Cancer Incidence in the Japanese Atomic-Bomb Survivors. Average ERR, and Average EAR of Thyroid Cancer with Increasing Age[a]

Category	Observed	Expected	Mean Dose (Sv)	PY	$AERR_{1Sv}$	AEAR 10^4 PYSv^{-1}
Male	22	14.9	0.27	307,167	1.80	0.87
Female	110	79.4	0.26	510,388	1.49	2.32
ATB[b] 0-9	24	7.6	0.21	185,507	10.25	4.21
ATB 10-19	35	14.6	0.31	190,087	4.50	3.46
ATB 20-29	18	17.5	0.28	132,738	0.10	0.13
ATB > 30	55	54.5	0.25	309,224	0.04	0.06
ATB All	132	94.3	0.26	817,600	1.5 (0.5-2.1)	1.8 (0.8-2.5)

[a]Modified from Table 17, UNSCEAR, 2000; page 408.
[b]Age at time of bomb.

TABLE 4.4 Risk of Thyroid Cancer After External Radiation Exposure in Children

Study	Observed	Expected	Mean Dose (Sv)	PY	AERR$_{1Sv}$	AEAR 10^4 PYSv^{-1}
Shore et al., 1993	37	2.7	1.4	82,204	9.5 (6.9-12.7)	3.0 (2.2-4.0)
Tucker et al., 1991	23	0.4	12.5	50,609	4.5 (301-6.4)	0.4 (0.2-0.5)
Lundell et al., 1994	17	7.5	0.26	40,6395	4.9 (1.3-10.2)	0.9 (0.2-1.9)
Lindberg et al., 1995	15	8.0	0.12	37,0517	7.5 (0.4-18.1)	1.6 (0.09-3.9)
Ron et al., 1989	43	10.7	0.10	27,4180	34 (23-47)	13 (9.0-18)
Shore, 1990	13	5.4	0.24	34,700	5.9 (1.8-11.8)	9.1 (2.7-18.3)
Schneider et al., 1993	309	110.4	0.60	88,101	3.0 (2.6-3.5)	37.6 (32-43)
Ron et al., 1995	436	NA	NA	NA	12.0 (6.6-20)	3.5 (2.0-5.9)

possible explanations include differences in case ascertainment and in surgical removal of suspected thyroid neoplasms. The relative risk between ^{131}I and external radiation is not well established. As in many of the other risk comparisons, the data on the atomic-bomb survivors provides the best information on risk as a function of age at exposure and dose. Ron et al. (1995) used a pooled analysis of data from seven cohort studies (atomic-bomb survivors, children treated for tinea capitis, two studies of children irradiated for enlarged tonsils, and infants irradiated for enlarged thymus), and two case-control studies of patients with cervical cancer and childhood cancer. The studies were conducted on almost 120,000 people about 58,000 exposed to a wide range of doses and 61,000 nonexposed subjects) and included nearly 700 thyroid cancers and 3,000,000 person years of followup. For persons exposed to radiation before the age of 15 years, a linear dose-response relationship best described the data down to 0.10 Gy. For childhood exposures, the pooled excess relative risk per Gy (ERR/Gy) was 7.7 (95% CI = 2.1-28.7) and the excess absolute risk per 10^4 PY per Gy (EAR/10^4 PY-Gy) was 4.4 (95% CI = 1.9-10.1). The ERR was greater (p = 0.07) for females than for males, but the findings from the individual studies were not consistent. The ERR began to decline about 30 years after exposure but was still increased at 40 years. Risk decreased significantly with increasing age at exposure; little risk was apparent after the age of 20 years. On the basis of the data, there was a suggestion that spreading dose over time (from a few days to over a year) may lower risk, possibly because of the opportunity for cellular repair mechanisms to operate. The thyroid gland in children has one of the highest risk coefficients of any organ, and there is convincing evidence of increased risk at 1.10 Gy (Ron et al., 1995).

A high proportion of the thyroid cancers in the atomic-bomb survivors are accounted for by exposure at young ages. Little (2002) reports that over 50% of the excess cases associated with either the atomic bomb radiation or natural background radiation are linked to exposures under the age of 20 years, irrespective of the assumed risk model or natural background dose rate. The excess risk is overwhelmingly concentrated among females, again irrespective of the assumed model or natural background dose rate. Depending on the assumed natural background dose rate (in the range 0.5-2.0 mSv/year) between 17.3 and 32.0% of the thyroid cancers in this cohort may be associated with natural background radiation if an absolute-risk model applies; between 4.2 and 17.1% of the thyroid cancers may be associated with natural background radiation if the relative-risk model applies. The proportion of the thyroid tumors attributed to the atomic bomb radiation is between 21.1 and 22.0% for the absolute risk model, and is between 18.7 and 19.1% for the relative-risk model, in both cases irrespective of the assumed background radiation dose. The proportion of thyroid cancers accounted for by natural background radiation progressively increases with attained age, from 0.3% of cancers among those under the age of 15 years to 30.5% for those over the age of 60 years, assuming that the absolute-risk model applies. There is a similar increase in this percentage if it assumed that the relative-risk model applies (Little, 2002).

Conclusion   The thyroid in children is highly sensitive to ionizing radiation from x rays and an increased incidence of thyroid cancer has been noted in some populations after doses as low as 0.1 Gy (Ron et al., 1995; Ron et al., 1989). The highest risk observed is in the youngest children, especially in females, and the increase in risk lasts for 40 years or more but at a decreasing rate in the later years.

Studies of Thyroid Cancer after Medical Administration of [131]I

Two studies of children given ^{131}I in diagnostic doses have reported small increases in the appearance of thyroid cancer. A Swedish study found 50 thyroid cancers when 39.4 were expected (SIR = 1.27, 95% CI = 0.94-1.71) after a mean dose of 0.5 Gy (range, 0-40.5) (Hall et al., 1996a). However, the study included a very small number of young children. No increase was noted when persons referred with the diagnosis of suspected tumors were excluded (Holm et al., 1988).

A study conducted, by the US Public Health Service, found five thyroid cancers when 2.53 were expected after a mean dose of 0.9 Gy (range = 0-20) (Hamilton et al., 1987). Three studies of mostly adults given ^{131}I in therapeutic doses have been reported from Sweden, England, and the United States. The Swedish incidence study found 18 thyroid-cancer cases when 13.9 were expected (SIR = 1.29; 95% CI = 0.76-2.03) after doses over 100 Gy (Holm et al., 1991).

The British study of cancer mortality after radioactive-iodine treatment of 7,417 thyrotoxic patients found decreased overall mortality (SIR= 0.83; 95% CI = 0.77-0.90) (634 observed/761 expected). Overall cancer mortality in the study was also decreased (SMR = 0.90, 95% CI = 0.82-0.98) (448 observed/499 expected). The thyroid cancer incidence, however was increased (SIR = 3.25, 95% CI = 1.69-6.25) (9 observed/2.8 expected) (Franklyn et al., 1999).

The US (NCI) mortality followup study found 24 thyroid cancer deaths when 6.09 were expected (OR=3.94, 95% CI = 2.52-5.86) after a mean dose of 50-70 Gy (Ron et al., 1998a). If one assumes a 5-year latent period and excludes deaths in the first 5 years after therapy, the OR falls to 2.6, which is of marginal statistical significance. Although doses to the thyroid could not be estimated adequately, no exposure-response was observed when administered activity was used as a proxy measure of dose.

Conclusions   The thyroid cancer risk after medical ^{131}I exposure is poorly documented, whereas the risk after exposure to external radiation is very well documented. The small number of children who received ^{131}I in medical studies, and the medical considerations for the procedure, complicate interpretation of the findings, so there is little confidence in the results of the studies. The ^{131}I risk coefficients derived from environmental-epidemiology studies are likely to be more reliable despite dosimetry uncertainties, which are being reduced by efforts to compute individual doses for Chornobyl and other populations exposed to [131]I.

Thyroid Nodules after [131]I Exposure

Thyroid nodules are common in the general population, and they increase in number with age (Tan and Gharib, 1997). Before the 1980s, most thyroid epidemiology studies report thyroid nodules on the basis of manual palpation; more recent studies report results based on thyroid ultrasonography. In both circumstances, the findings are buttressed by fine-needle aspiration or surgical biopsy, which provides the information needed to distinguish benign from malignant nodules. They are more prevalent in regions where the diet is low in iodine (Gembicki et al., 1997). When ultrasonography is used, the prevalence of nodules is about 60% in persons over 70 years old. Thyroid nodules are relatively rare in children under 18 years old (Mettler and Upton, 1995). The risk coefficients for thyroid nodules reported in heavily exposed (high-thyroid-dose) populations (Marshall Islanders) are up to 8 times higher than the risk coefficients for thyroid cancer.

A Food and Drug Administration-sponsored study of children who received diagnostic [131]I (mean dose, 0.9 Gy) reported an ERR/Gy of 2.0 (95% CI = –0.5-12.5) for thyroid nodules (Hamilton et al., 1987). The frequency of thyroid nodules in 1,005 women given ^{131}I for diagnostic function and imaging tests was compared with that in a comparison group of women (248) attending a mammography screening clinic in a Swedish study (Hall et al., 1996b). The mean dose to the thyroid was 0.54 Gy, and the average age was 26 years old. The study found an ERR of 0.9 per Gy (95% CI = 0.2-1.9), but no difference in the ERR between those exposed under the age of 20 years, and those exposed after the age of 20 years.

External Radiations: Japanese Atomic-Bomb Survivors

The frequency of nodules was assessed with ultrasonographic screening in the Nagasaki Adult Health Study in 2,587 persons (61% women) 40 years old after exposure to the atomic bomb. The average dose was 0.77 Sv, and thyroid nodules were detected in 39 men and 151 women. A statistically significant increase in solid nodules was found only in women, but the authors did not describe the power of the test, nor did they calculate the ERR/Sv from their findings (Nagataki et al., 1994).

Conclusion Thyroid nodules increase in frequency with age, restricted intake of iodine in the diet, and radiation exposure. Only a small fraction of thyroid nodules become malignant. Given our review of the data and the fact that screening for thyroid nodules was considered and rejected by the Institute of Medicine for use in irradiated populations, we find no basis for reversing that recommendation. Thyroid nodules that progress and are diagnosed as malignant are covered under RECA.

STUDIES OF POPULATIONS OCCUPATIONALLY EXPOSED TO RADIATION

The causal association between exposure to ionizing radiation and the appearance of late effects, primarily cancer, was first recognized among groups of early radiation workers. Populations at risk for occupational exposure to radiation since the early 1940s have been subject to increasingly stringent occupational radiation-protection standards (Jones, 2004). Epidemiologic studies of such populations continue to make important contributions to the understanding of radiation-induced disease, particularly of the risks of late effects after low doses (< ~200 mGy [< ~20 rad]) that often are of public concern. The epidemiologic strengths and weaknesses of such studies must be borne in mind when reviewing their findings.

The strengths of the occupational population studies include the availability of large numbers of people, many of whom were individually monitored for radiation exposure on the job and have long periods of followup. In many instances, records exist of individual workers' work and occupational-medicine histories and of the operations and processes. A weakness or limitation of the studies is that the worker populations have been predominantly healthy white adult males 18-65 year old, often with regular access to health care, and do not include representative numbers of people who are unemployed because of illness or other factors and other segments of the general population, such as women, children, and other races—all factors that can influence health. Thus, results of many of the low-dose studies cannot be extended to other segments of the general population. Also, the characteristics of worker populations are recognized as contributing favorably to their mortality and illness experiences relative to the general population, a bias described as the healthy-worker effect (McMichael, 1975). However, that effect is generally considered to be greater for noncancer death or disease rates among workers when they are compared with segments of the general population of similar age, sex, and race than when those rates are compared with cancer rates, because it is difficult to screen from the workforce those who might develop cancer in the future. The average cumulative radiation doses to individual workers generally are low with the uncertainties that are inherent in monitoring data on individual workers, so total population dose tends to be both low and poorly estimated. Those limitations diminish the statistical power of the worker studies to evaluate the risk of radiation-induction of disease, primarily radiogenic cancers, at low doses. Other factors that can limit evaluation of the risks of disease at low doses are exposures to multiple agents, including one or more additional types of ionizing radiation from external and internal sources, and to chemicals and other workplace hazards and individuals' lifestyle habits, often undocumented, such as smoking. Those factors limit the ability to detect and confirm a radiation-induced effect at low doses. Those issues have been discussed recently in more detail by Gilbert (2001) and Howe (2004).

In addition to uranium miners, millers, and ore transporters, several other groups of workers at risk of exposure to radiation have been followed over long periods to identify increases in causes of death relative to nonexposed comparison populations and to evaluate statistically significant relationships between such increases and occupational radiation exposures. The results have been increasingly available since the middle 1980s. They have particular relevance in considering the risks to the downwinders and onsite participant RECA populations because of the similarities between their radiation-exposure experiences and those of the occupationally exposed populations. Both these population groups were at risk of exposure to one or more types of radiation from external and internal sources at low doses and low dose rates over extended periods. They also probably had some similar non-occupational risk factors. Descriptions of most of the individual worker populations and findings published through the late 1990s are summarized in UN Scientific Committee on Effects of Atomic Radiation (UNSCEAR) reports (UNSCEAR 1994; 2000). For the purposes of this section, we discuss additional and updated epidemiologic studies of workers, other than uranium miners, millers, and ore transporters, who potentially were at risk of exposure to radiation on the job. The populations of interest are considered as five main but not always mutually exclusive groups; specifically:

1. Nuclear-Industry workers;
2. Commercial nuclear power-plant workers;
3. Nuclear shipyard workers;
4. Medical personnel; and
5. Military participants at nuclear-weapons test sites and US nuclear submariners.

To address the committee's charge regarding new epidemiologic information that might affect radiation risk estimates, the more recently reported findings of the major occupational epidemiologic studies, other than those involving uranium miners and millers, are summarized here.

Studies of Nuclear-Industry Workers

Epidemiologic studies involving nuclear-industry workers have been conducted or are in progress in several countries (UNSCEAR, 1994; 2000). Some of these studies have since been updated or have served as the basis of more focused studies of workers at the same or multiple sites and of specific cancers. Reports also are available of completed studies of workers at additional sites. They include studies of populations of civilians employed in the post-uranium-milling production and research and development operations of nuclear-energy and weapons-development programs at multiple facilities in the United States (US), United Kingdom (UK), Canada, the Russian Federation (formerly part of the Soviet Union), Japan, and France. In some of those countries, the operations began in the early 1940s.

Followup studies of mortality from all causes conducted during the 1970s and 1980s for a number of facility-specific populations of nuclear program workers in the US, the UK and Canada (NRC, 1990; UNSCEAR, 1994; UNSCEAR, 2000). The more robust of those studies established a basis for combined population studies in the individual countries (Gilbert and Marks, 1979; Smith and Douglas, 1986; Beral et al., 1985; Beral et al., 1988; Howe et al., 1987) and across all three countries (Cardis et al., 1995), external low LET radiation being the primary exposure of interest. Workers in nuclear industry operations in the UK, Canada, and Japan also are included in national registries of radiation workers.

Depending on the characteristics of their jobs, nuclear-industry workers were at risk for chronic exposure to low doses of various types of radiation primarily from external or internal sources, or both, and other potentially hazardous agents present in the workplace. Mortality from all and specific causes, including all types of cancers, generally is measured as the ratio of the number of deaths observed in the study population to a number expected in the comparison or "nonexposed" group (standardized mortality ratio [SMR]) and is the main result reported in most of the studies. In many of those and other previously reported studies of nuclear-industry workers, the SMRs for total mortality and noncancer system-specific diseases are less than unity, reflecting a healthy-worker effect. Cancer mortality typically is similar to that expected among the general population, although some statistically significant increases in various site- or type-specific cancer mortality are noted. However, with the exception of nonCLL leukemia, there is a lack of evidence of a consistent pattern of such increases (Telle-Lamberton et al., 2004) or of their attribution to occupational radiation exposure. A causal association between chronic exposure to low doses of low LET radiation and multiple myeloma remains equivocal. In some studies, nonexposed workers or workers in different dose or job groups at the same facility are used as internal controls when radiation dose-response relationships are examined for all or site-specific solid cancers and leukemia, thereby taking the healthy-worker effect into account. The major, more robust studies have provided risk estimates expressed as excess relative or absolute risks (ERR, EAR, respectively) for radiation induction of radiogenic cancers and nonCLL leukemia respectively. To date, only a few morbidity studies of nuclear-industry workers have been conducted. Those have tended to focus on subcohorts of workers at nuclear facilities, who also were at risk of exposure to specific radionuclides, such as plutonium or nonradioactive toxic metals or chemicals.

Some previously evaluated facility-specific studies of nuclear workers have since been updated or have served as the basis of more focused studies of workers at the same or multiple facilities and of workers with specific exposures or cancers. Reports on recent studies of workers at additional nuclear sites also are available in the peer-reviewed literature. In Table 4.5, we reference the major studies in those categories and summarize their significant findings with respect to a risk for radiation-induced cancers and nonCLL leukemia as available.

Studies of Commercial Nuclear Power-Plant Workers

Workers at commercial nuclear power plants (CNPPs) are primarily at risk of chronic exposure to low external doses of high-energy penetrating radiation (x and gamma), and to a lesser extent to neutrons externally and possibly of alpha-particle emitters (such as, uranium, radium, and radon) and low energy beta emitters (such as, tritium,^{3}H) internally. The findings of two large combined-population cohort mortality studies have recently been reported; both focused a priori on evaluation of relationships between radiation and the risk of solid cancers and leukemia (except CLL).

Zablotska et al. (2004) followed a cohort of 37,735 male and 7,733 female employed and monitored for at least 1 year at four Canadian nuclear power plants in 1957-1994 with a total of almost 608,000 person year at risk (individual mean = 13.4 year). Cumulative radiation exposures (equivalent doses) for individual workers ranged from 0 (31.6%) to 498.9 mSv (49.9 rem) (mean = 13.5 mSv [1.25 rem]). Compared with the Canadian general population, mortality in male and female workers combined from all causes (1,599 deaths; SMR = 0.63, 95%

TABLE 4.5 Risk Estimates for Mortality from Selected Cancer Groups Among Atomic-Bomb Survivors and Cohorts of Nuclear-Industry Workers (Not Including Uranium Miner or Millers)

Study	Cohort Size[a]	Mean dose (mSv)	No. of deaths, All Causes/ All Cancers	ERR:All Cancers Except Leukemia ERR/Sv	CI	ERR: Leukemia Except CLL ERR/Sv	CI	Significant Specific Cancer Increases[b] Other Than Leukemia
Atomic-bomb survivors (Pierce et al., 1996)	86,572	NA	NA/7,578	0.24[c]	0.12,0.4	2.2[c,d]	0.4,4.7	NA/NA
IARC[e] (Cardis et al., 1995)	95,673	40.2	15,825/3,976	−0.07	−0.4,0.3	2.2	0.13,5.7	Multiple myeloma, 44 deaths p = 0.037 (1-sided)
NDRC[f] (Ashmore et al., 1998)	206,620	6.3	5,426/1,632	3.0[g]	1.1,4.9	0.4	−4.9,5.7	0 (CI 90%)
NRRW[h] (Muirhead et al., 1999)	124,743	30.5	NA/3,596	0.09	−0.3,0.5	2.6	−0.03,7.2	0 (CI 90%)
JNIWR[i] (Iwasaki et al., 2003)	175,939	12.0	2,934/1,191	NA	NA	0.01	−10.0,10.0	0 (CI 95%)
SETCEA[j] (Telle-Lamberton et al., 2004)	58,320	NA	4,809/1,898	NA	NA	NA	NA	Pleural cancer, 28 deaths, SMR = 1.79 Melanoma, 24, SMR = 1.50

[a]Total number of men and women.
[b]Confidence intervals.
[c]Adjusted for the effects of time since exposure and for nonlinearity in dose.
[d]Based on male atomic-bomb survivors ages between 20 and 60 years at exposure. as presented by Murihead et al., 1999.
[e]International Agency for Research on Cancer.
[f]National Dose Registry of Canada.
[g]Estimate for men, estimate for all cancers in women = 1.5/Sv , (90% CI = −3.3. 6.3).
[h]National Registry of Radiation Workers, UK.
[i]Japanese Nuclear Workers Registry.
[j]Suivi Epidemiologique des Travaillersdu Commissariat a l'Energerie Atomique, France.
NA = Not available.
SOURCE: adopted from Gilbert, 2001.

CI = 0.60-0.66) and all cancers (531 deaths; SMR = 0.74, 95% CI = 0.68-0.80) demonstrated a "healthy-worker" effect typical of a relatively young workforce. Deaths due to leukemia except CLL, for males and females combined, were fewer than expected, but the deficit was not statistically significant (18 deaths; SMR = 0.80; 95% CI = 0.47-1.26). The RRs for nonCLL leukemia increased monotonically across the four dose groups (<1, 1-49, 50-99, and >100 mSv) on the basis of one death in each of the two highest dose groups. ERR/Sv for all solid cancers (2.80, 95% CI = −0.038-7.13), and nonCLL leukemia (52.5, 95% CI = 0.205-291) were higher than those for the atomic-bomb survivors (Little and Muirhead, 1998) and the International Association for Research on Cancer (IARC) combined analysis of nuclear workers (Cardis et al., 1995), but the authors considered that they could have been due to chance. Uncertainties associated with the relatively small numbers of deaths to date also could have contributed to the findings.

A companion study (Howe et al., 2004) of US CNPP workers evaluated noncancer and cancer mortality among a predominantly male cohort (N = 53,698) with individual radiation monitoring data for at least a year while they were employed at 52 facilities nationwide some time between 1979 and 1997. As in the study by Zablotska et al. (2004), a marked healthy-worker effect was seen for noncancer deaths (773 deaths; SMR = 0.34, 95% CI= 0.32-0.36) and all solid-cancers deaths (368 deaths; SMR = 0.65, 95% CI = 0.59-0.72) relative to the general US population. However, positive but not statistically significant associations with radiation dose were seen for nonCLL leukemia (26 deaths; ERR/Sv = 5.67, 95% CI = −2.56-30.41) and for all solid cancers (368 deaths; ERR/Sv = 0.506, 95% CI = −2.01 - 4.64). The finding of a high mortality risk, ERR/Sv 8.78 (90% CI = 2.19-20.0) from arteriosclerotic heart disease in the worker population is considerably higher than reported in the LSS where the authors also found a significantly increased heart disease mortality risk (ERR/Sv = 0.17, 90% CI = 0.08-0.26) p = 0.001 (Preston et al., 2003). In contrast, the incidence of heart disease in the adult health study for about 10,000 participants during the period 1958-1998 (Yamada et al., 2004) showed no significant relationship with radiation dose for any of the cardiovascular diseases. The high rate of heart disease observed by Howe et al. (2004) was noted as being out of line with other observations, and they advised that further attention to the issue was warranted. McGale and Darby (2005) systematically reviewed the published findings of studies of mortality (25 studies) and morbidity (one study) from circulatory disease among various populations, including some worker populations at risk for exposure to radiation doses between 0-5 Sv. The authors concluded that there is no clear epidemiologic evidence of a risk of circulatory diseases at 0-4 Sv, as was suggested by the study of atomic-bomb survivors (Preston et al., 2003).

Because the commercial nuclear-power industry was established somewhat later than the nuclear-energy and weapons-development programs, mortality and the years of followup available for CNPP workers are less than for workers in the nuclear-development programs, so the statistical power of these studies is more limited. However, CNPPs provide the opportunity for followup studies of large cohorts of individually monitored people at risk of exposure to radiation at occupationally low levels, and both cohorts are included in the combined analysis of mortality among nuclear workers in 15 countries that is being coordinated by IARC.

As illustrated by the summary reviews, there are benefits to being able to evaluate the human risks posed by exposure to low levels of radiation through direct observation and measurement of exposed populations. They allow evaluation of the cancer risks estimated with extrapolation from data on populations such as the atomic-bomb survivors, who were exposed at high dose rates over a much wider range of doses—from very low doses to several Gy—than those measured directly in the low-dose and low-dose-rate populations. Such comparisons can show whether the cancer risk estimates obtained by extrapolation for low exposure levels significantly underestimate or overestimate the risks obtained through direct measurement. Also, despite their limitations, the multiple low-level exposure studies contribute to the "weight of evidence" with respect to the validity of the cancer risk estimates obtained by extrapolation that contribute to the basis of current radiation-protection standards.

Studies of Nuclear Shipyard Workers

Between the early 1950s and 1970s civilians were employed in the US naval nuclear propulsion program at facilities nationwide where they were involved in building and overhaul of US nuclear naval vessels. In those activities, workers were at risk of external exposure to gamma radiation from cobalt-60 and other radionuclides deposited in the nuclear reactor systems and to asbestos and industrial chemicals.

A recent study by Silver et al. (2004) has updated through 1996 the mortality experience with respect to radiation status of an expanded cohort of 37,853 predominantly white civilian men and women employed at the Portsmouth Naval Shipyard (PNS), Kittery, Maine, some time between 1952 and 1992. This population originally was the subject of a proportional mortality analysis that found greater than expected proportions of leukemia and all cancers among the men (Najarian and Colton, 1978). Reports of those findings, attributed to methodologic shortcomings, contributed to the concerns raised in 1978 that led to the formulation and eventual enactment of RECA 1990 as described in Chapter 2.

More rigorous followup studies of mortality from all causes (Rinsky et al., 1981), leukemia (Stern et al., 1986), and lung cancer (Rinsky et al., 1988) in a cohort of almost 25,000 white civilian men employed at PNS some time between 1952 and 1977 found leukemia and all-cancer mortality within the range expected among a comparable component of the US population, an increased risk of lung cancer in workers with career doses of at least 1 rem (10 mSv) externally and at least 15 years after first exposure and an increased risk of nonCLL leukemia in workers in the same group exposed to at least 1 rem (10 mSv)—with no latent interval period. However, the increased lung-cancer risk with respect to radiation appeared to be smaller when exposures to asbestos and welding fumes were taken into account, and no statistically significant association was found between leukemia and radiation or solvent exposure, although there was an increased risk of leukemia in electricians and welders.

PNS employees also were included in a combined population study of mortality that involved almost 62,000 civilian workers at eight US naval shipyards that serviced nuclear powered vessels (Matanoski, 1991). This study was designed to determine whether there was an excess risk of leukemia or other cancers in the population that was associated with their occupational radiation exposure to low doses of gamma radiation. Three subcohorts were identified for comparison; they were: nonradiation workers; radiation workers with individual cumulative external doses of 0.5 rem (5 mSv) or less; and radiation workers with more than 0.5 rem (5 mSv). The overall mortality risks in all three groups were generally similar to those in the US general population but were highest for the nonradiation-worker group and significantly lower than expected for the > 0.5 rem subcohort. The risks of nonCLL leukemia and lymphoma in the radiation worker groups were lower than those for the general population and the non-radiation workers. However, the risk for the > 5 rem group was greater than that for the < 0.5 rem group. The lung cancer risk was higher in the nonradiation worker group relative to the general population and slightly, but not significantly higher in both groups of radiation workers. However, this increased risk appeared to be associated with the effects of workers' exposure to asbestos rather than to radiation.

In the updated study of the PNS cohort (Silver et al., 2004), the healthy-worker effect was less evident than previously observed by Rinsky et al. (1981); overall mortality in the full cohort was similar to that expected for the US population (12,393 deaths; SMR = 0.95, 95% CI = 0.93-0.96). Mortality from all cancers (3,192 death; SMR = 1.06, 95% CI = 1.02-1. 10) was statistically greater than expected, owing in large part to increased risks that were statistically significant for cancers of the trachea, bronchus, and lung in exposed radiation workers (monitored with > 0.0 mSv) and workers who were not monitored for radiation; confounding was associated with asbestos exposure in the radiation workers and smoking in the nonmonitored workers. Leukemia mortality in the full cohort, although slightly increased was similar to that in the general population (115 deaths; SMR = 1.01, 95% CI = 0.84-1.22) but was lower among the exposed and unexposed monitored subcohorts, whereas it was nonstatistically increased among the nonmonitored subcohort. However, a positive dose-response relationship was observed between leukemia and cumulative external radiation dose; the authors interpreted this as being consistent with the conclusions of other reviews of leukemia among nuclear workers (Schubauer-Berigan and Wenzl, 2001).

Studies of Medical Personnel

Radiologists and Radiotherapists

Since the discovery of x rays by Roentgen in 1896, many physicians, technologists and physicists have been occupationally exposed to radiation at dose levels that changed substantially over time. Before 1920, when the British Radiological Society was formed and procedural standards were formulated, high doses were received by practitioners. Therapists who used manually inserted radium needles received very high doses to the fingers, and many lost digits from overexposure. Diagnostic radiologists did not wear protective shields, and they received much higher doses than was the norm in later years. The doses to those radiation workers were fractionated, delivered over many years, and very high by current standards. The turning point with respect to the protection of radiation workers in the United States was a 1928 international meeting that led to the adoption of the roentgen as the radiation unit and the creation of the group that led to the formation of ICRP and the introduction of occupational radiation-protection standards in the United States.

All followup studies on radiologists are hampered by lack of individual dose data. During the 1920s and 1930s, doses to individual radiologists were estimated to be 1 Sv per year (Braestrup, 1957). Smith and Doll (1981) estimated annual doses to British radiologists at 0.1 Sv before the 1950s and perhaps 0.05 Sv in the early 1950s, and they declined to 0.5 mSv by 1993 (Hughes and O'Riordan, 1993). The total number of 8 cases of leukemia were observed among British radiologists who had registered with a radiological society after 1920. The fact that radiation pioneers received doses likely to have been higher than the pre-1950 annual average (0.1 Sv) is a potential source of bias.

A report on mortality among almost 2,700 British radiologists who practiced in 1897-1997 reveals a number of important findings regarding the particular tumors that were increased and the periods involved (Berrington et al., 2001). Although the number of cancer deaths among radiologists registered after 1920 was similar to that expected among all medical practitioners, there was a statistically significant trend (p = 0.002) toward increasing cancer mortality with time since entry into practice (registration with the British Radiological Society), so that those registered for more than 40 years after 1920 had a 41% excess risk of cancer mortality (SMR = 1.41, 95% CI = 1.03-1.90). Practitioners who entered practice after 1954 did not show increased mortality from cancer, so the trend was most likely due to the highest mortality risk in the earliest period. In those registered after 1920 when the first recommendations for radiological protection were published in Britain, the death rates from cancer in radiologists were not greater than the death rates in all other medical practitioners combined (SMR = 1.04, 95% CI = 0.89-1.21). No evidence was found of increased mortality other than from cancer even in the earliest radiologists despite the fact that the estimated dose they received has been associated with more than a twofold increase in death rate in the Japanese atomic-bomb survivors. The greatest proportional excess in the post-1920 radiologists was in leukemia. Mortality from noncancer causes was lower than in the comparison populations even in the pre-1921 radiologists, who would have received the highest doses. The latter finding is at odds with recent noncancer-mortality data on the Japanese atomic-bomb survivors.

Parallel studies of radiologists carried out in the United States reached many of the same conclusions (Seltser and Sartwell, 1965; Matanoski et al., 1975a, b, 1984 *cited in* Yoshinaga et al., 2004). The turning point in the United States was a 1928 international meeting that led to the adoption of the roentgen as the radiation unit and the creation of the group that led to the formation of ICRP. The SMRs for US radiologists compared with other specialists were 1.38 for all cancers and 2.01 and 1.0 for leukemia for those who entered the specialty during the periods 1920-1939, and 1940-1969 respectively. SMRs for leukemia in the UK radiologists were 2.5, 2.7, 2.29, and 1.16 for those who registered with the British Radiological Society in the periods 1897-1920, 1921-1935, 1936-1954, and 1955-1979, respectively. For all cancers, the UK radiologist' SMRs were 1.58, 1.04, 0.91, and 0.78 for the same period. Thus, similar trends in leukemia and all-cancer mortality were observed in both groups.

Matanoski et al. (1984) noted a nonstatistically significant relative risk (RR = 2.1) for multiple myeloma in the US radiologists who joined the Radiological Society of North America in 1940-1969 when compared with other physician specialty groups. No increase was seen among the earlier member cohort (1920-1939). Berrington et al. (2001) also noted a similar increase in the risk for multiple myeloma (2.32) on the basis of four deaths among the later UK radiologists (> 1940-1969) but not for the earlier cohort (>1920-1939). Several hypotheses have been offered for the finding that the risks for nonCLL leukemia for the later cohort at entry appear to decrease while the risk for multiple myeloma increased but to date it remains unexplained (Matanoski et al., 1984).

Radiological Technologists

There are many more radiologic technologists than radiologists, and they spend more hours in conducting procedures. It is estimated that there are 2.3 million medical-radiation workers worldwide. About 146,000 radiologic technologists were followed for mortality through 1990 (Doody et al., 1998, Mohan et al., 2003). The cohort included all technologists living in the US who were certified for at least 2 years between 1926 and 1982. The cohort was composed primarily of women, in contrast with the radiologist cohorts analyzed that were limited to men. A smaller cohort consisted of 6,500 male x-ray technologists trained by the army during World War II. They were followed through 1963 (Miller and Jablon, 1970) and 1974 (Jablon and Miller, 1978) with death certificates obtained through Veterans Administration files (See Tables 4.6 and 4.7).

TABLE 4.6   Mortality Risks Observed in Radiology Technologist Studies

		SMR/Number of Deaths Observed in Study Population		
Technologists	Sex	All causes	All cancer	Leukemia
US 1926-1982 Mohan et al., 2003	Female	0.76/7567	0.86/2558	$0.92^a/98^c$
US 1926-1982 Mohan et al., 2003	Male	0.76/5057	0.73/1137	$0.95^a/60^c$
US Army, 1946-1963 Jablon and Miller, 1978	Male	1.06/289	1.05/55	$1.25^b/8$

^{a}Comparison vs US population.
^{b}Comparison vs pharmacy and lab technologists.
^{c}Not including CLL.

Higher leukemia SMRs were found in Japanese male x-ray technologists (Mohan et al., 2003), and higher leukemia SIRs were found in Chinese male and female x-ray workers than in US radiation technologists (Sigurdson et al., 2003). The higher rates are presumed to reflect higher exposures. The SMRs and SIRs observed in radiologists and technologists working in the US, UK, Japan, China, and Denmark have been summarized by Yoshinaga et al. (2004). As was found for the radiologists studied, the most consistent observation was an increased leukemia risk in the early cohorts of medical-radiation workers. The lack of individual dose estimates in the years before personal dosimetry was in wide use compromises the ability to capture dose-response information from the historical studies. Future data should not be so limited, albeit with lower individual exposure doses.

TABLE 4.7   Standardized Incidence Ratios/Number of Persons Observed in US Radiation Technologist Studies

		SIR/Number of Cases Observed in Study Population		
Technologists	Sex	All causes	All cancer	Leukemia^a
US 1926-1982 Sigurdson et al., 2003	Female	NA	1.07/2408	1.12/48
US 1926-1982 Sigurdson et al., 2003	Male	NA	0.94/884	$1.04^b/27$

^{a}All types of leukemia including CLL.
^{b}Comparison vs US SEER Program.
NA = Not applicable.

Military Participants at Nuclear-Weapons Test Sites

More than 200,000 US military personnel participated in atmospheric nuclear-weapons tests between 1945 to 1963. In the late 1970s concern about the long-term health of the participants in those tests, particularly the risk of leukemia and other cancers, prompted a series of followup studies, primarily of mortality among groups of military personnel identified as having participated onsite at the 'SMOKY' (1957) or at least one of five other test series conducted at the Nevada (NTS) or Pacific (PTS) Test Sites between 1953 and 1956 (Caldwell et al., 1980; Caldwell et al., 1983; MFU, 1985). Those studies found no consistent disease patterns. Compared to the US population, leukemia incidence and mortality was significantly increased among the more than 3,000 SMOKY participants, based on small numbers of leukemia deaths, but not in some of the other groups. Average individual and the total population radiation doses accumulated during the test periods were generally low, although a few participants received > 50 mSv (5 rem) during the test year; dose-response relationships were not evaluated.

Subsequent studies by Watanabe et al. (1995b) and Johnson et al. (1996) updated and evaluated mortality among 8,554 and approximately 40,000 US Navy veterans who participated in HARDTACK I (PTS 1958) and CROSS-ROADS (Bikini 1946) tests, respectively. Followup for the HARDTACK 1 cohort was from September 1, 1958 through September 1, 1991; it was through December 1992 for the five series test study. Veterans who had not participated in the tests comprised the comparison groups for these studies. The median dose of gamma radiation was 3.88 mSv (388 mrem) among the HARDTACK I veterans. Comparing unadjusted mortality ratios for the HARDACK 1 veterans and their comparison group, Watanabe et al. (1995b) found a statistically significant increase in deaths from all causes (1,083 deaths, RR = 1.10, 95% CI = 1.02, 1.19), but a nonsignificant deficit in leukemia deaths (6 deaths; RR = 0.69, 95% CI = 0.27-1.78). Mortality from other cancer causes, except for cancers of the digestive organs, was similar in both groups; only digestive organ cancer mortality was statistically significantly increased (66 deaths, RR = 1.47, 95% CI = 1.06 - 2.04). The dosimetry data available for the CROSSROADS study were considered unsuitable for epidemiologic analyses, so mortality was compared for three surrogate exposure groups: veterans who boarded target ships and thought to be at highest risk of exposure; those who did not, and a special tasks group. In that study all causes mortality among participants was slightly increased (~5%) compared with the nonparticipant veteran group. Small increases seen in mortality from all cancers (1.4%) and leukemia (2.0%) were not statistically significant. Thus, results of those studies were equivocal with respect to a radiation effect.

Subsequently, a mortality study was designed to update the five test series study and to analyze the timing and causes of death of about 70,000 servicemen who participated in at least one of five selected nuclear-weapons test series in the

1950s; almost 65,000 comparable nonparticipants served as the comparison group. More than 5 million person/year of mortality followup information was obtained. Overall, the participants and the comparison group had similar risks of death and cancer except that the participants had a 14% higher risk of leukemia than the comparison group. That higher risk was not statistically significant and was possibly real but also could have been a chance finding (IOM, 2000).

A study was conducted of 1,010 US veterans who had received the highest gamma-radiation doses (50 mSv [5 rem] or more per year) during the 1958 HARDTACK I test series. Cancer rates were compared with those in a group of 2,870 participants in lower-dose tests. Mortality from all causes (RR = 1.22, CI = 1.04-1.44) and from all lymphopoietic cancers (RR = 3.72, CI = 1.28-10.83) was significantly higher in the high-dose cohort than in the low-dose controls (Dalager et al., 2000).

Mortality and incidence of cancer in participants in the United Kingdom's (UK) weapons tests held during the 1950s and 1960s at bases on islands in the Pacific Ocean and in Australia, was updated through 1991 for 21,358 service personnel and civilians and a control group of 22,333 nonparticipants (Darby et al., 1993a), most of whom were included in an earlier followup study (Darby et al., 1988). During the seven years of further followup, the number of deaths in the test participants were fewer than expected from national rates. SMRs for all causes were 0.86, all neoplasms 0.85, leukemia 0.57, and multiple myeloma 0.46. In the period more than 10 years after initial participation, relative risk in the participants was near unity for all causes (0.99 CI = 0.95-1.04) and all neoplasms (0.95 CI = 0.87-1.04). Leukemia mortality was equal to that expected from national rates but greater than in controls for both the followup period (1.75 CI = 1.01-3.06) and the period 2-25 years after the tests (3.38 CI = 1.45-8.25). The authors concluded that participation in nuclear weapons tests had no detectable effect on life expectancy or on the development of cancer. They attributed the apparent increase in leukemia in participants to an apparent deficit in rates observed in the controls, although a small risk of leukemia could not be excluded.

Mortality and cancer incidence between 1957 and 1987 were evaluated among a small cohort of New Zealand naval personnel (n = 528) who participated in UK nuclear weapons tests in 1957 and 1958 at Pacific island bases and 1504 nonparticipant naval controls (Pearce, 1996; Pearce et al. 1997). Mortality from all causes combined, all causes other than cancers and all cancers combined, though increased in the participants relative to the nonparticipants in some cases, was as expected. The RRs for cancer incidence overall (RR = 1.12, 90% CI = 0.78-1.60); and the incidence of cancers other than hematological malignancies (RR = 1.14, 90% CI = 0.69-1.83) were slightly but not statistically significantly increased. Hematological cancers accounted for seven deaths among the partici-pants (RR = 3.25, 90% CI = 1.12.-9.64) including four from leukemia (RR = 5.58, 90% CI = 1.04-41.6), one of which was the CLL type (Pearce et al., 1997). No cases of multiple myeloma were identified among the participants. The small numbers of leukemia and the wide confidence intervals do not provide strong evidence of a radiation effect although it cannot be ruled out.

The committee comes to the same conclusion as the authors of the different reports concerning the mortality experience of military participants at nuclear test sites: the lack of a consistent pattern of cancer excess, with the only statistically significant increased mortality for lymphopoietic cancers suggests that factors other than or in addition to radiation may have been involved.

Study of US Nuclear Submariners

Mortality rates for all-cause and specific-disease categories in a cohort of 76,160 men who served in US nuclear submarines between January 1, 1969 and June 30, 1982 were compared with those in the comparable segment of the US population (Charpentier et al., 1993). During the study period the cohort accumulated almost 595,800 person/year and about 32,000 person/rem (mean cumulative dose to individuals 1.70 mSv [.0.17 mrem]). Notably, a statistically significant deficit of mortality was seen for all causes of death (811 deaths, SMR = 0.62, 95% CI = 0.58-0.66) combined and all cancers (77deaths; SMR = 0.71, 95% CI = 0.56-0.88) combined. However, as the authors noted, the study findings should be interpreted cautiously, because in addition to a marked healthy-worker effect, the cohort was relatively young (mortality through the study period of just over 10%) and the periods of radiation exposure and followup were relatively short. The authors evaluated mortality with respect to several measures including radiation dose, but did not develop cancer risk estimates.

Conclusion   Except for leukemia in some but not in all studies reviewed, the epidemiologic studies of populations occupationally at risk from chronic exposure to low doses of ionizing radiation continue to show a lack of a consistent pattern of statistically-significant mortality excesses related to radiation dose for the radiogenic cancers that currently are compensable under RECA (Table 2.1). A causal association between chronic exposure to low doses of low LET radiation and multiple myeloma remains equivocal. Apparent excess mortality from other specific cancers identified among the various worker populations were described by the study authors as chance findings, or were attributed to small numbers of deaths, or to factors other than radiation, including exposure to other workplace hazards. To date, studies of these populations also fail to show any evidence of increased mortality risks related to dose for cancers identified in Table 3.6 that are not compensated under RECA.

Conclusions   The committee reviewed information about the long-term risks to human health posed by radiation exposure that has been published since the BEIR III report in 1980, which was the basis of the original RECA legislation in 1990. Our review focused on epidemiologic studies that we considered pertinent to the RECA populations with respect to the types of radiation to which they were potentially exposed, the duration and magnitude of their exposure, and the cancers and other diseases that are compensable under RECA. We included data on all the cancers for which a radiation risk increase has been documented. In some of the less common cancers, such as cancers of the salivary gland and the small intestine, the data were too sparse and no numerical risk estimates were found in our review of the literature.

We found no evidence that the results of epidemiologic studies of radiation-exposed populations reported since RECA was formulated, substantially change the current estimates of risk of radiation-induced diseases among the RECA populations.

We concluded that to date the risk estimates for radiation-induced cancers and nonCLL leukemia obtained from the more statistically-powerful occupational studies for exposure to chronic low doses generally are consistent with those estimated for the low dose range obtained by extrapolation from the atomic-bomb survivors' data.

While recognizing the limitations of the epidemiologic studies of populations occupationally at risk of chronic exposure to low doses of low LET radiation, our review of the studies of such populations has provided little evidence of increased risks for disease related to low radiation doses, particularly for most of the site- or type-specific cancers compensable under RECA. These findings suggest that it is unlikely that onsite participants and the downwinders, particularly those who may have been exposed as adults to fallout from US nuclear weapons operations, are at significantly increased risk for cancers that are currently compensable under RECA, except possibly for nonCLL.

RECENT DEVELOPMENTS IN RADIATION BIOLOGY

This section discusses recent findings in radiation biology that might have a direct or indirect effect on cancer risk coefficients. Such information might result in a reconsideration of populations and geographic regions that RECA covers.

The dose-response relationship for the induction of tumors by ionizing radiation is generally described as being a LNT response. The risk assessment for cancer is based on human tumor data and so, relies directly neither on the use of cellular- and molecular-biology data, nor on the frequencies of radiation-induced tumors and genetic effects in laboratory animals. However, such data are used as part of the evidence for the LNT hypothesis. Thus, the committee does consider factors that might influence the shape of tumor dose-response the effects of genetic variation. Those may become important if, as has been discussed by UNSCEAR (2001), for example, future dose-response approaches to risk assessment are more biologically based than is currently the case.

Cellular and molecular radiation-biology studies have been used extensively to provide support to the LNT approach for extrapolation of tumor responses from low or medium doses to those at very low doses (NCRP, 2001). Gene mutations and chromosomal alterations have been shown to be involved in the formation of tumors, including radiation-induced ones (reviewed in Meltzer et al., 2002 and NCRP, 2001). Many studies have demonstrated that gene mutations and chromosomal alterations increase with radiation dose and that they are induced in a LNT manner at low doses (under 5 mGy). However, in recent years, several radiation-induced processes have been described by which radiation might either increase or reduce the frequency of those genetic alterations at very low doses compared with the currently accepted LNT extrapolation from low or medium doses. No effect of those cellular processes on cancer risk has been established at this time (Morgan, 2003).

Bystander Effects

The bystander effect is described as a response in cells that are not directly traversed by a radiation-particle track. The majority of such responses have been described for high-LET exposures (such as to alpha particles) because it is possible, using a microbeam or specific dose, to define the cells traversed or the proportion of the cell population irradiated. For low LET radiation (such as x-rays and gamma rays), unless specific energy microbeams are used, all cells are traversed by multiple ionization tracks. Thus, no measurable bystander effect will occur.

The bystander effect has been observed in several experimental in vitro systems and a variety of mechanisms have been proposed to explain it (Mothersill and Seymour, 2001). The lack of consensus illustrates the degree of speculation that is involved in the interpretation of bystander experiments, and this in turn results in an inability to relate the phenomenon directly to risk. Whether a bystander effect can be induced after in vivo irradiation is still quite uncertain. Thus, a concern remains as to just how relevant the in vitro cellular results are for predicting in vivo responses and how pertinent they are to the process of tumor induction. Certainly, the organization of tissues, as compared to cell cultures, and the nature of cell-cell interactions in vivo vs those in vitro support an overall concern about the relevance of the in vitro studies.

Two recent studies support the view that the use of in vitro approaches does not necessarily predict in vivo outcomes. Weaver et al. (2002) showed in an elegant set of studies that tumor cells in a three-dimensional organization responded to apoptotic (programmed cell-death) signals differently from the same cells grown on flat tissue-culture substrates. The data show that analyzing cell interactions in a more natural three-dimensional setting provides a view that is closer to what happens in living organisms. Prise et al. (2002) lend support to that view. They showed that multicellular, tissue-based models provide evidence of competing bystander processes at low doses of high-LET radiation, both protective and adverse ones. Those outcomes are quite different from the responses described for in vitro cellular systems.

Genomic Instability

The development of widespread genomic instability is a hallmark of tumor development. Such instability is both a cause and a consequence of the cancer process. The type of genomic instability described after radiation exposure is different from and much more limited than that observed in tumors (Little, 1998). Most studies that have investigated genomic instability have involved irradiating cells in vitro and observing the appearance of de novo genetic changes in descendants of irradiated cells. A few studies have shown increased genetic damage in descendants of irradiated cells that have been transplanted into whole animals, but there is no substantial evidence of the effects being induced and transmitted in vivo.

Adaptive Response

An adaptive response to radiation exposure has been described for chromosomal alterations and mutations for both in vitro and in vivo exposure (UNSCEAR, 1994). The phenomenon is one whereby the frequency of chromosomal aberrations was found to be approximately 50% lower after a small priming dose (such as 10 mGy) followed by a challenge dose of 1 Gy or more compared to the frequency after a challenge dose of 1Gy or more without a priming dose.

A number of possible explanations of the adaptive-response phenomenon have been proffered, but none has convincingly explained it. In addition, the adaptive response is highly variable and depends on the cellular (or tissue) system used. For human cell studies, samples from some people show an adaptive response, and those from others do not. The induction of an adaptive response appears to be transitory, that is, the protective effect of the priming dose generally lasts for only a few hours. Furthermore, very small doses and dose rates, of the kind encountered environmentally, do not seem to induce an adaptive response. Having reviewed the literature on the induction of an adaptive response, National Council on Radiation Protection and Measurements (NCRP) Report 136 (NCRP, 2001) concluded that the data are generally interpreted not to exclude the LNT model and thus do not provide sufficient grounds for rejecting the LNT dose-response model as a basis for assessing the risks posed by low-level ionizing radiation in radiation protection.

Genetic Susceptibility

The evidence is clear that certain genotypes enhance susceptibility to cancer of different types. A subset of those genotypes can confer sensitivity to radiation-induced cancer. At the population level, these mutations are predicted to have little overall effect on cancer risk estimates, because their frequency in the population is very low (around 1 per 10,000 live births). That view is supported by computational modeling approaches conducted by the

International Commission on Radiological Protection (ICRP, 1998) for assessing the effect of autosomal dominant or recessive mutations on radiation-induced tumor frequencies.

However, at the individual level, persons in such susceptible groups face the potential of an increased risk at the individual level. ICRP (1998) in its report *Genetic Susceptibility to Cancer* concluded:

> The principal conclusion by the Commission is that, on current knowledge, the presence of familial cancer disorders does not impose unacceptable distortions in the distribution of radiation cancer risk in typical human populations. For individuals with familial cancer disorders, radiation cancer risks relative to baseline are judged by the Commission to be small at low doses and insufficient to form the basis of special precautions. It seems likely however those risks to those with familial cancer disorders will become important at the high doses received during radiotherapy.

NCRP in its Report 136 (NCRP, 2001, page 194) endorsed the ICRP statement on the effect of susceptibility mutations at the population level:

> The studies to date of the rare genetic mutations do not suggest they will have a major impact on total irradiated-population risk or on the shape of the dose-response.

Again, it is the effect at the individual level that would probably be influenced by mutations for susceptibility to radiation-induced cancer at the high doses received during therapy. No information is available on a specific sensitivity to cancer induction by low doses received occupationally, medically, or environmentally of people who carry susceptibility mutations. However, for people with such mutations, exposure to a given dose of radiation might be more likely to induce a cancer but the individual's baseline risk is also elevated.

In recent years, the approaches for identifying single-nucleotide polymorphisms in the human population have improved substantially (Carlson et al., 2004; Belmont and Gibbs, 2004). In addition, recent studies have provided evidence of links between specific polymorphisms and increased disease outcome (Houlston and Peto, 2004). Such studies do not include radiation-induced cancers. However, given the prevalence of polymorphisms in the population and their relative frequency (over 1% by definition), scientists and risk assessors need to follow the research in this field to determine whether specific genetic polymorphisms can enhance individual risks of radiation-induced tumors.

Minisatellite Alterations and Hereditary Risk

Minisatellites are variable regions of DNA characterized by a series of repeat nucleotide sequences that usually occur in noncoding regions of DNA. Mutations in minisatellite regions involve changes in the number of repeat sequences, and they are about 1,000 times more common than base-change mutations that occur in protein-coding genes. Because of their high mutability by ionizing radiation (for example, about 4% in the exposed people discussed in Dubrova et al., 2002a), minisatellite mutations have been proposed for use in measuring hereditary effects of radiation exposure.

Dubrova and colleagues have conducted several studies on populations exposed to fallout from the Chornobyl accident (Dubrova et al., 1996, 1997, 2002b) and on families living in the vicinity of the Semipalatinsk nuclear test site (Dubrova et al., 2002a). They have demonstrated a 1.6- to 2.0-fold increase in minisatellite mutations in the offspring of irradiated parents. However, not only does the increase appear to be independent of the dose received, but also no mechanism has been identified by which radiation could induce such changes in the number of repeats in a particular minisatellite region.

Using a similar technique, Weinberg et al. (2001) reported a 7-fold increase in repeat sequence mutations in people born to fathers who were involved in cleanup at the Chornobyl plant. However, Jeffreys and Dubrova (2001) responded by describing the method used by Weinberg et al. (2001) as unreliable and concluded that the mutants detected had to be validated. That has not been done, so the study by Weinberg et al. (2001) remains controversial.

Other studies of radiation-exposed populations have failed to demonstrate an increase in minsatellite mutations in the offspring of exposed fathers. They include two studies of Chornobyl cleanup workers (Livshits et al., 2001; Kiuru et al., 2001) and a study of the offspring of the Japanese atomic-bomb survivors (Kodaira et al., 1995). In addition, no evidence of increased minisatellite mutations was observed in the sperm of radiotherapy patients sampled at various times after treatment (May et al., 2000).

The UK National Radiological Protection Board (NRPB) has recently commented on the studies conducted at Semipalatinsk by Dubrova et al. (2002a), noting that although all other studies have had negative results or been methodologically flawed, Dubrova et al. (2002a) provides the most convincing demonstration to date of a radiation-induced effect on minisatellite mutation frequency (Bouffler and Lloyd, 2002). However, in concluding that, Bouffler and Lloyd (2002) noted Dubrova et al. (2002a) reported a 1.8-fold increase in minisatellite mutation frequency for doses cited as greater than 1 Sv. That value is broadly consistent with the genetic doubling dose of 1 Sv used by ICRP (1991) and UNSCEAR (2001).

In a recent comprehensive review of the basis and derivation of genetic risks, UNSCEAR (2001) discussed the work of Dubrova and colleagues and concluded that "minisatellite variations very rarely have phenotypic effects." UNSCEAR did not include data on minisatellite mutations in its genetic risk estimates. Where associations between minisatellite variations and phenotypic effects have been found, they have been for multifactorial diseases whose complex etiology involves multiple genes and interaction with environmental factors. Such diseases are far less responsive to an increase in mutation rate than those due to single gene mutations (UNSCEAR, 2001). Bouffler and Lloyd (2002) concluded that minisatellite mutations are unlikely to affect the incidence of heritable disease substantially.

We can conclude that no new evidence on radiation-induced minisatellite mutations has been published that requires revision of the human heritable risk posed by radiation exposure.

In summary, recent studies in cellular and molecular radiation biology are providing new insights into how radiation interacts with cellular components and how signals can be transferred from "hit" cells to "unhit" ones. The information should improve understanding of the underlying cellular changes that might be involved in the induction of mutations and how the changes are related to an excess risk of cancer or hereditary disorders after radiation exposure. In this context, radiation risk assessments and consequent risk estimates are disease-based. That is, they are derived from the findings of epidemiologic studies of exposed human populations buttressed by the results of experimental studies of irradiated laboratory animals. Thus, they do not rely directly on mechanistic considerations. Furthermore, risk-assessment approaches are supported by information on the dose-response relationships obtained for a variety of mutational end points known to be associated with carcinogenesis and hereditary effects. Reviews by various authoritative international and national scientific bodies of the risks to health arising from exposure to low doses of radiation have included knowledge of potential novel biologic mechanisms; they include the recent NCRP review that led to Report 136 (NCRP, 2001).

None of those reviews concluded that the epigenetic phenomena require modification of the LNT dose-response model that forms the basis of current risk estimates. A move to a more biologically based risk-assessment approach would require consideration of potentially confounding factors for low-dose response. With respect to genetic susceptibility to radiation-induced tumors, the current position of both ICRP and NCRP is that the effect of susceptibility mutations on population risk would be very small. For individual risk, there would be a minor effect of susceptibility mutations at low doses; they might have a much larger effect at the high doses received in therapy. The effect of single-nucleotide polymorphisms on sensitivity to tumor induction by radiation is not known.

Conclusions The committee concludes, on the basis of recent data on radiation-induced responses at the cellular and molecular levels discussed in this section, that current cancer risk estimates do not need revision. That conclusion is also based on the fact that current risk estimates are developed directly from human tumor frequencies.

The committee further concludes that continued monitoring of research in cellular and molecular radiation biology as related to radiation-induced cancer risk is needed.

RECENT DEVELOPMENTS IN RADIATION DOSIMETRY AND RADIATION DOSE AND RISK ASSESSMENT

Radiation Dosimetry

Estimates of health risks to exposed cohorts in the HRSA program have historically been obtained from dose assessment or retrospective dosimetry. This was necessary because many of the people, in particular downwinders, did not have personal dosimeters and there was a lack of comprehensive workplace or environmental monitoring. Reconstructing the external dose requires information on fallout deposition patterns, life styles, shielding by building materials and dose conversion factors. Reconstructing the dose from internal emitters involves detailed studies of the movement of the deposited radionuclides through the food chain into the body and the resultant organ doses obtained by using physiologically based pharmacokinetic (PBPK) models. Descriptions of these procedures for fallout are presented by Bouville et al. (2002) and Simon and Bouville (2002).

There are continuing efforts to update conversion factors relating radioactivity to dose for both internal and external exposures. Conversion coefficients for external radiation for use in radiologic protection have been revised by ICRP (ICRP, 1996) and the US Environmental Protection Agency (EPA) (Eckerman and Ryman, 1993). A summary of procedures for dose estimation from radionuclides in the environment has also been published by the International Commission on Radiation Units and Measurements (ICRU) (ICRU, 2002). Doses from internally deposited radionuclides were estimated with physiologically based pharmacokinetic (PBPK) models, such as those developed by ICRP (1979). Dose-conversion factors for internal deposition of radionuclides have been revised by EPA (Eckerman et al., 1988).

Tissue weighting factors, w_T are defined as the fractions of stochastic risk of carcinogenesis or hereditary effects resulting from radiation exposure of organ T, relative to the total risk posed by uniform exposure of the entire body (ICRP, 1991). The most recent accepted values are shown in Table 3.7. Modifications to w_T are being reviewed by ICRP on the basis of the latest assessment of cancer incidence from epidemiologic studies. The w_T for the gonads may be reduced by a factor of 5 to 0.04. The value of w_T may increase for breast cancer from 0.05 to 0.12. There should be no changes in the w_T proposed for thyroid cancer or respiratory cancers. There has been complete revision of the model for the human respiratory tract (ICRP, 1994) and of basic anatomic data on the skeleton (ICRP, 1995).

Collectively, the revisions in dose-conversion factors and other dosimetry measures should reduce uncertainty in estimates of dose, but they will not substantially change the general assessment of risk to cohorts in the HRSA program.

The revised PBPK model for the human respiratory tract does not include dosimetry for inhalation of radon or the short-lived descendants of radon that are referred to as radon daughters. Historically, the risks posed by radon have been related to the time-integrated concentration of potential alpha energy from short-lived radon daughters, usually expressed in working level month (WLM). The committee does not expect that practice to be revised. Any changes in risk estimates associated with radon will be related to radiation biology or observed cancer incidence rather than to a revised paradigm for dosimetry.

The most comprehensive database of risks associated with external exposure from ionizing radiation is the Life Span Study of Japanese atomic-bomb survivors conducted by the Radiation Effects Research Foundation. Previous estimates of risk were related to dose assessments for each person according to a system called DS86 (NRC, 1987). In 2001, a National Research Council report made recommendations regarding revisions to DS86 to reduce uncertainty in dose assessments (NRC, 2001). The revisions have been completed and will be published as DS02 in 2005. The protocol has been used to obtain revised estimates of dose for each person in the study. The new data indicate that cancer-mortality risk factors (relative risk per unit dose) will decrease by about 8% because of changes in dosimetry. That is principally because of an increase in the gamma-ray dose for both cities. There are, however, no changes in the apparent shape of the dose-response curve or the age and time-since-exposure patterns of risk. Efforts are under way to evaluate and reduce uncertainties in the risk estimates for mortality and to develop risk estimates for cancer incidence (Preston et al., 2004).

The risk of thyroid cancer in people exposed to ^{131}I has now been conclusively demonstrated as a result of the 1986 Chornobyl accident. The Institute of Medicine and National Research Council discussed the early skepticism that met reports of increased thyroid-cancer incidence at Chornobyl and the later findings showing that irradiation of the thyroid by ^{131}I is almost, if not equally, as effective as irradiation by external radiation (IOM-NRC, 1999). Most of the radiation-induced thyroid cancers incurred after that accident are papillary thyroid cancers, the latent period is short, and there are indications that they are more aggressive than usual. Recent findings on the dosimetry of ^{131}I exposure from Chornobyl and its related risk may clarify uncertainties in estimating their health effects.

Radiation Dose and Risk Assessment

National Cancer Institute 1997 [131]I Study

Since RECA was enacted in 1990, the National Cancer Institute (NCI) has completed a comprehensive study of radiation doses to the thyroid from ^{131}I released from tests at the Nevada Test Site (NTS) (NCI, 1997). The study uses fallout measurements, atmospheric modeling, and statistical analysis to estimate ^{131}I fallout deposition density in each county of the continental United States and the corresponding radiation doses to the thyroid for each atmospheric test at the NTS. NCI took into account the ages of those at risk of exposure and their consumption of milk and other foodstuffs. NCI presented its results in tables and in a series of maps showing the doses in all counties for four milk-consumption rates for people born in selected years from 1930 to 1962; NCI also produced maps showing doses for different test series.

NCI has provided the committee with updated versions of several of the maps, and we present them in this report. The maps show the radiation doses to people by county. They include the latest revisions to the dose calculations and show the doses based on contours. Thus, they offer a more accurate representation than the earlier NCI maps in that the doses are based on where the fallout was deposited and did not stop at county boundaries.

Figure 4.1 shows the estimate of the dose to the thyroid of a child born on January 1, 1951, for average milk consumption. Estimated doses range from less than 1 mGy to greater than 100 mGy. The map gives the total thyroid dose from both external and internal radiation. The great majority of the dose, however, is from the ingestion of ^{131}I in foodstuffs, particularly milk.

The dose to the thyroid from ^{131}I depended significantly on the age of the person when the exposure was received. Because of the relatively higher uptake of iodine in young children and the smaller thyroid, which resulted in a greater

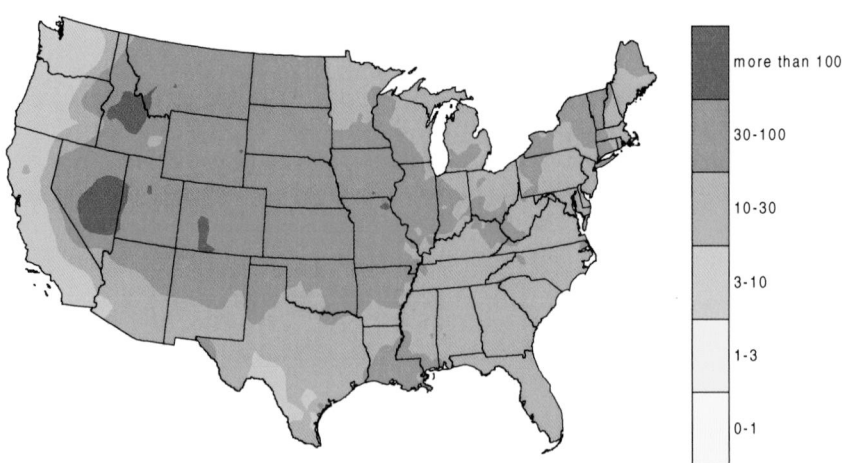

FIGURE 4.1   Geographic distribution of estimated total (external + internal) dose (mGy) from all NTS tests to the thyroid of children born on 1 January 1951 and who were average milk drinkers (map courtesy of National Cancer Institute).

iodine concentration, the thyroid dose in young children is higher than that in other age groups for the same amount of fallout and the same dietary intake of ^{131}I. The dependence of dose on age at exposure tends to decrease with age until adulthood when dose varies little with age.

Because of the dependence of dose on age at exposure in young children, maps of thyroid doses to people born at other times may differ from Figure 4.1. Such maps—for example, for a person born on January 1, 1954—would reflect the influence of ^{131}I deposited in fallout from tests occurring while that person was a small child.

For comparison, Figure 4.2 shows the thyroid doses from all tests at the NTS to people who were adults during the time of nuclear testing.

Figures 4.1 and 4.2 show that people living in many parts of the United States, not just those living near the NTS, received high thyroid doses as a result of nuclear tests. For example, for children born on January 1, 1951, thyroid doses in areas in Idaho, Montana, and Colorado, were also higher, and thyroid doses in other areas, such as the Midwest and up-state New York and Vermont, were elevated.

Much of the geographic distribution is due to the dynamics of ^{131}I. First, once away from the NTS, ^{131}I is deposited mainly through precipitation ("wet" deposition), so areas that receive precipitation when a fallout cloud is passing overhead are more likely to have high deposition. Second, once it is deposited, the main exposure pathway is ingestion of milk from cattle or goats that grazed on pasture that received the fallout. Consequently, thyroid doses tend to be

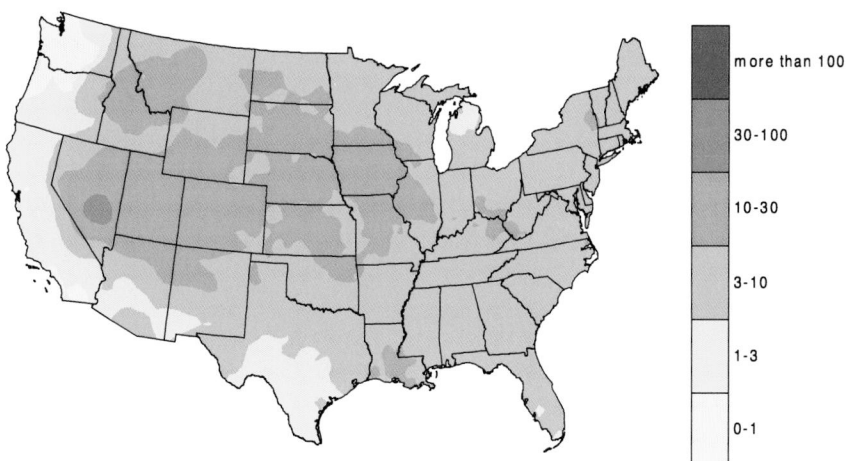

	more than 100
	30-100
	10-30
	3-10
	1-3
	0-1

FIGURE 4.2   Geographic distribution of estimated total (external + internal) dose (mGy) from all NTS tests to the thyroid of those of adult age at the time of exposure and who were average milk drinkers (map courtesy of National Cancer Institute).

elevated in areas receiving fallout through wet deposition where entry into the milk pathway is possible.

On the basis of the 1997 study (NCI, 1997), NCI developed a ^{131}I dose calculator that was published on the Web at http://ntsi131.nci.nih.gov/. The user supplies date of birth, sex, locations and dates of residence, and milk-consumption pattern. The calculator then uses the results of the 1997 study to estimate the thyroid dose from ^{131}I and its 90% credibility interval.

Institute of Medicine-National Research Council Review of 1997 National Cancer Institute [131]I Study

The NCI ^{131}I study was reviewed by an Institute of Medicine-National Research Council committee in 1999 (IOM-NRC, 1999). That committee stated that the NCI approach was generally reasonable, but found that the county-specific estimates of thyroid dose were too uncertain to be useful in estimating individual doses.

Individual doses depend strongly on specific variables, such as age at exposure and amount of milk consumed, which are not considered in the county doses. Estimating individual doses is possible but highly uncertain because important data are not available or are of questionable reliability. The committee also observed that there was little epidemiologic evidence of a widespread increase in thyroid cancer.

Centers for Disease Control and Prevention-National Cancer Institute 2001 Draft Feasibility Study

In 2001, the Centers for Disease Control and Prevention (CDC) and NCI published a draft feasibility study of the health consequences of nuclear-weapons testing on the American population (CDC-NCI, 2001). The report considered all radionuclides that contributed substantially to the radiation dose, and estimated the effective dose and the dose equivalent to the organs at risk. Both NTS fallout and global fallout were considered. Global fallout included not only the fallout from American tests but also the contribution from tests of other nations. The draft feasibility study concluded that a full dose assessment was possible but that it would be a major effort comparable with the NCI ^{131}I study discussed above.

The CDC-NCI study used the ^{131}I fallout deposition densities found in the 1997 NCI study as a starting point to calculate the deposition densities from NTS fallout of the 33 other radionuclides that contributed substantially to the radiation dose. The study then calculated the doses from both internal and external exposure to those radionuclides. Only the dose to the thyroid from ^{131}I resulted in an internal radiation dose that substantially exceeded the dose from external radiation. As a result, most organ doses (except to the thyroid) were roughly the same.

NCI has prepared updated maps showing the current best estimates of dose to various organs and made them available to the committee. The maps are used in this report rather than the original maps in the 2001 draft feasibility study. In Figures 4.3 and 4.4, the dose to the red bone marrow from nuclear tests at the NTS is shown as representative of the other organ doses for an adult and a child born on January 1, 1951.

Both the 1997 NCI study and the 2001 CDC-NCI draft feasibility study estimated doses to the thyroid from ^{131}I at the NTS. Although the study results are similar, they are not identical. The difference is discussed briefly in the 2001 report and attributed to differences in estimating the amount of fallout retained by vegetation. In addition, the 2001 results are preliminary in that they are for the draft feasibility study, and did not include uncertainties.

In addition to estimates of doses from NTS fallout, the 2001 draft feasibility study evaluated doses from global fallout. The doses to red bone marrow were found to be slightly higher from global fallout than from NTS fallout but less for the thyroid.

The 2001 CDC-NCI draft feasibility study presents maps similar to Figures 4.3 and 4.4 for global-fallout red marrow doses by county in the United States (CDC-NCI, 2001). The study did not estimate the ^{131}I doses to the thyroid from global fallout by county (although some ^{131}I was occasionally present), because of lack of data. The ^{131}I doses were given for the United States as a whole. Doses from ^{3}H and ^{14}C, which affect the hydrological and carbon cycles, respectively,

FIGURE 4.3   Geographic distribution of estimated total (external + internal) dose (mGy) from all NTS tests to the red bone marrow of children born on 1 January 1951 (map courtesy of National Cancer Institute).

FIGURE 4.4   Geographic distribution of estimated total (external + internal) dose (mGy) from all NTS tests to the red bone marrow of those of adult age at the time of exposure (map courtesy of National Cancer Institute).

were also estimated for the United States as a whole. The report notes that the proportion of global fallout due to US weapons testing can be roughly determined from the fission yield of the US tests relative to the total fission yield from all high-yield nuclear testing.

National Research Council Review of Centers for Disease Control and Prevention-National Cancer Institute 2001 Draft Feasibility Study

The study was reviewed by a National Research Council committee that published its report in 2003 (NRC, 2003c). Among its conclusions and recommendations, the National Research Council found the following:

• The ^{131}I fallout data and the resulting dose and thyroid-cancer risk should be reanalyzed to include the new dosimetry and risk estimates from Chornobyl to update the 1997 NCI report.

• On the basis of the results of the draft feasibility study, further work with fallout radionuclides other than ^{131}I would not be warranted, because of the very low levels of associated exposure and the uncertainties in their distribution over time and location.

• In agreement with the authors of the draft feasibility study, the dose and risk estimates that were presented were developed as population averages and should not be used to estimate risks to specific individuals.

• A program should be established to examine and archive fallout-related documents from sites operated by the Department of Energy and the Department of Defense and other relevant sites.

National Cancer Institute and Centers for Disease Control and Prevention Working Group 2003 Revision of NIH Radioepidemiology Tables

An NCI and CDC working group (NCI-CDC, 2003) reviewed and revised the 1985 National Institutes of Health radio-epidemiology tables (NIH, 1985). The revision was principally based on the 1958–1987 Life Span Study Tumor Registry data on the atomic-bomb survivors at Hiroshima and Nagasaki. The computer program Interactive Radio-Epidemiological Program (IREP, version 5.3) incorporated the results of this work to give probability of causation/assigned share values for individual radiation exposures.

Risk coefficients and associated PC/AS values in some cases have been substantially changed, both from the original NIH tables and from their 1988 revision by the Committee on Interagency Radiation Research and Policy Coordination (CIRRPC) (CIRRPC, 1988). The NCI-CDC report used cancer-incidence data on the atomic-bomb survivors, rather than the cancer mortality data on most cancers used for the 1985 report. For thyroid cancer, the NCI-CDC report used a compilation of seven studies (Ron et al., 1995), which was considerably more extensive than that used by the NIH report.

CIRRPC also assumed that, for a particular cancer, an applicant had a low baseline risk at the 10th percentile of the cancer risk distribution and that the ERR varied inversely with the baseline risk. The NCI-CDC revision did not use those assumptions, which had accounted for a factor of two increase in the ERR for most cancers (NCI-CDC, 2003).

CONCLUSION

This chapter has presented the results of recent studies in radiation epidemiology, biology, and dosimetry. The overall aim is to develop a database that forms part of the consideration of new populations or geographic areas for coverage by RECA. Chapters 5 and 6 consider the issue of additions to RECA.

5

Expanding RECA Eligibility: Scientific Issues

The committee was charged with making recommendations to the Health Resources and Services Administration (HRSA) that are based on scientific knowledge and principles:

> whether other classes of individuals or additional geographic areas should be covered under the Radiation Exposure Compensation Act (RECA) program.

The present criteria for downwinder eligibility include diagnosis of a compensable disease and proof of residence in selected counties of Arizona, Nevada, and Utah during the period of atmospheric testing at the Nevada Test site (NTS) (see Map 2.2). The committee reviewed the criteria and compared them with those of other compensation programs related to radiation exposure (Chapter 2, Table 2.4). This chapter describes a process for expanding geographic areas for coverage of people who may have been exposed to radiation from fallout from US nuclear-weapons testing that is based on diagnosis of disease and on scientific methods to determine the extent to which radiation was responsible for the disease. The objective is to suggest a process that would not take an inordinately long time to develop and implement.

In an attempt to identify an appropriate way to consider geographic areas for compensation of people who may have been exposed to radiation from fallout from US nuclear-weapons testing, the committee began evaluating the current system of eligibility on the basis of residence alone. One method that the committee used was to estimate the absorbed dose of radiation for populations living in affected areas surrounding the NTS. People can be exposed to radiation from fallout by many pathways including external radiation as a plume or cloud passes over the region, external exposure to radioactivity deposited on the ground and remaining there for extended periods, and internal exposure to radioactivity that accumulates in the body from inhalation or ingestion of plants, meat, and milk.

A collection of data is available to map exposure rates across the United States from radioactivity deposited on the ground as a result of fallout (NCI, 1997). Analysis of the data has revealed that doses from external radiation to radiosensitive tissues are small and, in all but a few cases, not significantly greater than those from natural background radiation (CDC-NCI, 2001).

The most important pathway is the ingestion of iodine through the consumption of milk (NCI, 1997). Iodine is absorbed rapidly in the gastrointestinal tract and almost completely transferred to the blood through the small intestine. Of the iodine in the blood, about 30% is transferred to the thyroid, and the remainder is eliminated by excretion. Stable iodine in the thyroid is reduced by 50% in approximately 120 days. The isotope iodine-131 (^{131}I) has a radioactive half-life of 8 days. Thus, more than 90% of the atoms of ^{131}I taken up in the thyroid will decay there. Beta particles from that decay deliver a substantial dose to the thyroid.

There has therefore been an extensive effort to determine the dose to populations from ^{131}I originating in the atmospheric weapons-testing program at the NTS. The decade-long effort has produced a dose calculator that can conveniently be used to estimate the dose to a person on the basis of age, location, and milk consumption (NCI, 1997) (http://ntsi131.nci.nih.gov/).

We have used the calculator to provide information that might be useful in identifying geographic areas in the continental United States that might be eligible for compensation through RECA. To begin the process, we created an imaginary person. The person was a male born on January 1, 1948, who remained in a single county throughout the testing period. He consistently drank one to three glasses of processed milk from local dairies. We computed the total dose to the person's thyroid. We repeated that process for every county in Utah. Several counties with sufficient fallout data were subdivided into regions, and we performed the dose calculation for each region separately. Figure 5.1 shows the results of our calculations. Each circle represents the computed dose for a specific county or sub-county region in Utah. The counties on the X axis are arranged from the highest thyroid dose to the lowest.

The variation in dose is large, ranging from 30 mGy for the lowest value in Iron County to 210 mGy in Washington County.

We assumed that the person was diagnosed with thyroid cancer some time after the testing at the NTS stopped. We then determined whether he was eligible for compensation through RECA. The solid circles in Figure 5.1 represent counties where he would be eligible for compensation, and the open circles represent counties in which he would not. Some counties with relatively low doses are

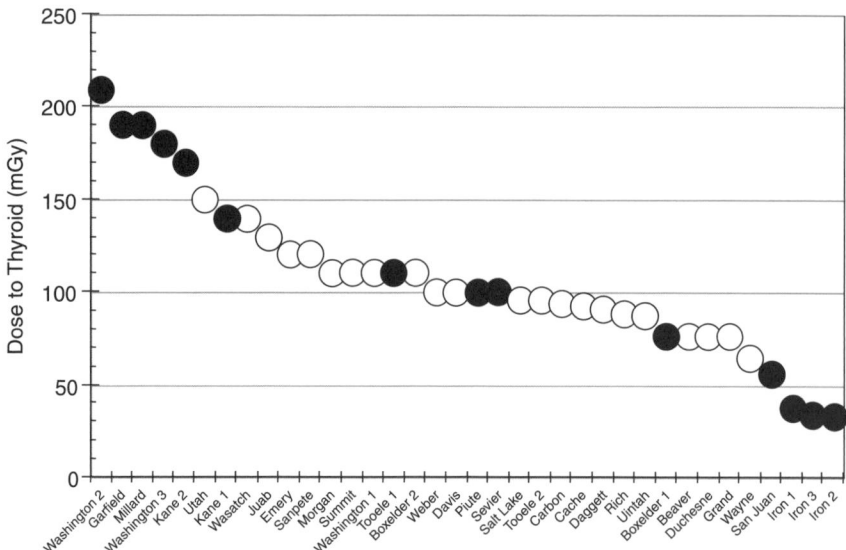

FIGURE 5.1   Calculated absorbed dose to the thyroid of a person born in 1948 who resided in same county in Utah for entire period of NTS testing.

included in RECA, and some with relatively high doses are not. That result leads to confusion and concern.

We repeated the computation with different persons of different ages and sexes. When plotted on a graph, the data were always similar to those in Figure 5.1. We then extended the computation to include counties is some states other than Utah. The results are shown in Figure 5.2. The solid circles represent counties in Utah where the person would be eligible for compensation through RECA. The open circles represent counties in the states listed.

This is not a comprehensive survey of all counties in the continental United States. It does however demonstrate that thyroid doses in other regions of the country can be greater that for counties currently eligible for compensation. We would expect similar results for other cancers, but have not made these computations because the dose calculators have not been thoroughly developed at this time.

That simple exercise does illustrate the problem one encounters when attempting to recommend extensions of RECA coverage for exposure to fallout. Doing so might well include counties throughout most of the United States. The committee also recognized that including absorbed dose in the determination of eligibility for compensation would not be sufficient because the risk of radiation-induced cancer depends on the age at exposure and age at diagnosis in addition to dose. A process based on risk would therefore use dose and the other criteria to

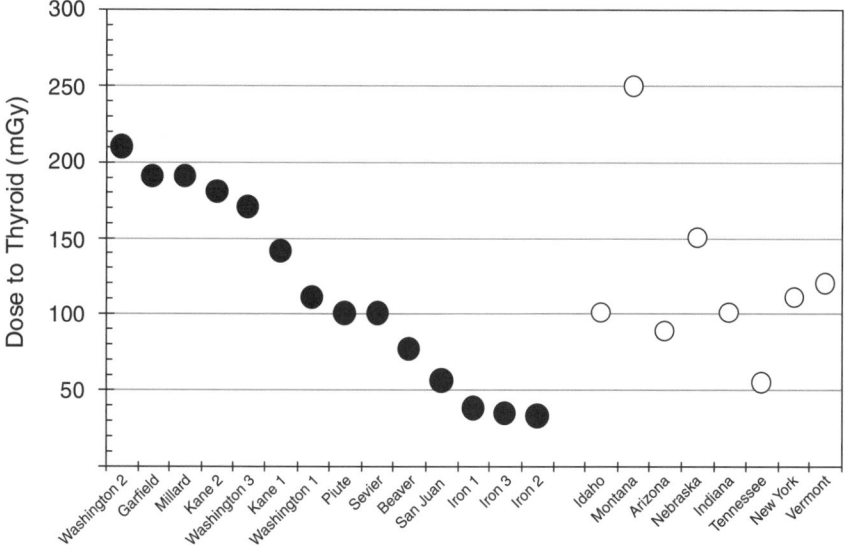

FIGURE 5.2   Calculated absorbed dose to the thyroid of a person born in 1948 who resided in same county for entire period of NTS testing. The solid circles are for counties in Utah that are currently eligible for compensation in RECA and the open circles are counties in states other than Utah.

determine the probability that an identified cancer was caused by radiation rather than by other agents.

The mandate before the committee involved scientific justification. One approach that is being used in other compensation programs, both in the United States and in the UK, is referred to as probability of causation (PC) or assigned share (AS). That approach has been reviewed extensively in the literature and has been used for tort litigation in many situations related to exposure to ionizing radiation. This chapter presents the history and implementation of PC in issues related to ionizing radiation. It defines PC in the context of compensation and presents examples that illustrate the process. Chapter 6 applies the PC process specifically to the task of expanding RECA eligibility.

PROBABILITY OF CAUSATION

Litigation for damages from ionizing radiation first became an issue after reports documented the hazards experienced by radium-dial painters (Martland, 1929). Many cases have since been adjudicated in the United States for medical, industrial, and environmental exposures to radiation. The problem faced by the legal system is that no specific form of cancer is caused only by radiation. Furthermore, features of a cancer caused by radiation cannot be distinguished from those of the same cancer caused by other mechanisms. Legal decisions were often based on the opinions of experts who determined whether radiation was likely to be the cause of cancer in a particular person. That combination of circumstances created considerable controversy. In 1983, Congress instructed the National Institutes of Health (NIH) to develop data that would provide a scientific basis for determining whether a cancer diagnosed in a person was caused by a documented dose of radiation. The data were ultimately published in the form of tables that were entitled the probability of causation.[1]

The tables were based on information from epidemiologic studies of many groups of people. Technically, PC is the fraction of a group of identical persons in whom a radiation-induced cancer would be expected to occur at some specified time after a dose of radiation was received. In that sense, PC is a prospective concept applied to a population. In practice, however, PC is applied to a single exposed person in whom cancer has already been diagnosed; that is, it is used retrospectively. It has been suggested that the term "assigned share" is more appropriate for that situation because the value of PC for a group has been "assigned" to an individual. It is therefore the individual's "share" of the total cancer risk due to radiation (NIH, 1985). PC and AS are numerically the same. The committee has adopted the nomenclature PC/AS for the purposes of this report. PC/AS is now widely used for claims against the government or its contractors in compensation programs such as Radiation-Exposed Veterans Compensation Act (REVCA) and Energy Employees Occupational Illness Compensation Program Act (EEOICPA).

Much of the information used to generate PC/AS for leukemias and solid tumors has been obtained from long-term followup of Japanese atomic-bomb survivors. Data on lung cancer resulting from exposure to radon decay products are obtained from studies of underground uranium miners. Estimates of the risk of thyroid cancer used for existing values of PC have been obtained, in part, from patients receiving diagnostic medical examinations involving the thyroid.

Risk-projection models are often required when information on specific doses is unavailable. In particular, that occurs for low doses of radiation when the spontaneous risk of cancer is at least as large as the risk of cancer from radiation. The models generally agree with a linear nonthreshold (LNT) hypothesis for radiation effects. They predict that there is some chance that a cancer will occur at any dose. The probability of occurrence is proportional to dose and does not reach zero until the dose is zero. Cigarette-smoking is the only other risk factor that is considered in the PC tables for radiation; it is applied only to the calculations for lung cancer.

---

[1]The original radioepidemiological tables (NIH, 1985) were mandated under Public Law 97-414 (known as the "Orphan Drug Act").

A significant issue that needs to be addressed is how to choose a value of PC/ AS that is accepted as "proof" that radiation was responsible for the diagnosed cancer in an individual. If the PC/AS has a value of exactly 0.5, it implies that it is as likely as not that the cancer was caused by radiation. For a PC/AS greater than 0.5, it is assumed that it is more likely than not that the cancer was caused by radiation. Most tort cases are based on a decision value of PC/AS equal to or greater than 0.5 (at least as likely as not that the cancer was caused by radiation). That criterion is discussed below, as are uncertainty in determining PC/AS and how uncertainty is incorporated into the decision-making process.

Definitions of PC/AS

If a person is exposed to radiation, the risk that a specific radiation-induced cancer will develop at a given age is referred to as R_{rad}. It depends on the absorbed dose to the organ, sex, age at the time of exposure, and age at the time of diagnosis. The risk that a specific cancer from all other causes will develop at the same age at diagnosis can be referred to as $R_{baseline}$. Probability of causation, PC/AS, is defined as

$$PC/AS = \frac{R_{rad}}{R_{rad} + R_{baseline}} \tag{1}$$

In most cases, the risks posed by radiation must be obtained from risk-projection models based on epidemiology. For example, when a relative-risk (RR) model is used,

$$R_{total} = R_{rad} + R_{baseline} = RR \cdot R_{baseline}, \tag{2}$$
$$R_{rad} = (RR - 1) \cdot R_{baseline}, \text{ and} \tag{3}$$
$$R_{rad} = ERR \cdot R_{baseline}, \tag{4}$$

where RR is relative risk factor and ERR is excess relative risk factor (see Chapter 3).[2] Useful relationships among PC/AS, RR, and ERR are

$$PC/AS = \frac{RR-1}{RR} \text{ and} \tag{5}$$

$$PC/AS = \frac{ERR}{1 + ERR} \tag{6}$$

The relationship between PC/AS and ERR is plotted in Figure 5.3.

---

[2]ERR is sometimes used only for a linear risk model. In this case, however, the committee is using it in its more general form, ERR = RR − 1, as discussed in Chapter 4.

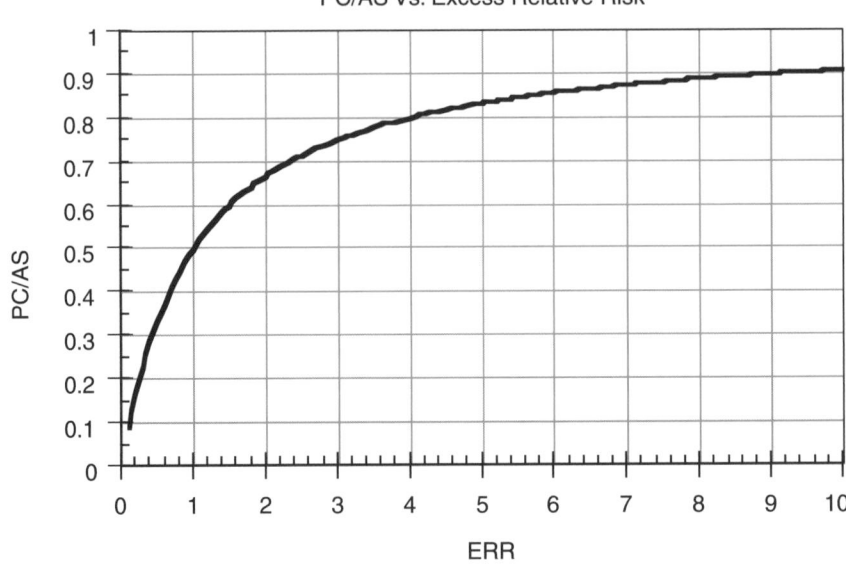

FIGURE 5.3   PC/AS as function of ERR. If excess risk from dose of radiation is 10 times baseline risk, PC is 0.91 (that is, 10/(10+1)).

For example, suppose that a person receives an absorbed dose to the thyroid and is later diagnosed with thyroid cancer. Presume that the risk of developing thyroid cancer from that absorbed dose and the person's age at exposure is found to be twice the baseline risk of thyroid cancer. Then

$$R_{total} = R_{rad} + R_{baseline} = 2R_{baseline},$$ (7)

$$RR = 2,$$ (8)

$$ERR = RR - 1 = 1, \text{ and}$$ (9)

$$PC / AS = \frac{ERR}{1 + ERR} = 0.5$$ (10)

That leads to the intuitive result that if the risk added by radiation exposure is just equal to the baseline risk (ERR = 1), then there is an equal chance (PC/AS = 0.5) that the diagnosed cancer was caused by the absorbed dose from radiation in comparison with all other factors.

The process of obtaining PC/AS is in effect a process of determining the ERR for a person exposed to radiation and diagnosed with cancer. The determination of ERR for a particular person must rely on dosimetry to determine dose and how the ERR depends on dose. Very often, the dose is not measured directly but is estimated through a dose-reconstruction process that relies on many assumptions. In addition, risk is modified by factors that depend on the type of radiation, the dose rate, age at time of exposure, and age when disease is diagnosed. Variations in baseline risk also depend on life style and socioeconomic factors.

All components used to estimate ERR have uncertainty. They are combined to form a probability distribution for ERR that reflects the combined uncertainty of all the components used to obtain ERR. Figure 5.4 is an example of a distribution of ERR.

On Figure 5.4, the median value of ERR is 1.0 (half the estimates of ERR are smaller than 1.0, and half are greater than 1.0). The average value of ERR is 1.8, and there is a small chance that the ERR could be as high as 10. That distribution can be converted to a probability distribution of PC/AS by using the relationship shown in Equation 6. For illustration, several examples of the combined uncertainty (credibility) of PC/AS are shown in Figure 5.5. Each plot, A-D, is the resulting distribution of PC/AS obtained from different hypothetical situations for dose and risk.

Figures 5.5 shows that for any given case, the range of possible values can be very large. The arrow represents the median, or central, value of PC/AS (half the

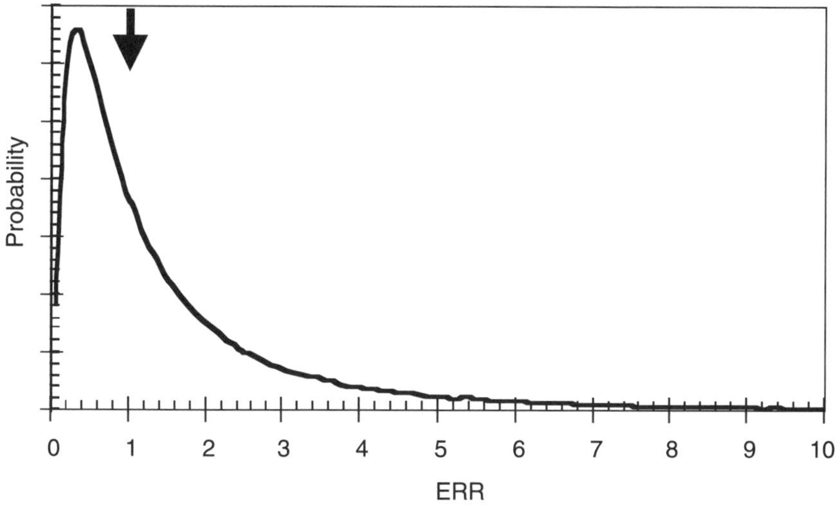

FIGURE 5.4   A probability distribution of excess relative risk, ERR resulting from uncertainties in estimates of the absorbed dose and associated cancer risks. Arrow indicates median value.

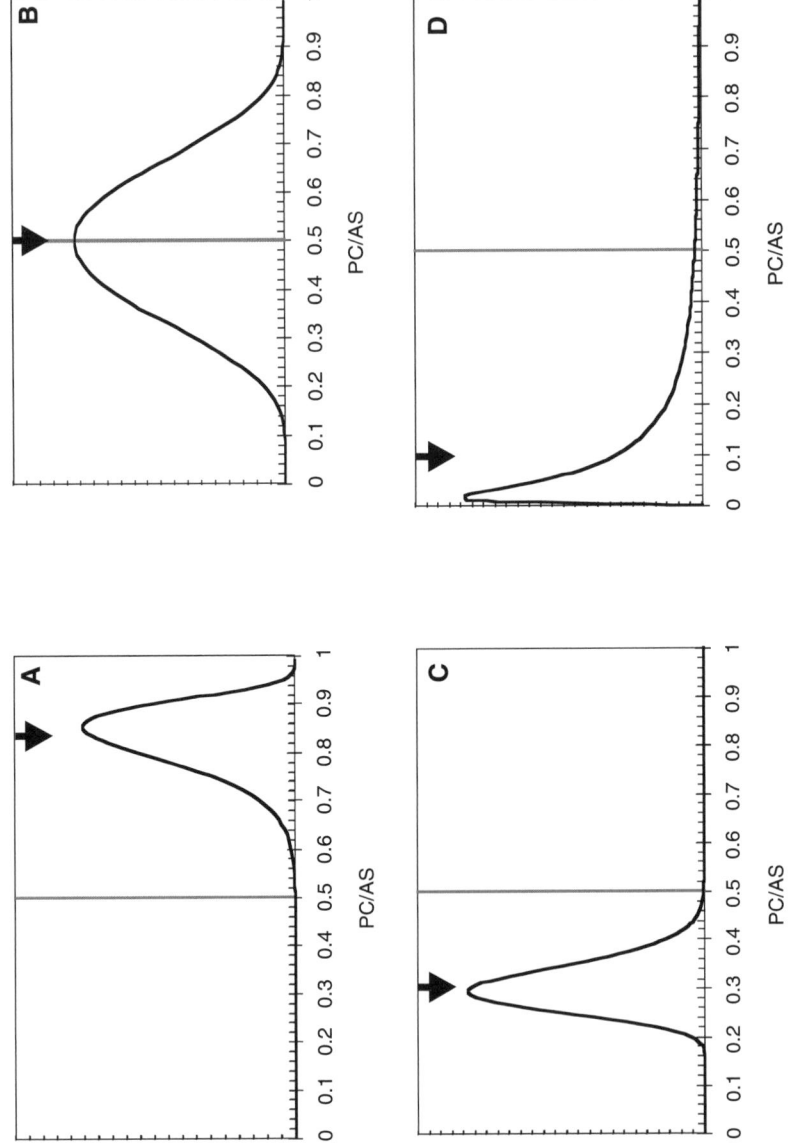

FIGURE 5.5   Distribution of PC/AS for individuals A, B, C, and D obtained from different estimates of dose and risk.

estimates are larger than this value, and half are smaller). The vertical line corresponds to a PC/AS of 0.5. If the true value of PC is greater than 0.5, it is more likely than not that the diagnosed cancer for this person was caused by radiation. The true value is not known exactly, so the graphs, including all the uncertainties, must be used to determine eligibility for compensation.

One possible criterion for awarding compensation under RECA is if the median value of the PC/AS distribution is equal to or greater than 0.5, that is, odds are at least 1 to 1 that the cancer was caused by radiation. For such a case, the persons represented by A and B in Figure 5.5 would be eligible, but the persons represented by C and D would not be eligible. Another possible criterion is a median value of the PC/AS distribution is equal to or greater than 0.2, that is, odds are at least 1 to 4 that the cancer was caused by radiation. For such a case, the persons represented by A, B, and C would be eligible, but the person represented by D would not be eligible.

A third possible criterion is whether some portion of the distribution exceeds 0.5. EEOICPA and the Department of Veterans Affairs (VA) award compensation if at least 1% of the distribution is greater than 0.5. In effect, only a small fraction of the right-hand tail of the distribution needs to exceed 0.5. In that case, the persons represented by A, B, and D would be eligible, but the person represented by C would not be eligible.

At first, one could conclude that using the 1% rule always gives the benefit of the doubt to the claimant. There are circumstances, as in D, where the average and the median values of PC/AS are small, but compensation is awarded because the uncertainty (the spread of the distribution) is very large. That often happens in the case of cancers where the association with radiation is not well established and the corresponding risk factors have large uncertainties. However, even though person C has a median value of PC/AS that is higher than person D, C would not receive compensation, because of the confidence that the true value of PC/AS is less than 0.5.

TOOLS FOR DETERMINING PC/AS

NIH Radioepidemiological Tables

The original radioepidemiological tables that ultimately served as a basis for determining PC/AS were mandated under Public Law 97-414, the Orphan Drug Act (NIH, 1985). A section of that law, pertaining to the development of radioepidemiological tables, was originally introduced by Senator Orrin Hatch (R-Utah) as a part of Senate bill S 1483, which became the Radiation Exposure Compensation Act. The tables were intended to provide a means for estimating the likelihood that a person who has or had any of several radiogenic cancers developed it as a result of exposure to ionizing radiation from the nuclear-weapons tests in Nevada.

The term radiogenic means that credible research has established a connection between exposure to radiation and an increased risk of the disease in human populations. In general, a radiation-induced cancer cannot be distinguished from the large number of similar cancers that are related to other causes. Public Law 97-414 required the US Public Health Service to develop radioepidemiological tables that set forth the relationships between PC/AS and radiation dose for various cancers. The main purpose of the tables has been to provide a mechanism for making decisions for compensation after exposure to ionizing radiation.

A committee of the National Research Council was formed to review the development of the radioepidemiological tables (NRC, 1984). The committee objected to the use of the term probability of causation and recommended the use of assigned share.[3]

An ad hoc working group of NIH developed the first radioepidemiological tables for estimating PC/AS in 1985. Its report (NIH, 1985) identified 13 different cancer sites that had statistical evidence of an association between absorbed dose and cancer in human populations.[4]

The final NIH report was issued in 1985 and has been used by VA as a guide to adjudicating compensation claims, through the Veterans Dioxin and Radiation Exposure Compensation Act, for cancers diagnosed in persons who were exposed during military service.

Tables Created by the Committee on Interagency Radiation Research and Policy Coordination

In 1988, the Committee on Interagency Radiation Research and Policy Coordination (CIRRPC) developed tables to be used in screening claims of radiation-induced cancer. The tables made extensive use of uncertainties in the values of PC/AS listed in the 1985 NIH report. A person passed the screening test when there was at least 1% probability that the estimated PC/AS exceeds 0.5. The CIRRPC report noted that

> this procedure is designed to insure that cases which have even a small chance of a true PC that is 0.5 (50%) or greater (i.e., that meet the "at least as likely as not" criterion), are developed for assessment of causality, yet will avoid detailed development of those cases for which there is virtually no chance that the true PC would be as large as 50%.

---

[3]The committee noted that the so-called probability of causation was strictly speaking not a probability, but rather an estimate of the *proportion* of cancers that were caused by exposure in a large (hypothetical) group of similarly exposed cases—an estimate that was then *assigned* to all members of the group.

[4]Cancer sites covered by the 1985 radioepidemiological tables are: bone and joint, breast, colon, esophagus, kidney and bladder, leukemia, liver, lung, pancreas, salivary gland, stomach, and thyroid gland.

Revision of the Radioepidemiological Tables

In 2003, a working group of the National Cancer Institute (NCI) and the Centers for Disease Control and Prevention (CDC) reviewed and updated the 1985 NIH radioepidemiological tables (NCI-CDC, 2003, available at http:// dceg.cancer.gov/docs/Report03.pdf, accessed February 22, 2005.). Rather than publish a new set of tables, the working group developed the computer program Interactive Radio-Epidemiological Program (IREP 5.3), which can be used to estimate PC/AS with the revised radioepidemiological tables. The program incorporates the following revisions for the process of estimating PC/AS:

- New incidence and mortality risk data.
- Calculation of risk and AS for all ages at exposure starting at birth.
- New cancer sites including some less strongly associated with radiation exposure.
- New analytic approaches and ways to summarize data.
- More attention to uncertainty and presentation of risk.
- Use of organ-specific equivalent dose, in sievert (Sv).

The 2003 revision is based on two working assumptions: most types of cancer can, in principle, be induced by radiation; and the most important question is the magnitude of the risk associated with a particular exposure. In all, more than 25 cancers and groups of cancers are considered, including several cancer types not strongly associated with an absorbed dose of ionizing radiation. A list is shown in Table 5.1. The revised treatment of uncertainty included information from the original report and a more recent information (NCRP, 1996, 1997). Essentially, the method involves the calculation of a distribution of ERR similar to that shown in Figure 5.2 for a person in whom cancer has been diagnosed and who had been exposed to ionizing radiation.

The 2003 working group had access to expanded cancer-incidence data from atomic-bomb survivors participating in the Life Span Study being conducted by the Radiation Effects Research Foundation (RERF). Those data are not only more recent but based on more timely and accurate diagnoses of disease than were previously available from death certificates. Incidence data are also more relevant than mortality to compensation claims for cancers of delayed or low fatality. Direct access to RERF data allowed the working group to conduct its own analyses directed at the needs of that report, including modeling of dose-response modifiers, such as age at exposure, and including cancer types not strongly associated with radiation exposure.

The 2003 report is based on linear dose-response models for all solid cancers, which included dose and dose-rate reduction effectiveness factors (DDREFs) to allow for the possibility that risk per unit dose decreases with decreasing dose and dose rate. The 2003 report also treats the relative biological effectiveness

TABLE 5.1   Cancer Sites Covered by 2003 Radioepidemiological Program

Solid-cancer sites

1. Oral cavity and pharynx

2. Digestive system
   - Esophagus
   - Stomach
   - Colon
   - Rectum
   - Liver
   - Gallbladder
   - Pancreas
   - Other digestive

3. Respiratory system
   - Trachea, bronchus, and lung
   - Other respiratory cancers

4. Bone

5. Skin
   - Basal cell carcinoma
   - Other non-melanoma skin cancer

6. Female breast

7. Female genital
   - Ovary
   - Other female genital

8. Male genital
   - Prostate gland
   - Other male genital

9. Urinary system
   - Bladder
   - Kidney and residual urinary organs

10. Nervous system

11. Thyroid gland

12. Other and ill-defined sites (residual solid cancers)

Hematopoietic cancers

1. Leukemia, all types (except chronic lymphocytic leukemia)
2. Acute myelogenous leukemia
3. Acute lymphocytic leukemia
4. Chronic myelogenous leukemia
5. Lymphoma and multiple myeloma

(RBE) of densely compared with sparsely ionizing radiation as an uncertain quantity. The 2003 report does not include chronic lymphocytic leukemia (CLL) as a radiogenic disease, because data at the time were insufficient to formulate a relationship. And it does not address the health consequences of in utero exposure to ionizing radiation.

The PC/AS for lung cancer associated with radon exposure is given separately from lung cancer associated with other types of exposure. The estimates of risk for radon-related exposures are based on an analysis of data from a report to the US Department of Justice in 1996 (DOJ, 1996). A more comprehensive analysis based on risk estimates published by the National Research Council (BEIR VI), was available at the time (NRC, 1999), but implementation of BEIR VI into a PC/AS framework would have required more computational and staff resources than had been available to the NCI-CDC working group.

Interactive Radio-Epidemiological Program (IREP version 5.3)

The 2003 NCI-CDC working group replaced the tabular format for probability of causation with IREP version 5.3. That program eliminates nearly all the computational labor of estimating PC/AS values and their uncertainties, but permits a more detailed and comprehensive expression of the various components of the calculation and their uncertainties.

IREP version 5.3 includes the updates in the 2003 NCI-CDC report. That version is accessible at the NCI Web site at http://irep.nci.nih.gov (accessed February 22, 2005). It enables the user to calculate the PC/AS for a particular cancer in a person exposed to a given dose of radiation and certain modifying factors such as the person's age and sex. The PC/AS calculations are based on cancer-incidence data from the study of the Japanese atomic-bomb survivors and radon-induced lung cancer. The computer code allows consideration of uncertainty in an estimated dose and uncertainties in all the other items that enter into a calculation of PC/AS. The code computes the median (central estimate) of PC/AS value for a specified dose and the upper 95% and 99% percentiles of PC/AS, taking into account all uncertainties.

Implementation of IREP in Compensation Programs

The National Institute for Occupational Safety and Health (NIOSH) had adopted the 1985 NIH tables and added a few cancers to the original cancers, and it was using the results of earlier versions of IREP. It has now adopted the results of the 2003 NCI-CDC working group and is using the program NIOSH-IREP version 5.3, a modified version of the NCI-CDC code.

EEOICPA is using the modified version of the program prepared by NIOSH. The NIOSH version is accessible on the internet at http://www.niosh-irep.com/

irep_niosh/, accessed February 22, 2005. The original NCI-CDC version is maintained at http://irep.nci.nih.gov for archival and research purposes.

USE OF PC/AS IN ADJUDICATION

PC or AS has been used to adjudicate claims in the United States and the UK. The following sections summarize some of the situations in which it is used.

United Kingdom

In 1982, British Nuclear Fuels Ltd. (BNFL) and trade unions representing employees initiated a compensation program based on PC and cancer mortality. In 1987, the program was extended to cancer incidence. The program, which is voluntary and offered as an alternative to litigation, is now known as the Compensation Scheme for Radiation-Linked Diseases (CSRLD) and has been extended to UK nuclear electricity generators, the Ministry of Defense, and the Atomic Weapons Establishment (Wakeford et al., 1998).

Values of PC for the CSRLD are adapted from risk-projection models developed by BEIR V (NRC, 1990), which are based principally on the Life Span Study of the Japanese atomic-bomb survivors. The CSRLD covers all cancers except CLL, malignant melanoma of the skin, and Hodgkin's lymphoma. Those cancers are excluded because there was no evidence of an association with radiation exposure. The CSRLD also excludes mesothelioma because it is associated almost exclusively with exposure to asbestos rather than radiation. Breast cancer and cancer of the digestive tract are included but are treated separately. There was a concern related to transferring risk estimates between populations in Japan and the UK for cancers that have different background rates in the two countries.

The CSRLD uses a sliding scale for compensation. Cancer victims with estimated values of PC/AS that are 0.5 or higher receive full awards. Persons whose estimates are between 0.2 and 0.5 receive graduated partial awards. The sliding scale is shown in Figure 5.6. Information posted on the CSRLD indicates that over 1,000 claims have been assessed. Compensation has been awarded to 97 claimants. Over half of the awards were for a PC/AS less than 0.5 (http://www.csrld.org.uk/html/scheme_history.php, accessed February 22, 2005).

Department of Veterans Affairs

In 1988, Congress passed PL 100-321, the Radiation-Exposed Veterans Compensation Act. Under that law, it is presumed that a veteran's disease was caused by radiation if the veteran was present during a nuclear detonation regardless of the dose received. Thirteen radiogenic cancers were included in the list of presumptive diseases. A second regulation (38 CFR 3.311) governs diseases that are

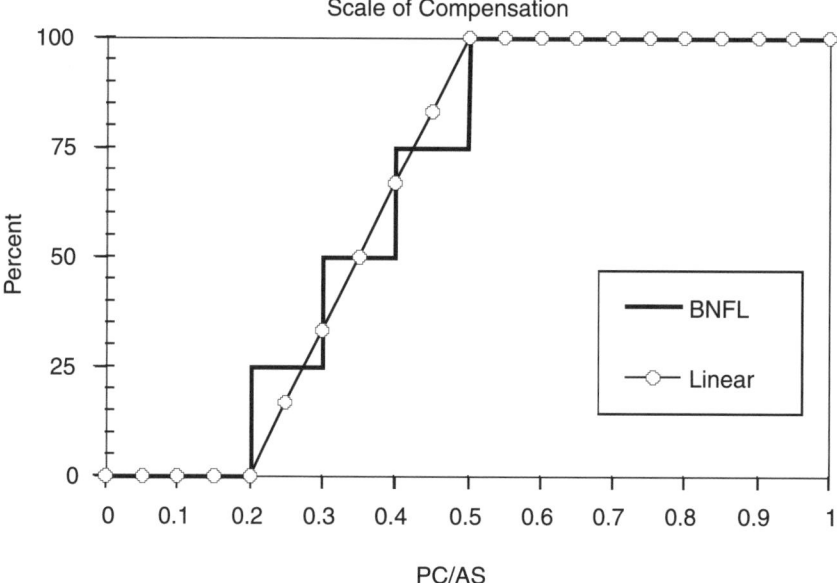

FIGURE 5.6   Percentage of full compensation based on PC/AS, (a) as used in the BNFL Compensation Scheme for Radiation-Linked Diseases (solid line) and (b) for a smooth linear scale (circles).

not presumed to be caused by radiation but could be associated with radiation if the dose was large enough.

VA has used the process developed by CIRRPC to evaluate claims (NRC, 2003b). CIRRPC produced tables that give the dose that would produce a PC/AS of 0.5 for various organs. To give the benefit of the doubt to the veteran, VA has chosen to use a threshold dose associated with a 99% credibility limit. At that dose, there is 99% confidence that the true PC is not greater than 0.5.

The PC/AS process is one of several tools in the medical decision process. It is not the only factor used in deciding whether a veteran with a nonpresumptive radiogenic disease receives compensation for service-connected exposures to radiation. However, the VA does not deny a claim if the PC/AS criterion is satisfied as long as all the other necessary conditions are fulfilled.

Department of Labor

Congress passed the EEOICPA in 2000 for persons who have worked for the Department of Energy or at any other facility involved in the development of nuclear weapons. In May 2002, the Department of Health and Human Services published its final rule on the guidelines to be used by the Department of Labor

(DOL) for determining whether a cancer included in an EEOICPA claim was caused by occupational exposure to radiation during nuclear-weapons production. DOL was advised to make such a determination by using a dose reconstruction provided by NIOSH and an estimate of PC/AS based on it. A person is eligible for compensation if DOL determines that the PC/AS was at least 0.5 (that is, at least as likely as not). Compensation is denied when there is a 99% chance that the estimate of PC/AS is less than 0.5 (that is, there is less than a 1% chance that PC/AS is at least 0.5).

APPLICATION OF PC/AS FOR FALLOUT FROM ATMOSPHERIC WEAPONS TESTING

This section presents a hypothetical example to illustrate the capabilities for determining PC/AS with existing tools and methods for a situation related to RECA. It involves thyroid cancer diagnosed in a person who was exposed to ^{131}I from fallout. It shows how PC/AS is determined with user-friendly computer programs available on the Internet and developed by NCI. The process consists of two steps. The first step is a computation of the absorbed dose to the thyroid, and the second step is estimation of PC/AS on the basis of the dose and a diagnosis of cancer. The next section shows a sample calculation with Monte Carlo methods that demonstrate the procedures embedded in IREP.

For the hypothetical example, we consider a male born on January 1, 1951, in an area that received little fallout. On January 1, 1952, he moved to and lived in a location that received a large amount of fallout. He then returned to the original area on January 1, 1953. While in the high-fallout area, he consumed 1-3 glasses of milk a day.[5] Thyroid cancer was diagnosed in 2004.

NCI Dose Calculator

The NCI calculator was used to estimate the absorbed dose to the thyroid from ^{131}I (http://ntsi131.nci.nih.gov/, accessed February 22, 2005). The result was 0.32 Gy (32 rad), with a range of 0.032-5.2 Gy (3.2-520 rad). That 95% confidence interval can be approximated using a lognormal distribution with a geometric mean of 0.32 Gy (32 rad) and a geometric standard deviation of 4.0.

The next step was to determine PC/AS with IREP Version 5.3 at http://www.irep.nci.nih.gov/, accessed February 22, 2005. For simplicity, the small radiation doses incurred during residence at the location where the fallout was low have been ignored. Other input values supplied to the program are

- An exposure in 1952.

---

[5]The actual locations used were San Diego County, California, for the lower fallout area, and Custer County, Idaho, for the higher fallout area.

- Chronic exposure rate.
- Radiation type—electrons with energy greater than 15 keV.
- Organ dose lognormally distributed with a mean dose of 0.32 Gy (32 rad) and geometric standard deviation of 4.0.

The median value of PC/AS was 0.67, and this was assigned to the person. That value indicates that it is more probable than not that the thyroid cancer was due to radiation exposure.

Age at exposure is important in determining both the dose to the thyroid and the PC/AS of thyroid cancer from radiation exposure. In the above example, if the person had been born in 1931 rather than 1951 and all other information were the same (residence during 1952 and milk consumption of 1-3 glasses per day), the NCI dose calculator would yield a dose of 0.036 Gy (3.6 rad) with a range of 0.0037-0.65 Gy (0.37-65 rad). That result was considered to be a lognormal distribution of dose with a geometric mean of 0.036 Gy (3.6 rad) and a geometric standard deviation of 4.2. IREP would compute a median value of PC/AS of 0.04; it would be very unlikely that the thyroid cancer was related to the dose received from ^{131}I in 1952.

We have also made an independent computation of PC/AS for the first scenario. We used data obtained directly from the NCI-CDC working group to compute ERR and used a Monte Carlo process to estimate the uncertainty (credibility limits). Details of the computation are presented in an annex to this chapter.

CONCERNS WITH USING PC/AS IN COMPENSATION PROGRAMS

Several authors, including those of the National Research Council 1984 committee and 2000 subcommittee reports, have pointed out various inequities that can arise in a compensation scheme that provides full payment to claimants with an estimated PC/AS greater than 0.5 but no payment to those with PC/AS less than 0.5 (NRC, 1984; NRC, 2000). For example, two people with cancer could have almost identical likelihood that their cancer was caused by exposure to radiation, but one applicant might have a PC/AS of 0.51 and the other a PC/AS of 0.49. If 0.5 were used as a bright line cut point, the first person would receive full compensation and the second none. A large difference in compensation at a single cut point does not recognize the large degree of uncertainty in the determination of PC/AS, and it assumes differences in PC/AS where statistically none exists.

Figure 5.5-D illustrates a situation where the estimated PC/AS has a wide credibility interval because of large uncertainties in estimated dose and the corresponding risk for a particular organ. For these situations, the true PC/AS might be very low, but the upper tail of the distribution could exceed established eligibility criteria.

A compensation scheme based on a sliding scale avoids that problem to some extent, in that people with PC/AS slightly below the value for full compensation would receive at least partial compensation. The 1984 National Research Council committee reviewed several alternative compensation schemes, including one in which compensation is proportional to PC/AS and another scheme in which full compensation would be received for PC/AS at least 0.5, no compensation for PC/AS is less than 0.1, and compensation scaled to PC/AS for values between 0.1 and 0.5 (NRC, 1984).

As noted above, the UK CSRLD has used a sliding compensation scale since 1982. That scale is not related linearly to PC/AS but has discontinuities at 0.2, 0.3, 0.4, and 0.5 (see Figure 5.6).

Robins and Greenland (1991) examined compensation procedures in which compensation is paid in proportion to PC/AS and compared them with a procedure in which compensation is proportional to years of life lost (YLL). They found shortcomings in both methods, but the YLL procedure had the advantages of being "robust" to model misspecification and "economically rational." Payment schemes can be "potentially optimal," and a scheme based on actual but unknown, biologic mechanisms and population heterogeneity is "truly optimal." They define a payment scheme as robust if the average payment to a worker harmed by a radiation exposure equals the average payment under a truly optimal scheme. Schemes are economically rational if the total damages that a defendant pays equal the total damages assessed under a truly optimal scheme. Robins and Greenland (1991) concluded, however, that "in certain settings, payments in proportion to probability of causation might be preferred to payments in proportion to years of life lost for social, ethical and/or legal reasons."

Robins re-emphasized several issues about PC/AS in a recent paper (Robins, 2004):

- The PC/AS cannot be unambiguously calculated with epidemiologic data, because of the lack of knowledge of the underlying biological mechanisms of cancer.
- PC/AS compensates only for the occurrence of a cancer; it does not completely account for the harm done to a person. For example, a person who dies from a compensable cancer at the age of 45 years suffers a greater loss than an individual dying from the same cancer at 70, but both would receive the same compensation.
- PC/AS encourages counterproductive public-health practices. In the example of radiation workers, if the radiation dose and resulting risk is spread out over many people rather than over just a few, PC/AS values could be kept small, and no one would receive compensation. The collective dose to a group of workers could remain the same and in many cases increase. With a linear risk model, the total number of cancers may be the same or larger, even though none are compensable.

Robins suggested a two-step compensation scheme based on the total YLL of the group, which is independent of the actual distribution of YLL among the people in the group. The program would first determine the total YLL of the group and set the compensation amount in proportion to that number. Then the distribution of compensation among the members of the group would be negotiated in some manner because it is not possible to determine the YLL for each member. That proposal could address the second and third issues noted above and optimize the response to the first one.

Other programs, as noted in previous sections, award compensation when there is at least a 1% chance (99% credibility limit) that PC/AS is greater than 0.5. That takes into account the uncertainty associated with PC/AS. The 2000 National Research Council subcommittee report criticized such an approach as favoring compensation for cancers whose connection with radiation is not well established (where the estimates of PC/AS have large uncertainties) over compensation for cancers that are clearly associated with radiation (NRC, 2000).

A similar concern could be applied to the estimation of dose. The absorbed dose to a person living far from a fallout monitoring station would typically have greater uncertainty than that of a person living closer to a station (NCI, 1997). If they contracted the same cancer and all risk factors were equal, the PC/AS of one might be 0.4 with 99% credibility limits of (0.35, 0.45), and that of the other might be 0.4 with 99% credibility limits of (0.25, 0.55). The first person would not be compensated, and the second would be because the 99% credibility limits included 0.5.

The National Research Council 2000 subcommittee recommended that the full probability distribution be taken into account. Although it did not recommend any particular compensation scheme, it suggested as an example consideration of schemes in which compensation is scaled according to the probability that PC/AS exceeds 0.5 or according to a direct function of the central value of PC/AS. The subcommittee also noted that the PC/AS method does not take degree of harm into account, in contrast with such a method as compensating according to YLL. It recognized, however, "that the current policy is grounded on the notion of AS, and we do not wish to discredit the efforts of the working group by implying that its estimates are not useful as a basis for resolving such claims, even in the face of fundamental problems of interpretation" (NRC, 2000).

CONCLUSION

The development of PC/AS was mandated by Congress to provide a more scientific basis for awarding compensation to persons who have been exposed to radiation and later have a diagnosed radiogenic disease. The concept is used in many compensation programs that involve radiation exposure. There is a well-documented infrastructure of tools and methods that has taken many years to develop.

The National Council on Radiation Protection and Measurements (NCRP) statement *The Probability That a Particular Malignancy May Have Been Caused by a Specific Irradiation* (NCRP, 1992) concluded that PC/AS provided a logical procedure for addressing the issue of radiation-induced malignancy that had no other scientific or medical solution. It had the advantage that the process could be cast into a formula that would yield a figure of merit for the presumption of causation. A recognized limitation of the PC/AS approach was that population-based data were applied to individuals. That implies that a person has average susceptibility to cancer. PC/AS thus provides a probability, rather than proof of causation. It can, however, be tailored to individuals by using dose, sex, age at exposure, age at diagnosis, and type of malignancy.

The committee does not make specific recommendations on establishing a threshold value of PC/AS for compensation. However, we have compared several approaches to determine compensation on the basis of both point estimates of PC/AS and the entire distribution of PC/AS as illustrated in Figure 5.5 (see Table 5.2).

The first five schemes base compensation on point estimates of PC/AS. The term, PC/AS$_{med}$, represents the median, or central, value of the distribution. PC/AS$_{0.99}$ represents the upper tail of the distribution where there is a 1% chance that PC/AS exceeds this value. PC/AS$_{dist}$ represents the entire distribution estimated for a person. The last scheme uses the linear scale shown in Figure 5.6 weighted by PC/AS$_{dist}$. That is to say, you compute the fraction of compensation for each value of PC/AS multiplied the probability that the person's PC/AS might be this value and then sum up over all possible values of PC/AS from 0 to 1. This avoids the problem of selecting a single payment threshold value of PC/AS (vertical line in Figure 5.5) and eliminates the need to base the decision on a single point estimate of PC/AS for a person.

The committee has considered criticisms of the PC/AS approach. Alternative schemes also have serious shortcomings. They would require time and resources to develop for RECA. Implementation could lead to confusion and

TABLE 5.2   Examples of Compensation on the Basis of PC/AS Using Several Criteria with Distributions Illustrated in Figure 5.5

	A (0.83)	B (0.5)	C (0.26)	D (0.10)
PC/AS$_{med}$[a] ≥ 0.5	100%[b]	100%	0	0
PC/AS$_{med}$ ≥ 0.2	100%	100%	100%	0
PC/AS$_{0.99}$ ≥ 0.5 (EEOICPA)	100%	100%	0	100%
PC/AS$_{med}$ with BNFL scale (Figure 5.6)	100%	100%	25%	0
PC/AS$_{med}$ with linear scale (Figure 5.6)	100%	100%	20%	0
Fraction of PC$_{dist}$ ≥ 0.5	100%	50%	0	2%
Integrate PC/AS$_{dist}$ with linear scale	100%	81%	23%	8%

[a]PC/AS$_{med}$ refers to the median value for each distribution, shown in parentheses at the head of each column.

[b]The numbers represent a percent of the maximum award for each scheme.

possible conflicts with the other compensations programs. It is not obvious that the alternative schemes would be more favorable or equitable to claimants. A compensation scheme based on PC/AS has sufficient flexibility to provide a scientifically based process for including persons who are not in geographic areas that are now eligible for compensation under RECA.

ANNEX

Calculation of PC/AS for Thyroid Cancer with Monte Carlo Method

To understand the process of determining PC/AS in more detail, the estimate of risk posed by exposure to ^{131}I was obtained with an independent calculation. A Monte Carlo calculation was performed with the computer code Crystal Ball® 2000.2, which was developed and distributed by Decisioneering (2001).

The NCI-CDC working group (2003) used the following expression to determine the ERR of solid tumors caused by low-dose low-dose-rate radiation exposure:

$$ERR = REF_L \times \frac{R_{\gamma,H}}{DDREF_\gamma} \times D \tag{11}$$

where

REF_L is the radiation effectiveness factor for the radiation type and cancer type,

$R_{\gamma,H}$ is the ERR per Gy at high doses and high dose rates of the reference high-energy gamma radiation,

L and H refer to low and high doses and dose rates,

$DDREF_\gamma$ is the dose and dose-rate effectiveness factor, and

D is the estimated absorbed dose (Gy).

$R_{\gamma,H}$ must be modified for the exposure scenario being considered. In this example, for low-dose low-dose-rate (chronic) radiation primarily from electrons (beta particles from the ^{131}I decay), that was done by using REF_L and $DDREF_\gamma$. For clarity, the committee followed the NCI-CDC working group and used the terminology "$R_{\gamma,H}$" for the uncorrected ERR per Gy, and "ERR per Gy" for the ERR per Gy properly adjusted for the dose, dose rate, and radiation effectiveness factor. Then

$$ERR \text{ per } Gy = \frac{REF_L \times R_{\gamma,H}}{DDREF_\gamma} \tag{12}$$

Values for the four terms in Equation 11 all have associated uncertainties. In practice, a distribution of values is assigned to each term, depending on supporting scientific evidence, and the ERR is calculated by choosing values for each of the four terms that depend on the assigned distribution. Typically, the ERR is then calculated a number of times using a Monte Carlo sampling method, and the PC computed from each calculated value of the ERR. This process then determines the distribution of the PC.

The four terms in Equation 11 are as follows:

$REF_L = 1$ (NCI-CDC, 2003, p. 65) for electrons with energies greater than 15 keV.

$R_{\gamma,H}$ is the ERR per rad at high doses and high dose rates for high-energy gamma radiation. The NCI-CDC working group based its risk estimates for thyroid cancer on an analysis of seven studies by Ron et al. (1995). The resulting risk values depend on age at exposure, taken to be lognormally distributed, and are shown in Table 5.3 and Figure 5.7.

$DDREF_\gamma$, the dose and dose-rate effectiveness factor for gamma radiation, for this example of chronic exposure, is a discrete probability distribution given in Table 5.4. In this context, chronic exposure to radiation occurs when the dose rate is less than 6 mGy/h (600 mrad/h) averaged over the first few hours (NCI-CDC, 2003, p. 32). The 6 mGy/hr dose rate is higher than what would occur for the ^{131}I thyroid doses considered here as long as the total dose from one exposure is less than about 150 rad.

D, the absorbed dose, is the result of the NCI dose calculator, which is lognormally distributed with a geometric mean of 0.32 Gy (32 rad) and geometric standard deviation (GSD) of 4.

TABLE 5.3   Excess Relative Risk per Absorbed Dose (Gy) for Thyroid Cancer, High-Dose and High-Dose-Rate Gamma Radiation[a]

Age at Exposure, Year	$R_{\gamma,H}$/rad (Geometric Mean)	Geometric Standard Deviation
0	9.463	2.183
5	6.262	1.924
10	4.136	1.976
15	2.732	2.16
20	1.804	2.301
25	1.192	2.367
30	0.788	2.365
35	0.521	2.379
40	0.345	2.732
45	0.228	3.14
50	0.151	3.611

[a]SOURCE: NCI-CDC, 2003, p. 55.

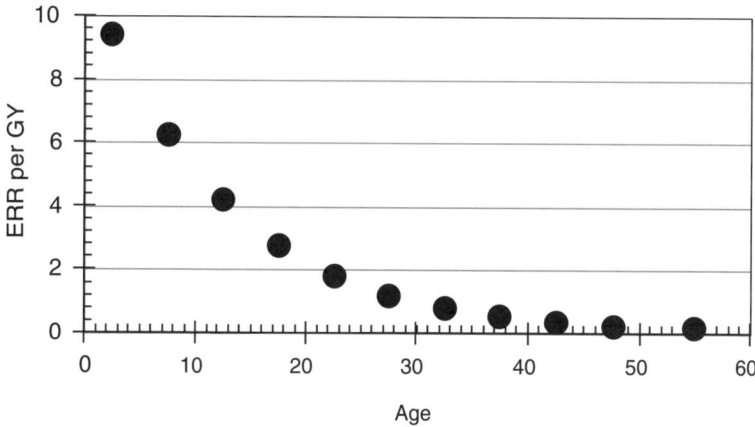

FIGURE 5.7   Excess relative risk per absorbed dose for thyroid cancer at high doses and high dose rates for gamma radiation (from NCI-CDC, 2003, p. 55).

The calculation was repeated 500,000 times for this person, each time with a different value of each of the four terms in Equation 11, randomly chosen according to its statistical distribution. Results of the Monte Carlo calculation are shown in Figures 5.8a, 5.8b, and 5.8c. The computations were repeated for the case in which the person was born in 1931 and was thus 20 years old at the time of exposure. The PC/AS for that scenario is also shown in Figure 5.8c.

Figure 5.8c displays the results as a complementary cumulative distribution. The vertical axis represents the probability that PC/AS is greater than the value on the horizontal axis. For example, the median value of PC/AS (the point where the probability is 0.5 on the vertical axis) for exposure at the age of 20 is 0.04, compared with 0.67 at the age of 1 year. Drawing a vertical line that intersects the horizontal axis at 0.5 shows that there is a 1% chance that PC/AS is greater than 0.5 at the age of 20 years and a 65% chance that PC/AS is greater than 0.5 at the age of 1 year.

TABLE 5.4   Probability Distribution for the Dose and Dose-Rate Effectiveness Factor[a]

$DDREF_\gamma$	Probability
0.5	0.01
0.7	0.04
1.0	0.35
1.5	0.23
2.0	0.23
3.0	0.10
4.0	0.04

[a]SOURCE: NCI-CDC, 2003, p. 60.

Dose for Thyroid Exposure to ^{131}I at Age = 1

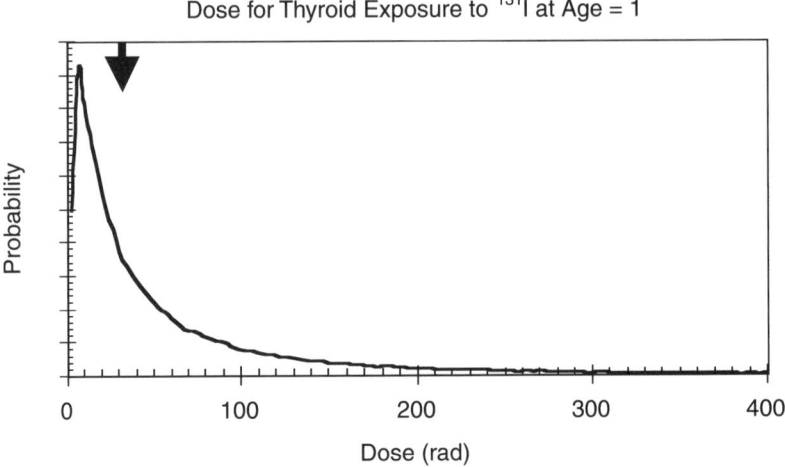

FIGURE 5.8a   Distribution of absorbed dose to thyroid from exposure to ^{131}I in fallout at age of 1. Median value of distribution is 32 rad, as indicated by arrow.

ERR for Thyroid Exposure to ^{131}I at Age = 1

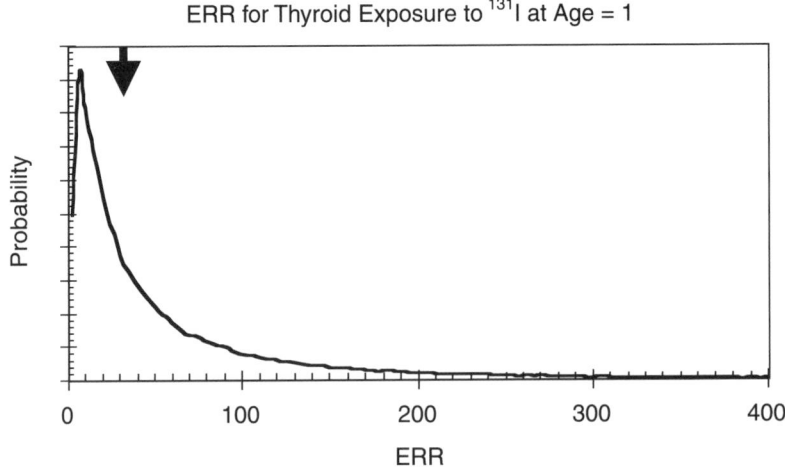

FIGURE 5.8b   Distribution of excess relative risk of thyroid cancer from exposure to ^{131}I in fallout at age of 1. Median value of distribution is 2.0 as shown by arrow.

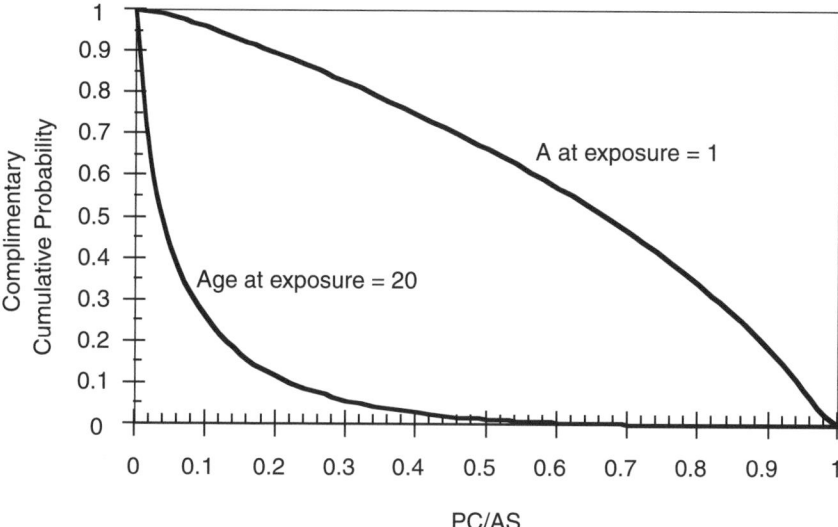

FIGURE 5.8c   Complementary cumulative distribution of PC/AS for exposure of thyroid to ^{131}I at age of 1 and age of 20. Curves show probability that PC/AC is greater than given value on X axis.

6

Expanding RECA Eligibility: Implementation

The committee was charged with making recommendations to the Health Resources and Services Administration (HRSA) as to whether the Radiation Exposure Compensation Act (RECA) should cover additional geographic areas. This chapter discusses some of the issues related to the current RECA downwinder areas, which include only people living in specific counties near the Nevada Test Site (NTS). It confirms the need to include additional geographic areas, and it proposes a procedure based on risk and probability of causation/assigned share (PC/AS) to determine which people would be eligible for compensation under RECA.

Eligibility for compensation through RECA depends on many factors. The committee believes that compensation for loss of health caused by exposure to radiation resulting from the United States government's nuclear-weapons program should be the first priority of any amendment to modify eligibility for compensation under RECA. Thus, we determined that epidemiologically based methods should be developed to identify additional populations and geographic areas that could be considered by Congress for inclusion in RECA.

PROBABILITY OF CAUSATION/ASSIGNED SHARE

Background

An important health detriment after exposure to ionizing radiation is the increased risk of several types of cancers, referred to as radiogenic, that can be caused by radiation. Such radiogenic cancers can also be caused by other agents or factors. No symptom or marker has been found that can identify a specific cancer as having been caused by radiation. That creates a problem when one is trying to establish eligibility for compensation for radiogenic cancer.

A method using probability of causation/assigned share (PC/AS) has been developed to address this issue. PC/AS was introduced and discussed in general terms in Chapter 5. We consider here how PC/AS could be used in determining eligibility under RECA to avoid some of the problems described in Chapter 5.

In Chapter 5, we noted several possible schemes for dealing with compensation that differ from the classical PC/AS or use a PC/AS with various adjustments with reference to the threshold for compensation and credibility interval. These include compensation based on years of life lost (YLL), compensation proportioned according to the posterior probability that PC/AS exceeds 0.5 or some other cutoff value, and compensation based on the values of the payment schedule weighted by the probability of having each value of PC/AS as determined with respect to the credibility distribution.

The committee recognizes the merit of those alternatives and how they address shortcomings in the classical PC/AS approach. After considerable discussion, however, the committee crafted its recommendations in terms of solely PC/AS (including credibility interval). Two positive factors affecting the committee's decision to adopt PC/AS are its widespread use in current compensation programs and the availability of user-friendly tools designed to implement the PC/AS approach. In addition, a compensation scheme based on YLL would neglect most individuals suffering from radiation-induced papillary thyroid cancer, the primary disease shown to be related to ^{131}I dose, because of thyroid cancer's small effect on longevity. A modified YLL method based on incidence rather than mortality, such as quality-adjusted years of life lost, was also considered and may be more reasonable. A compensation scheme based on YLL would be difficult to implement because the calculation of an individual's YLL would be a very uncertain projection. Compensation proportional to the area of the upper tail of the PC/AS distribution and compensation weighted by the PC/AS distribution are also attractive alternatives. Such approaches fall within the bounds of the committee's suggestion about PC/AS-based eligibility and merely adjust the amount that an individual would be awarded in compensation, which is not, in itself, a scientific decision.

The committee notes that compensation schemes based on other approaches may ultimately be preferable if the infrastructure is developed to support them. The committee would endorse such approaches if they provided for a more equitable distribution of compensation than would be possible with a PC/AS system.

In Chapter 1, we noted that the values of the threshold for compensation and the associated credibility interval result from societal decisions rather than scientific decisions. Later in this chapter, we recommend that Congress establish criteria for awarding compensation on the basis of computed distributions of PC/AS for any persons making such a claim. This chapter presents several examples using different thresholds for compensation; these are only examples and are not meant to suggest any specific value for the threshold for compensation or the credibility interval.

Use of PC/AS to Identify Geographic Areas for RECA Compensation

RECA explicitly defines the geographic areas in which people must have lived to be eligible for compensation for a cancer that could have been caused by their exposure to radiation as downwinders relative to nuclear tests at the NTS. It specifies portions of Utah, Nevada, and Arizona as areas whose residents may be eligible for compensation.

The committee considered whether other geographic areas should be added to the previously defined areas on the basis that residents had been at similar or higher risks from exposure to fallout from United States nuclear tests. We considered a range of possible expansions of the downwinder geographic areas currently included under RECA. In the next sections we will develop and present our recommendation that compensation be fundamentally based on a consistent PC/AS-based process rather than a priori solely on residence in specific geographic areas. We will propose that PC/AS criteria be used for both: 1) a pre-assessment to guide potential claimants and the implementing agency regarding which diseases, groups of individuals and geographic areas may satisfy RECA eligibility and 2) subsequently, a determination of individual claimant eligibility for RECA compensation. We then describe the underlying reasons that led the committee to this conclusion.

We arrived at this conclusion, in part, on the basis of the results of the 1997 National Cancer Institute's (NCI's) iodine-131 (^{131}I) report and the 2001 Centers for Disease Control and Prevention (CDC)-NCI draft feasibility study discussed in Chapter 4. The NCI ^{131}I report shows that estimated doses and associated risks for radiation-induced thyroid cancer from ingestion of ^{131}I depend not only on location but also on diet and age at exposure. Areas in which estimated thyroid doses were increased are not only near the NTS but also in such distant locations in the United States as the Midwest and upstate New York and Vermont (primarily the result of precipitation-borne fallout after test Simon). The distribution of those estimated doses is difficult to define geographically, especially because they also depend on diet and age at exposure. In addition, cancer risks from exposure to radiation depend not only on dose but also on other factors, including on the risk coefficient (see Chapter 5), which itself also depends on age at exposure. The risks to two individuals of different ages living in the same county would differ not only because their estimated doses may be different but also because their risk coefficients may be different.

Similarly, the 2001 draft feasibility study by the CDC-NCI shows the importance of factors other than location in the estimate of doses that individuals may have received from other radionuclides in fallout. The associated radiation risks of many radiogenic diseases also depend on such factors as age at exposure and age at diagnosis. Simple expansion of RECA eligibility based on location alone would not be equitable in that for some cancers it may fail to compensate higher-risk people in ineligible areas, such as those exposed to ^{131}I in fallout as newborns, and may compensate lower-risk people in eligible areas, such as those exposed to fallout at an advanced age.

It should also be recognized that, with the currently estimated levels of radiation and associated risk coefficients for cancer, it is unlikely that a very large number of individuals with cancer, even thyroid cancer, would be newly eligible for compensation. The actual number will depend on the threshold criteria established by Congress.

EXPOSURE TO FALLOUT RADIATION

Exposure of the Thyroid to NTS Fallout Radiation

Chapter 5 showed the distribution of the estimated thyroid dose from ^{131}I in fallout across all counties in Utah and showed which counties are compensated under RECA 2000 from NTS operations. As an example, the committee considered a male in each Utah County, born on January 1, 1948, residing in the same county during the nuclear-test period, and consuming an average amount of milk daily. That example showed that persons in one county (for example Joab County), who later developed thyroid cancer, would currently not receive compensation under RECA, whereas persons in other counties (such as Iron County) would receive compensation, even though the estimated doses in some of these other counties were substantially smaller (see Figure 5.1).

This section discusses additional variations in a person's risk of radiogenic thyroid cancer that the current RECA scheme does not take into account. Estimated thyroid doses vary not only between counties but also can vary considerably within a given county. The risk of thyroid cancer will vary even more markedly within a county than the thyroid dose, because the thyroid risk coefficient also varies with age at exposure and other factors.

For the RECA-compensable cancers, the current PC/AS tables and calculators do not take coincident factors into account except for lung cancer (smoking). As noted earlier, the committee has chosen to use the current PC/AS approach because it is available and practical; it recognizes that improvements and refinements could be and perhaps should be made.

The 1997 NCI ^{131}I report contains an extensive series of maps showing all counties in the continental United States and the estimated thyroid doses that would be received by county residents born from January 1, 1930, to January 1, 1962, from ^{131}I intake from NTS weapons tests given four milk-consumption scenarios (no, average, and high cow's milk intake from store-purchased milk, and high milk intake from a backyard cow). The Institute of Medicine-National

Research Council committee that reviewed the NCI ^{131}I report noted that the within-county variability in many cases exceeds the between-county variability. Age at exposure, sources of milk, and amount of milk ingested are particularly important risk factors expected to influence the occurrence of thyroid cancer in a person exposed to radiation.

As an example, the present committee considered the thyroid doses from ^{131}I for a person living in Custer County, Idaho, one of the counties identified in the 1997 NCI report as having a relatively high per capita thyroid dose from NTS operations. Although that county does not have the highest per capita thyroid dose (the county with the highest estimated per capita dose is Meagher County, Montana), it represents the general issue. Our committee noted that people who lived only in counties in Idaho or Montana during the nuclear-testing era are not eligible for compensation under RECA, even though some of these counties were among the ones that received the highest per capita thyroid dose.

Figure 6.1 shows the best estimate and upper and lower 90% bounds of the thyroid doses from ingestion of ^{131}I from all NTS tests for a male living in Custer County. The figure shows the estimated doses by year of birth and for average milk consumption (the upper 90% bound was over 1 Gy for birth years 1932-1953 and is not shown in the figure). We calculated the thyroid doses for male residents of this county with NCI's on-line thyroid-dose calculator, assuming that a person was born on January 1 of each year (doses for persons born on other days of the year may be different) and resided in the same county during 1951-1971. For example, the 0.2 Gy estimated dose shown for a person born on January 1, 1953, would be the estimated dose received from the aboveground tests conducted from 1953 through 1962, when aboveground testing at the NTS ended. The estimated dose also includes a small component from in utero exposure to fallout from tests in 1952, which would affect individuals born in the early part of 1953.

Residents of Custer County received most of their total estimated dose in 1952 from the Tumbler-Snapper series of tests that were conducted that year at the NTS, when the county was in the path of the fallout. As seen in the figure, the highest estimated dose for this set of birth dates occurred to people born on January 1, 1952, who were newborns during that test series. People born in 1948-1951 were infants and young children in 1952 and therefore had higher estimated thyroid doses than those born before 1948 or after 1952. Estimated dose decreases as the year of birth moves earlier from 1952; people born in 1930 (or before) would receive an estimated dose less than 10% that received by people born in 1952, even though they were residing in the same county. The estimated dose is also sharply lower for people born after 1952 because the major fallout events for this particular county occurred before their births. That illustrates further a problem in defining a geographic area for future RECA compensation: any such area would include people who received estimated doses that varied over a wide range, some of which would not be significant.

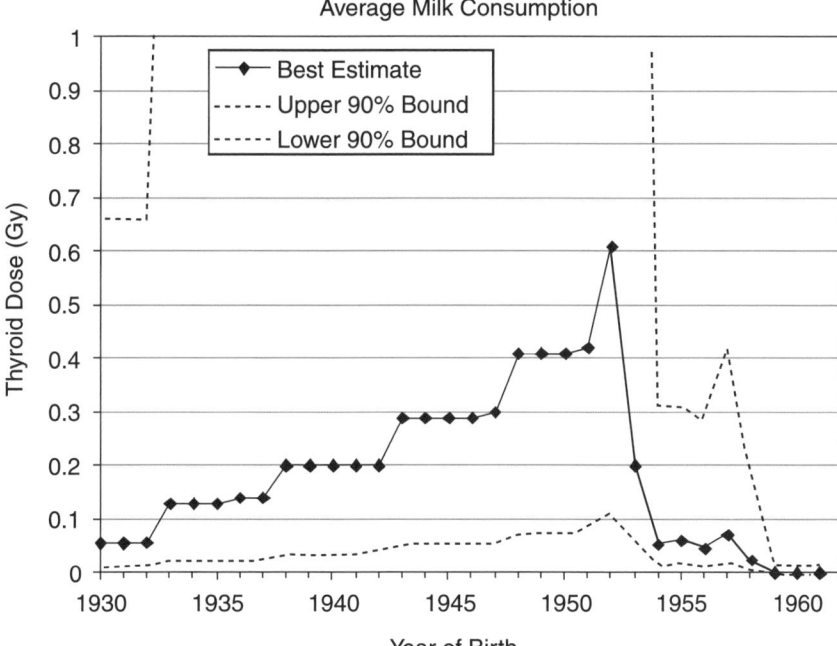

FIGURE 6.1 Thyroid dose (Gy) that the NCI dose calculator estimates a male resident of Custer County, Idaho, born on January 1, would have received from all NTS tests vs. year of birth for average milk consumption. The "step-like" structure of the graph is due to age-interval-based assumptions about variables, such as food consumption, and dose-conversion factors in dose calculator.

The highest thyroid dose in this set of people was estimated by the NCI thyroid dose calculator to be 0.61 Gy, for a person born on January 1, 1952. The uncertainties in the dose estimates are large; the 90% uncertainty limits are 0.11 to 8.4 Gy. By comparison, most residents of the United States would receive a thyroid dose from background radiation of about 0.001 Gy per year (estimated from NCRP, 1987). In the roughly eight years covered by atmospheric nuclear testing, they would receive a cumulative dose of 0.008 Gy to the thyroid from background radiation.

Several factors combine to increase the estimated ^{131}I thyroid dose to infants and young children, including the relatively higher uptake of ingested iodine because of increased thyroid metabolism at those ages and the smaller size of the thyroid, which increased iodine concentration relative to older children and adults (NCI, 1997).

As noted earlier in this chapter, the variability of fallout-related cancer risks among residents of the same county reflects the variability of the estimated thyroid doses, which depend on age at exposure and diet, and also the variability of the thyroid cancer risk coefficients, which also depend on age at exposure. As discussed in Chapter 5, the risk of thyroid cancer posed by a given thyroid dose is highest in newborns and decreases until the risk is small after the age of 20 years. As an example, Figure 6.2 reproduces from Figure 6.1 the estimated thyroid doses that residents of Custer County received from fallout from atmospheric tests. Assuming that a person developed thyroid cancer that was diagnosed in 2000, PC/AS can be found for that person's cancer depending on his or her estimated thyroid dose and age at exposure. Figure 6.2 shows these PC/AS values, and also the PC/AS compensation thresholds of 0.5 and 0.3 (two of the values used in other compensation schemes) and the effect on eligibility for compensation. The PC/AS curve in Figure 6.2 is the 50th percentile value of PC/

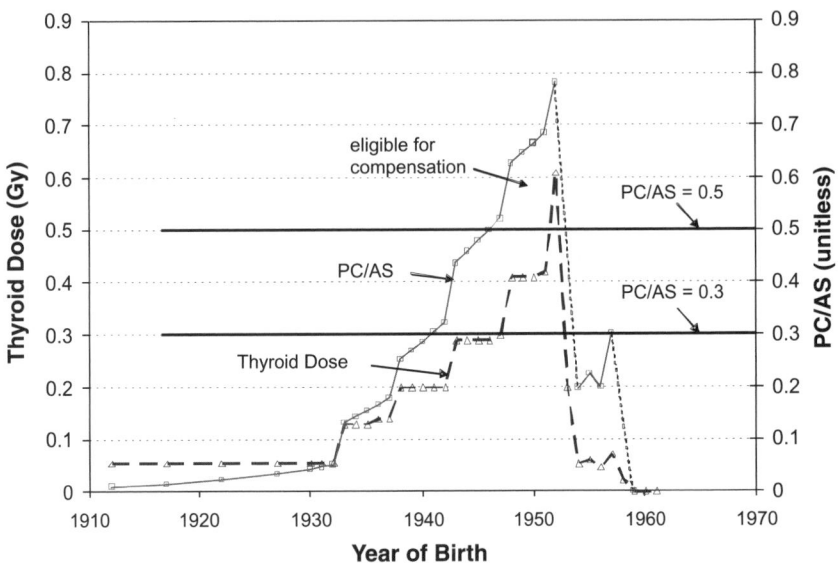

FIGURE 6.2   Estimated thyroid absorbed dose (Gy) from all NTS tests, and 50th percentile PC/AS, for a person living in Custer County, Idaho, born on January 1, by year of birth. Assumptions include average milk consumption, diagnosis of thyroid cancer in 2000, and exposures from NTS operations. The "step-like" structure of graph is due to age-interval-based assumptions about variables, such as food consumption, and dose-conversion factors in dose calculator. NOTE: PC/AS is not calculated for 1953 and 1958. Dose to individuals born on January 1 in those years included dose from the previous year because of in utero exposure. IREP is not currently designed to evaluate doses from in utero exposure.

AS by year of birth. (The 50th percentile value of the PC/AS value of 0.5 or 0.3 was chosen for convenience. Other percentile values, such as 99th percentile used in other compensation programs, can also be used.) The figure shows that PC/AS values for people born before the maximum dose period in 1952 decrease even more rapidly than dose with age at exposure.

PC/AS and estimated thyroid dose are highest for a newborn and decrease with age at exposure. If, for example, the compensation criterion is that the diagnosed thyroid cancer has to be "as likely as not" due to fallout radiation exposure, PC/AS would have to be at least 0.5. The figure shows that if a PC/AS threshold of 0.5 were used for compensation, people who were born in 1946-1952 would be eligible for compensation. Others would not be eligible. The age-at-exposure dependence of the thyroid-cancer risk coefficient complicates further the use of geographic areas as the sole criteria for RECA compensation.

Different groups of people may be eligible for compensation if other PC/AS thresholds for compensation are used. For example, if a scheme similar to that used by British Nuclear Fuels Ltd. (BNFL) were adopted, partial compensation would occur at PC/AS of 0.2, 0.3 (shown in Figure 6.2), and 0.4.

The above scenario represented by Figure 6.2 uses consumption of purchased cow's milk to illustrate how the general outline of the dose distribution within a county depends on age at exposure. Other scenarios—such as using goat's milk, obtaining milk from a backyard cow, or consuming no fresh milk—are possible; they would give higher or lower estimated doses depending on which scenario is examined.

Another perspective on the relation of age at exposure and PC/AS is shown in Figure 6.3. The committee used NCI's Interactive RadioEpidemiological Program (IREP) version 5.3 to estimate the thyroid doses that gave a PC/AS of 0.5 at the 50th and 95th percentiles for geometric standard deviations (GSD) of the thyroid dose of 2 and 4 (the GSD is used because dose distributions are found to be skewed to the right and assumed to be lognormal). Those doses, which depend strongly on age at exposure, are shown in Figure 6.3. The curves for GSD equal to 2 and 4 for the 50th percentile coincide, because the 50th percentile value, or median, of the distribution is insensitive to the spread of the distribution given by the GSD. The figure shows that a person would have a PC/AS = 0.5 at the 50th percentile if the person received 0.15 Gy as a newborn, or 0.35 Gy as a 10-year old. If the person only received 0.15 Gy as a 10-year-old, the PC/AS value would not reach 0.5.

For the 95th percentile, the GSD = 4 curve lies below the GSD = 2 curve. As discussed in Chapter 5 and illustrated in Figure 5.5D, the wider spread of the distribution for GSD = 4 causes the tail of the distribution to cross the 0.5 PC/AS value for lower values of the estimated dose than for GSD = 2. Consequently, as seen from the figure, the PC/AS of a 30-year old with an estimated thyroid dose of 0.15 Gy and GSD = 4 would exceed 0.5 for the 95th percentile, while that of a 30-year old with the same estimated dose but a GSD = 2 would not. An estimated

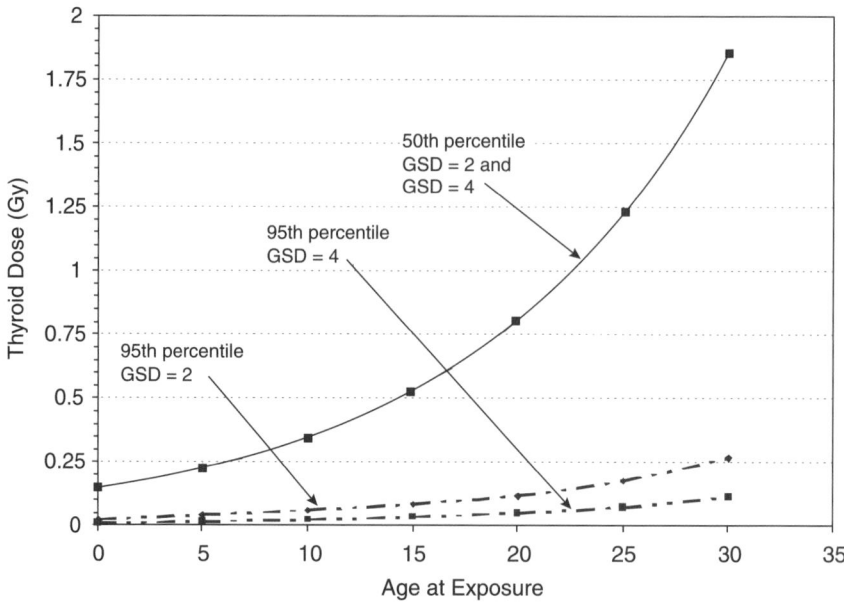

FIGURE 6.3. Estimated thyroid dose (Gy) for which PC/AS = 0.5 for different ages at exposure. Dose is given for 50th percentile and 95th percentile. The geometric standard deviation was assumed to be 2 and 4. Curves for the 50th percentile GSD = 2 and GSD = 4 coincide with each other.

dose of about 0.26 Gy would be needed for the PC/AS of the second person to exceed 0.5.

As an example of a possible PC/AS-based approach, to determine PC/AS for a particular person born in Custer County, Idaho, on January 1, 1950, the NCI dose calculator would be used to estimate the thyroid dose received from ^{131}I each year. This illustrative use of the calculator yielded the doses shown in Table 6.1.

The thyroid doses for the years not shown in the table are equal to zero. If the person had a diagnosed thyroid cancer, PC/AS could be calculated. The estimated doses are then entered into the NCI IREP program with the years of exposures, the date of birth, and the year of diagnosis to give the results presented in Table 6.2.

The 50th percentile value of PC/AS is 0.67 (66.55%), indicating that it is more likely than not that the person's cancer was due to exposure to ^{131}I in fallout. If the threshold for compensation were set at the 95th or 99th percentile, rather than the 50th percentile, the person would of course also qualify for compensation.

If one wants to ensure that people whose thyroid dose from NTS operations contributed substantially to the development of their thyroid cancer are properly compensated, a possible mechanism is to determine the estimated thyroid dose to a

TABLE 6.1 Estimated Annual Thyroid Doses Estimated Using the NCI Dose Calculator for a Person Born on January 1, 1950, and Living in Custer County, Idaho[a]

| Year of Exposure | Estimated Thyroid Dose (Gy) | | |
| | Best Estimate | Uncertainty Range[b] | |
		Low Estimate	High Estimate
1951	0.0014	0.000093	0.028
1952	0.33	0.028	5.2
1953	0.0048	0.0011	0.028
1955	0.0024	0.00057	0.023
1957	0.03	0.0074	0.18
1962	0.000062	0.0000012	0.0037
1965	0.000018	0.00000019	0.0024

[a]Doses do not add to total dose shown in Figure 6.1; best estimates are medians of lognormal distributions and are not additive.
[b]Uncertainty range given by NCI dose calculator corresponds to "credibility intervals" used in this report.

person who has a diagnosed thyroid cancer on the basis of the person's date of birth, milk-consumption pattern, and place and time of residence. The estimated thyroid dose and age at exposure would be used to determine PC/AS. If a designated PC/AS percentile equals or exceeds a threshold for compensation (such as 0.5), the person would be compensated. That procedure could be used for any person in the United States, and not be limited to particular geographic areas.

Exposure of Other Organs or Tissues to NTS Fallout Radiation

The committee also examined PC/AS for radiogenic cancers in addition to thyroid cancer for people in areas receiving NTS fallout. We used NCI's IREP version 5.3. For estimating doses, we relied on the information in the CDC-NCI draft feasibility study (CDC-NCI, 2001). The committee agrees with the authors of the draft feasibility study that the dose estimates would have to be defined

TABLE 6.2 PC/AS for Person Who Received Thyroid Doses Shown in Table 6.1 and Had a Thyroid Cancer Diagnosed in 2000[a]

Percentile	PC/AS
1st	15.00%
5th	26.83%
50th	66.55%
95th	92.10%
99th	95.67%

[a]For example, the 50th percentile shows the PC/AS value (66.55%) that exceeds 50% of the calculated values of the PC/AS distribution.

better before they could be used for individuals, so we are using the information here only for illustration. If the PC/AS approach is adopted, the dose estimates for other organs and tissues, for example, red bone marrow, will need to be completed and finalized.

An important difference between the thyroid and other cancer sites with respect to the radiation dose from fallout is that the thyroid is the only organ exposed to radiation from NTS tests for which the NTS-derived internal dose exceeds the external dose (CDC-NCI, 2001). For all other organs and tissues, the external dose from fallout is higher than the internal dose. The external dose is relatively uniformly distributed over the body. The most important organs for consideration are those most susceptible to cancer induction by evenly distributed whole body radiation.

The committee specifically examined the estimated dose to red bone marrow because of its susceptibility to induction by radiation exposure of all types of leukemia except of chronic lymphocytic leukemia (CLL).

The draft feasibility study concluded that the maximum total (internal + external) estimated dose from NTS fallout to red bone marrow was 3-10 mGy (CDC-NCI, 2001) (see Figures 4.3 and 4.4 of the present report). CDC-NCI noted, however, that differences in such variables as rainfall in a county might have resulted in higher estimated doses. Because of the importance of external radiation from fallout deposited on the ground, the time a person spent outdoors would also affect the estimated dose. In the draft feasibility study, the uncertainties in estimated dose were not thoroughly studied, but we considered a higher-upper bound dose to investigate the possible range of PC/AS for leukemia. We only considered median values of PC/AS in this example because the uncertainty intervals for estimated dose were not determined.

To estimate the possible higher values of PC/AS for leukemia, the committee used the NCI's IREP version 5.3 calculator to estimate PC/AS for the highest estimated dose to red bone marrow (10 mGy), as determined by the draft feasibility study, and to estimate an upper bound on the maximum estimated dose (40 mGy). For simplicity, the committee assumed that the entire dose occurred in one year (1952), although in actuality people would have incurred doses from the longer-lived radionuclides well beyond that date. Assuming that all the dose occurs in one year maximizes PC/AS because of its dependence on age at exposure.

In contrast to radiation-induced thyroid cancer, the age at diagnosis of radiation-induced leukemia is an important factor in addition to age at exposure and dose. PC/AS values for a newborn and a 10-year-old are shown in Figure 6.4 for different years of diagnosis.

As noted above, the PC/AS values were calculated for two different values of the estimated dose to the red bone marrow, 10 mGy and 40 mGy. Figures 4.3, 4.4, and 6.4 indicate that people who received the typical maximum estimated red bone marrow dose of 10 mGy would not have a PC/AS over 0.5. However, individual variations may exist within the average values used for the dose as-

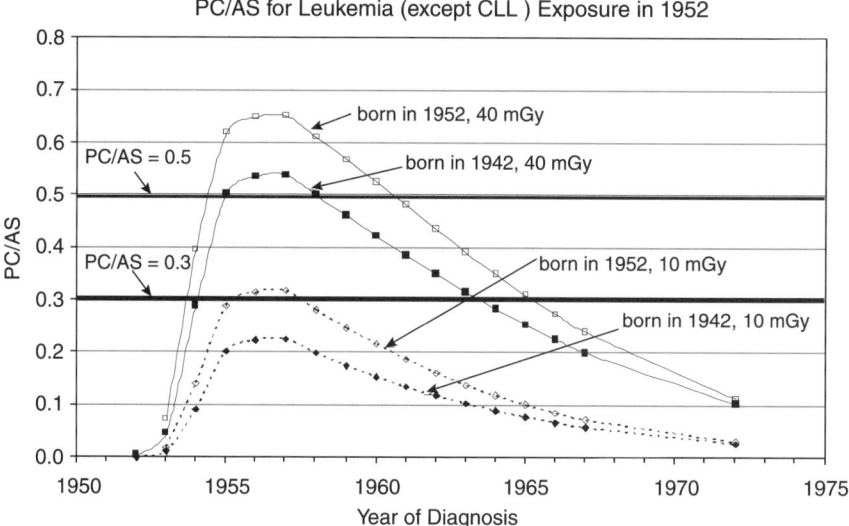

FIGURE 6.4   PC/AS for leukemia (except CLL) for estimated bone marrow doses of 10 and 40 mGy in 1952.

sessment; such variations may increase the estimate of the dose and were discussed but not specifically included in the dose estimates in the draft feasibility study. A person born in 1952 who received an estimated bone marrow dose of 40 mGy would have a PC/AS greater than 0.5 if the disease were diagnosed within 3-9 years of the exposure, whereas for a person born in 1942, the PC/AS would exceed 0.5 if the disease were diagnosed 3-6 years after exposure. If a compensation scheme like that used by BNFL were chosen, partial compensation would occur at other values of PC/AS, such as 0.3, as shown in Figure 6.4; people born in 1952 who received 10 mGy and developed leukemia would receive some compensation if the disease was diagnosed in 1956 or 1957.

For perspective, those estimated bone marrow doses can be compared with the dose a person could receive from natural background radiation. The type of radiation received from fallout is dominated by external gamma radiation, for which 10 mGy corresponds to 10 mSv. As pointed out in Table 3.2, a person living in the United States typically receives an effective dose of 0.6 mSv/year from external radiation from natural background (cosmic radiation and terrestrial radioactivity). The effective dose from external background radiation can be as much as 1.2 mSv/year for a person living in the mountains. This effective dose can be taken to equal the bone marrow dose from natural background external radiation. In the roughly 8 years of atmospheric testing at the NTS, the total bone marrow dose from natural background was 4.8-9.6 mSv. Consequently, the radiation dose of 10 mSv received by organs other than the thyroid from NTS fallout is roughly of the same order of magnitude as that received from natural background in the same time frame.

Because of the high sensitivity of bone marrow to radiation-induced malignancy, the PC/AS values for other, less sensitive organs or tissues would not be expected to exceed 0.5 unless individual variability results in a higher dose. PC/AS values for other high-sensitivity organs, however, such as lung and female breast, may also exceed 0.5 in some cases but these were not thoroughly evaluated by the committee.

In some circumstances that the committee expects to be infrequent, PC/AS values for cancers other than thyroid cancer may exceed the compensation criteria. That would typically correspond to special maximal-exposure situations in which specific organ doses were higher than those found in the draft feasibility study but within the range noted in that report. No excess cancers of those types have been found epidemiologically, but compensation may be warranted in some circumstances.

Later in this chapter, the committee discusses the need for a preimplementation assessment to identify classes of people at higher risk for radiation-induced cancers on the basis of such factors as age at exposure, age at diagnosis, diet, and geographic area who would be eligible for compensation. The pre-assessment would use PC/AS-based compensation criteria set by Congress. It would provide guidance to potential claimants and the implementing agency that would focus the compensation program on those classes of higher-risk people. This would educate and inform affected parties and, ultimately, simplify implementation for all concerned.

Exposure from Global Fallout

The previous discussion considered exposure from fallout created by tests conducted at the NTS. In addition to those tests, the United States detonated weapons at several sites in the Pacific, in the south Atlantic, and in the United States outside the NTS (DOE, 2000).

Global fallout resulted principally from debris injected into the stratosphere from high-yield nuclear tests, such as many of those conducted by the United States in the Pacific. That material was then mixed with stratospheric debris from high-yield nuclear tests of other nations. Residence times of the material in the stratosphere are relatively long, typically around a year. As a result, the shorter-lived radionuclides, such as ^{131}I, generally will decay before returning to the earth's surface. In contrast with NTS fallout, global fallout consists primarily of longer lived radionuclides, such as ^{90}Sr ($T_{1/2}$ = 29 years), ^{137}Cs ($T_{1/2}$ = 30 years), and ^{95}Zr ($T_{1/2}$ = 64 days)–^{95}Nb($T_{1/2}$ = 35 days but in equilibrium with ^{95}Zr).

The CDC-NCI draft feasibility study reviewed the health consequences for the US population of global fallout as well as NTS fallout (CDC-NCI, 2001).

The study included estimates of organ doses and cancer risks from those materials. As noted earlier in the case of dose estimates to organs and tissues other than thyroid from NTS fallout, the global-fallout dose estimates, although they have been released in a predecisional draft for peer review and public comment, are tentative because the report has not been finalized. In addition, as the authors of the draft feasibility study pointed out, the actual dose estimates themselves, even when they become finalized, are preliminary and support a feasibility study that shows that more detailed and accurate dose estimates are possible, but still must be developed (CDC-NCI, 2001). Considerable work would need to be performed to obtain dose estimates suitable for individual PC/AS calculations.

For example, the authors of the feasibility study noted that only a "crude model was developed to describe the geographical variation in Sr-90 deposition density . . ." Determination of the ^{90}Sr deposition density in each county across the United States, which was calculated from each county's precipitation and its estimated ^{90}Sr concentration, was essential in estimating the radiation dose to residents of those counties from most of the other fallout radionuclides. As a result, the study's authors stated that "the specific county estimates or estimates for years prior to 1958... may be quite uncertain and should be used with discretion" (CDC-NCI 2001). They also had serious reservations about current dose estimates from the small amount of ^{131}I that does return to the earth's surface in global fallout and the estimated doses from global fallout of ^{3}H and ^{14}C due to their entry into the hydrological and carbon cycles, respectively.

Once debris from a particular test enters the upper atmosphere, it mixes with debris from some of the other tests conducted by the US and other nations. Global fallout is a mixture of debris from all those tests. The United Nations Scientific Committee on the Effects of Atomic Radiation (UNSCEAR) has estimated the fraction of ^{90}Sr deposition due to the nuclear testing programs of each country, including the United States, on the basis of their fission yields (UNSCEAR, 2000). The CDC-NCI draft feasibility study also indicated that the relative contribution of a country's tests to global fallout could be roughly determined from the relative fission yield of its tests compared with that from tests conducted by other countries. In principle, then, the fraction of global fallout due to the US tests can be estimated, albeit roughly, from the fission yields of its tests.

Because RECA was established to compensate people for harm incurred as a result of the US nuclear-weapons program, radiation doses and associated risks from exposure to the US fraction of global fallout should be included in RECA. However, the discussion above indicates that the information and methods needed to estimate individual doses and risks from global fallout are still in a preliminary stage, and a more detailed understanding of global fallout needs to be developed before they can be determined. Consequently, the committee will recommend in the next section that a detailed preassessment of fallout doses be performed. That includes determining and finalizing the global fallout doses, and identifying geographic areas in the US in which the estimated doses from fallout for people with diagnosed RECA-specified cancer could result in a PC/AS value high enough to qualify for compensation on the basis of PC/AS criteria established by Congress.

Recommendations for Expanding Eligibility for People Exposed to Radiation from Fallout

The committee recommends that Congress establish a process using probability of causation/assigned share (PC/AS) to determine the eligibility of any new claim for compensation for a specified RECA-compensable disease in people who may have been exposed to radiation from fallout from US nuclear-weapons testing. The committee further recommends that Congress establish criteria for awarding compensation on the basis of computed distributions of PC/AS for any persons making such a claim.

• Prior to implementation of the revised compensation program, the National Cancer Institute (NCI) or other appropriate agencies should perform a population-based preassessment of all radiogenic diseases using PC/AS to provide guidance to individuals who might apply for compensation by determining the likelihood any individuals in a given population of being compensated. This analysis would be determined by disease identified, places of residence at the time of exposure, ages at the time of exposure and at diagnosis, and other demographic factors using the PC/AS criteria (including consideration of the upper credibility intervals) established by Congress. The calculation would use data for the maximal doses that such individuals may have received from fallout. In settings where variability is important in evaluating risk, there may be several such defined populations, and each would be evaluated on its own merits. The criteria for evaluating such population-based preassessments should be the same as those established by Congress for compensation of claims under RECA.[1, 2] The preassessments should be made for the following two purposes:

---

[1]Although agreeing with the general PC/AS and preassessment approach, committee member Kathleen N. Lohr wishes to emphasize one aspect of the committee's suggested plans for the preassessment activities of particular concern to her. The acceptability of the PC/AS approach and the proposed equivalency of RECA compensation criteria and population-based preassessment criteria rests on the assumption that the preassessments will be done using values for variables in the calculations that will in fact mimic "worst-case" scenarios and produce the highest possible dose estimates for the population groups in question. In this way, possible bias against individuals with specific characteristics different from the majority of the population group under consideration can be minimized.

[2]Although concurring in principle with the PC/AS approach and with the concept of population-based preassessment to enhance efficiency, committee member Stephen G. Pauker points out that if the criteria for evaluating population-based preassessments are quantitatively the same as the criteria for compensation, a bias may be introduced against some individuals whose PC/AS is substantially greater than the majority of the population. Rather, the criteria for evaluating preassessments might need to be quantitatively less strict than the compensation criteria to avoid such bias.

a. To provide guidance to potential claimants and the implementing agency as to which diseases may satisfy the compensation criteria established by Congress.

b. To provide guidance to potential claimants and the implementing agency as to which population groups or geographic areas may satisfy the compensation criteria established by Congress.

• The recommendation applies to residents of the continental US, Alaska, Hawaii, and overseas US territories who have been diagnosed with one of the specified RECA-compensable diseases and who may have been exposed, including exposure in utero, to radiation from US nuclear-weapons testing fallout. Both Nevada Test Site fallout and the US fraction of global fallout should be considered.

• PC/AS for any individual should be obtained from an estimate of the radiation dose resulting from US nuclear-weapons testing and the risk estimate associated with such dose.

• Uncertainties in PC/AS cannot be avoided and may be part of the compensation decision process. Because of substantial gaps in the existing data, the uncertainties in estimated doses[3] incurred by people exposed to radiation from fallout, and consequently the uncertainties in the associated PC/AS estimate, are large. This emphasizes the need to choose compensation criteria carefully. For example, a PC/AS value associated with a high percentile of uncertainty could exceed the criteria for compensation even for some very small median doses. The challenge Congress faces will be to decide if it is best to define criteria that avoid rewarding compensation in cases in which there is very low risk, but the uncertainty associated with its PC/AS is very large, because the connection of these cancers with radiation is not well established or the estimated doses are not well known.

• To support the use of the PC/AS process for compensation, The Centers for Disease Control and Prevention (CDC) and the NCI or other appropriate agencies should complete dose estimates for all significant radionuclides in fallout from US nuclear weapons testing to the population groups identified above. This should include all the major sources of dose related to US nuclear-weapons tests considered to have potential health consequences that the CDC-NCI 2001 draft feasibility study described.

• An updated dose calculator, similar to the existing NCI dose calculator for ^{131}I, should be developed for determining dose to the thyroid and other important

---

[3]The dose estimates depend on the measured deposition of radionuclides taken at the time of the nuclear weapons tests. Given the very small number of monitoring stations, most estimates represent interpolations over very large areas. Among the 3000 plus counties in the continental United States, fallout monitoring in areas other than a limited region in Nevada and its neighboring states occurred at never more than 95 stations through the years of aboveground US nuclear-weapons testing. See Chapter 5 and sections of this chapter for further discussion.

organs from fallout. Such an updated dose calculator should be directly coupled to a risk calculator similar to IREP Version 5.3 that can compute PC/AS and propagate uncertainties for establishing credibility intervals.

• NCI or other appropriate agencies should maintain and revise the parameters in the models or calculators for estimating PC/AS based on risk estimates recommended by the National Research Council Committee on Biological Effects of Ionizing Radiation, report number 7 (BEIR VII). Over time, the agency should update the PC/AS calculators with the latest risk parameters.

IMPLEMENTATION AND ANTICIPATED IMPACT OF THE RECOMMENDATION TO EXPAND THE SCOPE OF RECA TO ADDITIONAL GEOGRAPHIC AREAS

The recommendation would remove the a priori requirements based solely on geographic location (but would continue to require the presence of a RECA-specified disease) to establish eligibility for compensation. However, it is critically important to provide procedures to ensure that claims that are likely to be eligible for compensation could be processed efficiently and rapidly. Thus, before this modification to RECA is implemented, the NCI or other appropriate agencies is expected to perform population-based preassessments of PC/AS for different diseases, geographic areas and population groups in those areas. A population-based dose reconstruction is considered part of a detailed preassessment in this sense. The preassessments would provide information to potential claimants and the implementing agency as to whether a person might be eligible for RECA compensation under the new criteria that Congress establishes.

For example, the previous discussion noted the strong dependence of thyroid-cancer risk on age at exposure. The preassessment may show that people diagnosed with thyroid cancer who were children in some areas of the United States when testing occurred have PC/AS values eligible for compensation but that persons with the same diagnosis who were adults in those same areas during the same time period do not. RECA implementation would then encourage claims from people who were living in those areas while they were children. Thus, the preassessment would provide information about conditions based on age at exposure that would encourage submission of claims that have a chance of success while discouraging other claims.

That conclusion is especially true for diseases other than thyroid cancer. As discussed earlier, the CDC-NCI draft feasibility study found that estimated radiation doses to organs and tissues other than the thyroid were less than 10 mGy in a typical exposure scenario. If a person who received that dose developed a cancer other than thyroid cancer, the PC/AS value would be low no matter where the person lived at the time of exposure. Although the person's specific eligibility for RECA compensation would depend on the criteria for compensation that Congress establishes, the radiation risks of almost all the nonthyroid cancers at those low doses suggest that eligibility for RECA compensation for these cancers would be infrequent.

For example, the previous discussion in this chapter showed that even for exposures of highly radiosensitive bone marrow, doses considerably greater than 10 mGy were needed for PC/AS values to satisfy the PC/AS criterion of 0.5 or above at the 50th percentile for induction of leukemia and only if the disease had been diagnosed within a few years after exposure. The preassessment, based on these criteria, would demonstrate that there could be large regions of the country where people could not have received doses (with their associated uncertainties, as may be applicable under the criteria) sufficient to be eligible for compensation. Preassessments would, therefore, greatly streamline implementation of the modified RECA program.

The committee also evaluated several other cancer types. We used, as an example, a PC/AS of 0.5 at the 50th percentile. (Other criteria would lead to different results.) As noted in the paragraph above, bone marrow is highly radiosensitive, and a dose greater than 10 mGy is needed for the PC/AS to exceed 0.5 for leukemia. Because other organs and tissues are not as radiosensitive as bone marrow, that finding suggests that other organ doses must be considerably greater than 10 mGy for the PC/AS in those cases to exceed 0.5. If the criterion for compensation were a PC/AS of 0.5 at the 50th percentile, applying for compensation for the cancers that may have resulted from these other organ doses would be unwarranted. But, of course, if Congress established the criterion to be based on the upper limit of the 99th percentile credibility interval, compensation of some populations for some other radiogenic cancers might be possible.

Population-based preassessments would greatly simplify implementation of the RECA program in those areas. The simplification would be essential for the PC/AS procedure to be useful for determining eligibility for RECA compensation. Without it, a person diagnosed with a RECA-compensable disease would not know whether he or she was eligible for compensation until his or her application was reviewed and the person's PC/AS determined. The preassessment is intended to encourage potential claimants to submit only viable applications for compensation. This will improve the efficiency of the administrative process and reduce delays in awarding compensation to people who are eligible. Hence, preassessment information would be communicated in advance by the implementing agency to the affected public.

The outcome of the preassessment could be presented in several ways. One could be a set of tables for each compensable disease, population group or geographic area that would be likely to satisfy the criteria for compensation. Another possibility would be a series of maps that show areas of eligibility, similar to those presented in the NCI report on [131]I. It might also be useful to develop a web-based program, similar to the NCI dose calculator or IREP that would allow a potential claimant or agency staff member to enter variables from the claimant's history. The results would then indicate if an applicant should consider submitting a formal application.

Uncertainties in Dose and PC/AS Estimates

The committee notes an important difference between the precision of the dose determination of people exposed to radiation from fallout with the precision of dose determination in other compensation programs using PC/AS, such as the compensation programs that apply to military veterans and the Energy Employees Occupational Illness Compensation Program (EEOICPA). These programs award compensation if a person's PC/AS exceeds 0.5 at the upper 99th percentile level.

The NCI and CDC determined doses for the people exposed to fallout from dose-assessment models based on environmental measurements taken at the time of the tests. In some cases, the estimated doses were based only on atmospheric-transport models, which tracked a fallout cloud from the point of detonation to the location of the downwind resident and included no environmental measurements. Estimates of deposition of fallout based on these models have considerable uncertainty. In addition, even when measurements had been made, fallout deposition in large areas of the United States was estimated by interpolation among up to 95 fallout monitoring stations (Beck, 1980). Estimating fallout across the entire United States with such a small number of stations also involves large uncertainties.

Dose estimates based on such methods are not as precise as those based on measurements more directly associated with the person. Data from measurements are routinely available for many of the claimants in the other compensation programs. In the EEOICPA program, for example, many workers wore personal dosimeters from which to estimate the external radiation dose and participated in bioassay programs to estimate the internal dose. People exposed to fallout rarely had similar personal dosimetry.

As a result, the uncertainties in the estimated doses incurred by people exposed to radiation from fallout, and consequently the associated PC/AS estimates, are generally much larger than many of those for the other programs. Geometric standard deviations (GSD) of 4 are not unusual for some of the fallout dose estimates; thus, for example, an estimated dose of 10 mGy could range from 0.6 to 160 mGy at the 95th percentile level (i.e., 2 standard deviations). If Congress were to adopt compensation criteria for RECA that include some high percentile of the uncertainty, as in the other compensation programs, the PC/AS value at this high percentile value could exceed the criteria even for some very small estimated doses. That obviously could make the compensation program very large and difficult to administer. This emphasizes the importance of choosing compensation criteria carefully.

We have chosen an example to demonstrate the combined effects of uncertainty in the dose estimated and the selected criteria for compensation. It is meant to illustrate some consequences of implementing PC/AS, but should not be interpreted as an endorsement of any specific criterion for compensation. For this case, a person at age 25 is exposed to ^{131}I resulting from fallout. The person is diagnosed with thyroid cancer at age 35. A dose assessment is made and yields a geometric mean estimated dose of 80 mGy (8 rads) that is distributed log normally. The distribution of PC/AS is determined using the methodology described in IREP and presented in Chapter 5. Figure 6.5 shows the results of this computation in terms of the value of PC/AS corresponding to the median (central value) of the distribution, as well as the 75th percentile, 95th percentile and 99th percentile, as a function of increasing uncertainty based on the geometric standard deviation, GSD, in the estimate of thyroid dose. The PC/AS associated with a true dose of 80 mGy is 0.06. This corresponds to the median value of PC/AS shown in Figure 6.5.

The median value remains constant at 0.06 even when the GSD in the estimate of dose increases from 1.5 to 4.0. The 75th percentile for PC/AS increases with GSD, but remains below 0.2. The 95th percentile for PC/AS increases as the uncertainty of the estimated dose increases and eventually exceeds 0.5 when the

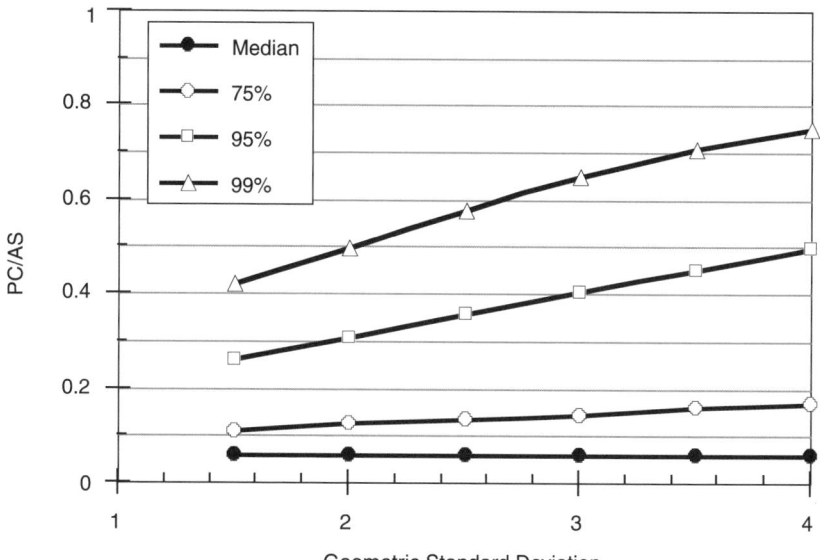

FIGURE 6.5   Values of PC/AS based on the median, 75th percentile, 95th percentile, and 99th percentile of the distribution for PC/AS as a function of the uncertainty based on the geometric standard deviation for a distribution of dose having a geometric mean of 80 mGy (8 rads).

GSD reaches 4.0. The 99th percentile of PS/AS is even more sensitive to uncertainty in the estimate of dose; it exceeds 0.5 when the GSD reaches 2.0. If the latter case were selected as the criteria for compensation, many claims would satisfy the criteria because of uncertainty in estimated dose rather than the central estimate of dose. Other examples using fixed cut-off criteria, as well as sliding scales for compensation, were presented in Table 5.2 using the complete distributions of PC/AS shown in Figure 5.5.

Uncertainties in PC/AS cannot be avoided and will be part of the compensation decision process. The challenge Congress faces will be to decide if it is best to define criteria that avoid rewarding compensation in cases in which there is very low risk, but the uncertainty associated with its PC/AS is very large, because the connection of these cancers with radiation is not well established or the estimated doses are not well known.

URANIUM MINERS, MILLERS, AND ORE TRANSPORTERS IN OTHER GEOGRAPHIC AREAS

The committee considered whether any circumstances warrant the extension of RECA to include workers who were employed in uranium mining and milling in geographic areas not now covered by RECA. The committee noted that RECA already covers uranium miners, millers, and ore transporters who worked in the RECA-designated uranium mining areas in 1942-1971, regardless of where they are now employed (see Figure 2.1).

RECA has a provision that allows individual states not covered under the uranium mining and milling sections to apply for inclusion in RECA if uranium mining occurred in the state during the period January 1, 1942-December 31, 1971.

The committee recommends that the provision which allows individual states not currently covered under Section 5 of the Radiation Exposure Compensation Act to apply for inclusion under RECA if uranium mining occurred in the state during the January 1, 1942 to December 31, 1971 period be expanded to include not only uranium mining but also uranium milling and ore transportation occurring during that period in support of the US nuclear-weapons program.

Probability of Causation/Assigned Share for White and Navajo Uranium Miners

The 2002 amendment to RECA added the duration-of-exposure option to the criteria for compensation. Uranium miners may now be compensated if they have accumulated 40 WLM or worked one year or more in uranium mines. The committee investigated how that decision affected the PC/AS for lung cancer associated with exposure to radon daughters in the mines. The committee used the relative-risk model developed in BEIR VI (NRC, 1999), which included a factor for duration of exposure. Data from the National Institute for Occupational Safety and Health study of 4,102 underground uranium miners were used to obtain all the estimates that study included 3,347 white miners and 755 Navajo miners who worked in the mines during the period 1950 to about 1970 (Hornung et al., 1998; Roscoe, 1997). The *BEIR VI* model was applied to four strata defined by Navajos and whites, each divided by duration of mining (less than 1 year vs. 1 year or longer). The *BEIR VI* model provides different estimates of relative risk for lung cancer for five increasing durations of exposure. Relative risk appears to increase for a fixed level of WLM as duration of exposure increases. The model is similar to that developed specifically for the Colorado Plateau uranium miners (Hornung et al., 1998). Accordingly, miners who have worked one year or more are estimated to have higher relative risk per WLM than those working less than one year.

Table 6.3 presents the results of the *BEIR VI* lung cancer model applied to the four strata described above. Results are provided for the median and the 5th, and 95th percentiles of the exposure distribution in each stratum. The exposure distributions are similar for Navajo and white miners. It is clear from examination of the table that most of the miners who worked more than one year had PC/AS greater than 0.50. The typical miner who worked less than one year (as defined by the median exposure) had PC/AS less than 0.50. However, about 10% of the short-duration miners had PC/AS over 0.50. By contrast, PC/AS for the RECA requirement of 40 WLM for compensation is 0.18 for short-duration miners and 0.70 for workers who accumulated 40 WLM in working more than five years underground.

CONCLUSION

The committee calculated thyroid doses and associated PC/AS values by using the on-line NCI dose calculator and IREP programs. The committee found that estimated thyroid doses and associated radiation risks and PC/AS values varied substantially both among different counties and among individuals in the same county. Age at exposure and diet were important factors affecting the within-county variability. Identification of RECA eligibility solely on the basis of geographic area would be scientifically problematic because a given area would contain residents having both relatively high and virtually nonexistent risks of radiation-induced thyroid cancer, depending on such factors as age at exposure. Consequently, the committee has recommended that compensation be based on diagnosis of thyroid cancer in an individual and the associated PC/AS, which takes those factors into account.

The committee also recommended extending the PC/AS-based compensation system to other eligible cancers in downwinders. However, performing an

TABLE 6.3   PC/AS of Diagnosed Lung Cancer in White and Navajo Uranium Miners in Colorado Plateau

| | White miners | | |
| | Duration: < 1 year | N = 515 | |
	Exposure (WLM)	RR	PC/AS
Median	46.5	1.26	0.20
5th percentile	2.1	1.01	0.01
95th percentile	262.4	2.44	0.59
	Duration: ≥ 1 year	N = 2,832	
	Exposure (WLM)	RR	PC/AS
Median	563.5	11.40	0.91
5th percentile	67.9	3.8	0.74
95th percentile	3059.8	49.6	0.98
	Navajo miners		
	Duration: < 1 year	N = 101	
	Exposure (WLM)	RR	PC/AS
Median	43.7	1.24	0.19
5th percentile	2.1	1.01	0.01
95th percentile	271.0	2.49	0.60
	Duration: ≥ 1 year	N = 654	
	Exposure (WLM)	RR	PC/AS
Median	513.9	10.64	0.91
5th percentile	66.3	3.79	0.74
95th percentile	2604.7	42.6	0.98

initial population-based PC/AS preassessment is essential if the approach is to be efficient and timely. The population-based preassessment would provide guidance concerning the extent of the problem and identify locations and population groups in which compensation would be viable. Estimated radiation doses to organs other than thyroid and the corresponding PC/AS values for the tissues at risk are substantially lower than those for the thyroid. Because of the low radiation doses estimated for organs other than the thyroid, the committee suggests that the preassessment would enable the implementing agency to focus on selected diseases, areas and population groups. The results of the preassessment, however, would depend greatly on the PC/AS criterion for compensation that is eventually adopted.

We presented an example showing the importance of the PC/AS criterion, especially the influence of the percentile of the PC/AS distribution that is specified for awarding compensation. If Congress chooses a 95th or 99th percentile, compensation would be awarded for very low values of estimated dose and PC/AS. This outcome is due to the large uncertainties in the dose estimates, which generally depend on an environmental measurement grid that had a limited number of fallout sampling stations to cover the majority of the United States. If the 50th percentile is chosen as the criterion, compensation would be based on the median value of the PC/AS distribution. As a consequence, the decision to award compensation would not be dominated by the large uncertainties in the measurements.

The committee also reviewed the geographic areas and population groups that RECA now covers for uranium mining, milling, and ore transporting. The committee recommended that the RECA provision allowing states not currently covered for uranium mining to apply for coverage be expanded to include uranium milling and ore transporting as well. The committee evaluated the PC/AS values for uranium miners meeting current RECA eligibility requirements and found that the miners who worked for more than 1 year generally would have PC/AS values greater than 0.5. The PC/AS values for miners having exposure of 40 WLM would depend heavily on the duration of the exposure.

7

Diseases, Populations, and Other Issues of Public Concern

In oral and written testimony presented to the committee at or in response to its information-gathering meetings, members of the public voiced their concerns and raised questions about diseases or conditions for which individuals in one or more of the populations designated in the Radiation Exposure Compensation Act (RECA) of 1990, and as later amended, are not entitled to compensation. These included some cancers and benign neoplasms, and several diseases with known or suspect autoimmune basis, and we have reviewed the literature on those conditions. The public also drew the committee's attention to groups of people that may have been occupationally or environmentally at risk for radiation exposure from nuclear-weapons development programs but who are ineligible for RECA compensation because they do not otherwise meet certain eligibility criteria. Such groups include

• US civilians residing or working in areas overseas that are not designated by RECA but that may have been contaminated by fallout from US atmospheric nuclear tests conducted in the region.

• People living in the vicinity of mill tailings or in dwellings built with mill tailing or other mill or mine residues.

• People who were at risk for radiation exposure fell outside RECA-designated periods or who were ineligible based on their failure to meet the relevant residence requirement in the defined time interval between exposure and disease.

The committee reviewed epidemiologic and other pertinent scientific and clinical literature to identify new information about the specific diseases of interest, including their characteristics in the general or "nonexposed" population and their known or suspected causes. We looked particularly for new information about relationships between exposure to ionizing radiation and other potential hazards during activities similar to those experienced by the designated RECA population and the individual diseases and conditions identified by the public. We reviewed new information about the effects of in utero exposure to radiation and the psychologic consequences of exposure or suspected exposure to radiation, and reviewed information related to the potential for exposure of other populations not eligible for RECA compensation but with circumstances and potential for exposure that may have been similar to those of RECA-eligible populations, such as those who were in utero and at risk during the nuclear-testing period. We also address other issues regarding eligibility for RECA compensation which the public brought to the committee's attention. These include concerns in two areas about claimants failed attempts to use affidavits in establishing proof of eligibility.

Our findings are reported here for diseases not currently compensable under RECA for some RECA-eligible populations and for specific populations that are not listed among those eligible for compensation. The diseases are categorized as malignant (or cancerous) and nonmalignant (or noncancerous).

MALIGNANT DISEASES

Leukemia

Members of the public drew the committee's attention to an apparent inconsistency in compensation under RECA: all types of leukemia except chronic lymphocytic leukemia (CLL) are compensable as radiogenic diseases for downwinders and onsite participants but not for uranium miners, millers, or ore transporters. Non CLL leukemia is compensable in a number of other exposed populations.

Leukemia in Uranium Miners

As reported in Chapter 4, a strong association between radiation and all types of leukemia other than CLL has been identified in several populations exposed to external penetrating radiation (gamma and x rays) at doses generally greater than 200 mSv and at high dose rates. However, as noted in Chapter 4, uranium miners are exposed primarily to alpha-particle radiation from inhaled radon. No exposure pathway has been established whereby immature blood cells are exposed to radiation from radon daughters suspended in air. However, a pathway has been proposed on the basis of the transfer of radon gas from the pulmonary region of the lung into the blood and from the blood to fat cells distributed in the bone marrow (Eatough and Henshaw, 1993). Allen and colleagues (1995) explored that proposed pathway using marrow fat content in a sample of 20 human ribs. They concluded that the bone marrow fat fraction is the important variable related to the alpha-particle radiation dose from radon in fat. Beta particles from lead-214 (^{214}Pb) and bismuth-214 (^{214}Bi) have sufficient range to deliver a dose of radiation to neighboring bone marrow cells, and this is the proposed pathway of the induction of leukemia in underground uranium miners.

According to a recent report by the National Research Council Committee on Biological Effects of Ionizing Radiation, *BEIR VI* (NRC, 1999), no significant excess of leukemia cases has been found among mining cohorts. As discussed in Chapter 5, the most definitive study of cancers other than the lung was the pooled study of 11 mining cohorts reported by Darby and colleagues (Darby et al., 1995). They found that the relative risk (RR) of all lung cancer deaths combined (n = 1,179) was close to expected (RR = 1.01, 95% CI = 0.95-1.07). For leukemia, there was a statistically significant increase in the RR for miners employed in the mines for less than 10 years (21 deaths; RR = 1.93, 95% CI = 1.19-2.95), but there was no increase in the RR for workers employed more than 10 years (48 deaths; RR = 0.99, 95% CI = 0.73-1.31). The overall RR was not significantly increased (69 deaths; RR = 1.16, 95% CI = 0.90-1.47). Darby et al. (1995) thus concluded that protection standards for radon exposure should continue to focus on lung cancer alone. Mortality from leukemia in radon-exposed mining populations has also been summarized (UNSCEAR, 2000). The trend in mortality from leukemia with dose (mean Working Level Month [WLM] = 155) was not statistically significant. Roscoe (1997) reported a nonsignificant increase in the RR of leukemia mortality in uranium miners (13 deaths; standardized mortality ratio [SMR] = 1.6, 95% CI = 0.8-2.7).

In addition to the studies of uranium miners, several ecologic studies have been done of the relationship between leukemia and indoor radon. Their results have been equivocal; some indicate a significant association, and others do not show a relationship. A review of 19 ecologic studies, six miner studies, and eight case-control studies (Laurier et al., 2001) concluded that the overall epidemiologic results do not provide evidence of an association between indoor radon exposure and leukemia.

Leukemia in Millers and Ore Transporters

Uranium millers and ore transporters were at risk for exposure primarily to soluble uranium dusts of low radioactivity, most of which is rapidly cleared from the body. Thus, the probability of radiation-induced leukemia was low in those populations. Pinkerton et al. (2004) show fewer than the expected number of deaths among the cohort of Colorado Plateau uranium mill workers from all types of leukemia, including CLL, compared with the US general population (5 deaths, SMR = 0.66, 95% CI = 0.21-1.53). The committee is unaware of any studies of leukemia in ore transporters and assumes that their risks would be similar to or less than those of uranium millers.

Conclusion On the basis of the epidemiologic evidence, the committee concludes that the uranium miner and miller populations now identified by RECA are not at significantly greater risk of dying from leukemia than males of similar age in the US general population. The leukemia risk to ore transporters is assumed to be similar to or less than the risk to millers. Thus, there is no epidemiologic basis for designating leukemia as a RECA-compensable disease for those populations.

Myelodysplastic Syndrome

The terms *myelodysplasia* and *myelodysplastic syndromes* (MDS) identify an imprecisely defined group of clonal hematologic disorders characterized by abnormal appearance of the bone marrow hematopoietic progenitor cells, ineffective hematopoiesis, peripheral blood cytopenias, and frequent evolution into the complete clinical picture of acute myeloid leukemia (AML). The disorders, which include aplastic anemia, may occur de novo without apparent cause at any age but mostly at the age of 60-80 years. Approximately 7,000-12,000 new cases of MDS are diagnosed each year in the United States. Secondary forms may occur at any age after the use of intensive chemotherapy with alkylating agents or the combination of chemotherapy and radiation therapy. The frequent progression of both de novo and secondary or therapy-induced MDS to clinically apparent AML suggests that the syndrome represents various steps in the continuous process of malignant transformation to leukemia. The natural progression of MDS syndrome is such that when the number of typical immature blast cells increases above a defined point in the peripheral blood, the diagnosis is changed from MDS to AML.

Leukemia studies in the Japanese atomic-bomb survivors have identified cases of MDS, but results of autopsy-based case studies could not be converted into a dose-response projection (Finch, 2004).

Conclusion The committee concludes that radiation exposures of downwinders and onsite participants are likely to be well below the doses at which MDS has been noted in prior studies and that MDS itself should not be included as an additional compensable disease under RECA. If leukemia is diagnosed, it may be compensable on those grounds based on PC/AS criteria adopted by RECA.

Multiple Myeloma

Multiple myeloma is compensated under RECA except in uranium miners, uranium millers, and ore transporters, and the committee was asked to reevaluate the justification for this exclusion. Multiple myeloma is a malignant disease of the blood characterized by an abnormality in plasma cells and possibly in the bone marrow stroma. An increased incidence of and mortality from multiple myeloma have been reported in the Japanese atomic-bomb survivors (Preston et al., 1994; Pierce et al., 1996). A similar conclusion was reached by the National Research Council *BEIR V* committee (NRC, 1990).

Nuclear-industry workers exposed to low LET radiations at lower doses and lower dose rates than the atomic-bomb survivors have not shown statistically significant increases although some of the findings have been of borderline statistical significance (Gilbert et al., 1979; Gilbert et al., 1993).

Estimated risks of multiple myeloma among persons exposed to alpha-particle radiation have generally been negative. The most recent update of the Colorado Plateau uranium miners study (Roscoe, 1997) showed a nonsignificant increase in RR of death from multiple myeloma (6 deaths; SMR = 1.8, 95% CI = 0.6-3.8), with no exposure-response trend. Similarly, the pooled study of 11 miner cohorts (Darby et al., 1995) found a nonsignificant RR of death from multiple myeloma (26 deaths; SMR = 1.30, 95% CI = 0.85-1.90). The most recent study of uranium millers (Pinkerton et al., 2004) found the mortality from multiple myeloma to be about equal to that expected in the US population (3 deaths; SMR = 1.02, 95% CI = 0.21-3.00).

Conclusion   Multiple myeloma is classified as having very low or no susceptibility to radiation induction. Incidence is related to low LET dose in some studies, and it is compensated in downwinders and onsite participants. However, there is no convincing epidemiologic evidence to warrant inclusion of miners, millers, or ore transporters for RECA compensation from multiple myeloma.

Renal Cancer

Under RECA, uranium millers and ore transporters are compensated for renal cancer, but miners, downwinder and onsite participants are not (see Table 2.1).

Renal cancer accounts for about 3% of all cancer deaths annually in males in the US general population. It is about 3 times more frequent in men than in women and generally occurs in people in their 60s and 70s. Of all renal cancers in adults, 85-95% are renal cell carcinomas, which develop from the epithelial cells lining the kidney tubules. A strong association has been observed between cigarette-, pipe-, and cigar-smoking and death from renal cancer (Bennington, 1973).

Although the atomic-bomb survivors and some populations exposed medically to penetrating low LET radiation at high doses and high dose rates have been found to be at increased risk for renal cancer (NRC, 1990), susceptibility to induction of renal cancer by such radiation is considered to be low (see Chapter 4, Table 4.1 in Mettler and Upton, 1995).

Renal Cancer in Miners

The most recent update of the Colorado Plateau uranium miners study (Roscoe, 1997) found a deficit of deaths due to renal cancer (SMR = 0.4, 95% CI = 0.05-1.4). In their collaborative analysis of mortality in 11 cohorts of underground miners, Darby et al. (1995) also found no statistically significant deficit of deaths due to renal cancer (44 deaths, 9 deaths; SMR = 0.91, 95% CI = 0.66-1.22).

Studies of mortality among several cohorts of uranium processors in the US nuclear industry whose occupational risk of exposure to uranium was primarily to insoluble uranium compounds with low to high enrichment (< 5% to > 90%) also failed to find any significantly increased risk of renal cancer (Polednak and Frome, 1981; Checkoway et al., 1988; Loomis et al., 1996; Ritz et al., 1999a and 1999b). None of those negative studies provided dose-response estimates for radiation-induced renal cancer.

Conclusion   The committee concludes that there is no epidemiologic evidence that indicates that uranium miners are at greater risk for death from renal cancer than men of similar ages in the United States. Therefore, there is no epidemiologic justification for adding renal cancer to the list of cancers for which uranium miners are compensated under RECA.

Renal Cancer Among Downwinders and Onsite Participants

In addition to the risk of exposure to ^{131}I, downwinders and onsite participants also were at risk for external exposure to low LET radiation from fallout, albeit at lower levels. A study of the incidence of solid cancers between 1958 and 1987 among 79,972 people in the Life-Span Study (LSS) cohort of atomic-bomb survivors found no significant radiation effect of cancer of the kidney (Thompson et al., 1994). On the basis of 73 incident renal cancer cases (0.8% of all the solid cancer cases ascertained in this study), radiation exposure was estimated to be associated with an excess relative risk (ERR) of 0.71/Sv (95% CI = 0.21-2.25), and an excess attributable risk (EAR) of $0.29/10^4$ person-years Sv (95% CI = 0.50-0.79), with an AR of 15.2% (95% CI = 2.6-41.3). For renal cancer, similar ERRs were obtained by using either attained age or age at exposure in the model.

In a more recent study of mortality from solid cancers between 1950 and 1997 among 86,572 members of the LSS cohort of atomic-bomb survivors, 60% of whom had radiation doses of at least 5 mSv (0.5 rem), Preston et al. (2003) did not specifically report risk estimates for renal cancer mortality, but in an example they gave the estimated risk of renal cancer at the age of 70 years in a woman who was 30 years old at exposure as an ERR of 0.97/ Sv (90% CI = <3-40), and an EAR of 0.14 (90% CI = < –0.1-0.4), with an AR of 14% (90% CI = <3-42).

Probably more pertinent to evaluating the risk of renal cancer among the downwinders and onsite participants in the absence of site-specific cancer studies of these populations are the studies, primarily of mortality, among several cohorts of military personnel who participated onsite in nuclear-weapons tests conducted by the United States and the UK at test sites in the United States and Pacific islands and in New Zealand, respectively (Darby et al., 1993a, and 1993 b; Pearce, 1996; Pearce et al., 1997, Watanabe, 1995a, and 1995b; Johnson et al., 1996; IOM, 1999). In none of those studies was mortality due to renal cancer statistically significantly greater than that in the comparison group.

Conclusion   On the basis of current epidemiologic evidence, the downwinders and civilian onsite participants covered by RECA are probably not at increased risk for death from renal cancer.

Skin Cancer

None of the RECA populations identified by RECA is eligible for compensation for any type of skin cancer. Nonetheless, the committee was asked to reevaluate the justification for this exclusion. Two histologic types of skin cancer are described here: nonmelanoma skin cancer (NMSC) and melanoma.

NMSC is the commonest type of cancer found among Caucasians worldwide, among whom the incidence is increasing because the risk increases with age. NMSC now accounts for an estimated 1 million new cases of skin cancer each year in the United States. The fatality rate is extremely low, and typically neither incidence nor mortality data are reported by cancer registries. NMSC is somewhat more common in men than in women. It comprises two main pathologic conditions: basal cell carcinoma (BCC) and squamous cell carcinoma (SCC), which occur in a ratio of about 4:1. Exposed areas of the face and head are the common sites of both types, but BCC occurs on the trunk and legs more frequently than SCC. Several environmental and host risk factors have been described for both types; the major ones are exposure to ultraviolet (UV) radiation from the sun (which is nonionizing), genetic predisposition, and lack of much skin pigmentation (Schottenfeld and Winawer, 1996).

NMSC is considered to be moderately susceptible to induction by ionizing radiation at high doses and high dose rates, and it was the first type of cancer to be causally associated with exposure to ionizing radiation. Relationships between NMSC and exposure to ionizing radiations have been extensively reviewed by the International Commission on Radiological Protection (ICRP, 1991) and the United Scientific Committee on the Effects of Atomic Radiation (UNSCEAR, 2000). The latent period varies widely and averages just over 20 years. Epidemiologic studies have suggested a synergistic interaction between UV and ionizing radiation that is more than additive (Shore, 1990), but this was not confirmed in a more recent study of skin cancer risk among atomic-bomb survivors (Ron et al., 1998b).

Melanoma or malignant melanoma is a relatively rare type of cancer—about 55,000 new cases (4% of cancers) in US males and females are predicted in 2004 (Jemal et al., 2004)—but its incidence is increasing more rapidly than that of any other kind of cancer around the world, particularly among people of Caucasian origin. It has a much higher fatality rate than NMSC. Epidemiologic studies indicate that exposure to sunlight (UV radiation) is the major causative factor (Schottenfeld and Winawer, 1996).

Melanoma is generally regarded as nonradiogenic with respect to ionizing radiation; this conclusion is based on limited information (UNSCEAR, 2000). However, followup by Reynolds and Austin (1985) of an earlier finding (Austin et al., 1981) of a 4-fold excess incidence of melanoma among employees at the Lawrence Livermore National Laboratory (LLNL), Livermore, California, found statistically significant increases in the incidence of melanoma among males (21 cases observed, 6.46 cases expected) and females (7 cases observed/ 1.35 expected). No other increase was found in incidence of groups of cancers recognized as being susceptible to induction by ionizing radiation. A more recent followup study (Austin and Reynolds, 1997) compared 31 melanoma cases diagnosed in LLNL employees between 1969 and 1980 with 110 individually matched controls with respect to established or suspected several risk factors, including occupational exposure history. Several occupational risk factors, including working around radiation sources (odds ratio [OR] = 3.7), were strongly associated with the cases. Multivariate analyses of the risk associations suggested that these factors did not act as confounders to the radiation/melanoma relationship, that is, they were independent risk factors. After adjustment for constitutional and occupational risk factors of interest, the OR for the association with radiation status remained increased (OR = 2.3, 95% CI = 1.0-7.6. However, the finding of increased melanoma risk at LLNL remains equivocal since none of the several epidemiologic studies of workers at other US federal nuclear plants or national laboratory facilities has found increases in the risk of melanoma. A nonsignificant increase in melanoma found among US radiologic technologists employed for > 5 years before 1950 (Freedman et al., 2003) suggests a need for further studies to evaluate the role, if any, of ionizing radiation in its induction.

Skin Cancer in Uranium Miners, Millers, and Ore Transporters

Tomasek et al. (1993) and Denman et al. (2003) suggested that ambient conditions in underground mines might cause cancers at sites other than the lung. Specifically, they suggested that radon in ambient air might be responsible for excess risks of skin cancer and leukemia. The committee has reviewed the pertinent literature to assess the scientific strength of that suggestion.

The risk of NMSC in radon-rich atmospheres is not associated with inhalation or other intake of radioactivity (see Appendix B) but is related to the plate out of radon daughters on the skin. The dose to the germinal layer of cells in the skin depends on the concentration of radioactivity on the skin surface and the ability of alpha particles from polonium-218 (^{218}Po) and polonium-214 (^{214}Po) to penetrate down to the location of sensitive cells. Previous studies of the process have used a nominal value of 70 µm as the depth of the basal cell layer at the base of the epidermis. Recent measurements have indicated large variation in the depth of those cells, and this could lead to higher radiation doses to the basal cell layer than were previously expected (Eatough, 1997).

The incidence of skin cancer was reported (Sevcova et al., 1978) to be higher in Czech uranium miners than in nonminers in the Czech general population (observed = 28.6 per 10,000 workers, [95% CI = 14.4-51.2], expected = 6.3 per 10,000). Most of the excess was found in miners employed underground for more than 10 years. In all cases, the tumors were surgically removed, and there was no sign of recurrence. In a followup of the same cohort, Sevcova and colleagues (1978) again found a significantly higher incidence of basal cell carcinoma, with an attributable annual risk of 1 per 10,000 workers per year per 1 Sv. Not unexpectedly, because of its low fatality rate, no statistically significant increase in the RR of death from skin cancer was found in either the Czech study (Sevcova et al., 1978) or in the pooled study of 11 uranium-miner cohorts (Darby et al., 1995). Similarly, the most recent update of the Colorado Plateau uranium-miner mortality study (Roscoe, 1997) found a nonsignificant excess of skin cancer deaths among miners compared with the US population (2 deaths, SMR = 2.5, 95% CI = 0.3-9.2). The pooled study by Darby and colleagues (Darby et al., 1995) found a deficit of malignant-melanoma deaths (18 deaths; SMR = 0.92, 95% CI = 0.54-1.45). The committee found no reported risk estimates for skin cancer among uranium millers or ore transporters.

The lifetime risk of death from skin cancer after whole-body exposure to ionizing radiation is 2×10^{-4}/Sv (ICRP, 1991, p 139). That translates to a risk of 1×10^{-7}/WLM. The lifetime risk of lung cancer from inhalation of radon daughters is 2×10^{-4}/WLM (ICRP, 1991, p 139). Thus, the risk of death from skin cancer is about 1/2,000 the risk of lung cancer for the same exposure to radon daughters (see Appendix B).

Conclusion   The committee found no convincing epidemiologic or dosimetric evidence that uranium miners are at increased risk for death from basal-cell or other type of NMSC. However, the one study that examined the incidence of basal-cell cancer in miners (Sevcova et al., 1978) found an excess incidence among those more heavily exposed to radon decay products. Basal cell cancer is very rarely fatal, is the most common of all cancers, is caused primarily by exposure to sunlight, and causes no change in quality of life; the committee therefore found insufficient evidence to add basal-cell carcinoma to the list of RECA-covered.

Skin Cancer Among Downwinders and Onsite Participants

A study of cancer incidence among the extended LSS cohort of atomic-bomb survivors (n = 79,972) found an increased risk of NMSC in the high-dose group (over 1 Sv [100 rem], mean = 2.22 Sv) compared with the referent (less than 0.01 Sv) group and low dose group (0.01-0.99 Sv, mean = 0.18 Sv); in this study, the ERR for NMSC at 1 Sv was estimated to be 1.0 (95% CI = 0.41-1.89) (Thompson et al., 1994). The more recent study by Ron et al. (1998b) of skin tumor risk among the same cohort of atomic-bomb survivors found a strong nonlinear dose-response relationship for basal-cell carcinoma (80 cases ascertained) (ERR at 1 Sv = 1.9, 90% CI = 0.83-3.3) but not for squamous-cell carcinoma (69 cases) or melanoma (10 cases). As in earlier studies, age at exposure was found to be predictive of skin cancer development; the risk was highest among those exposed when young.

A recent study of NMSC among US radiologic technologists (Yoshinaga et al., 2005) found small increases in risk for basal cell carcinoma but not squamous cell carcinoma among technologists who first worked before 1960 compared with those who first worked after 1960. The effects were greater among those with light complexions compared with those with darker hair and eye color. The findings suggest that further studies are needed to evaluate the role of chronic exposure to low to moderate doses of ionizing radiation in the induction of basal call carcinoma and modification of the risk by pigmentation levels.

Conclusion The committee concludes that currently no epidemiologic or dosimetric evidence that the radiation doses received by either the downwinders or the onsite participants increased their risk of NMSC or melanoma relative to that of the US general population. Thus, there is no epidemiologic justification for adding NMSC or melanoma to the list of compensable diseases under RECA.

Hodgkin's Disease

Hodgkin's disease (HD), sometimes called Hodgkin's lymphoma, is not compensated under RECA and the committee was asked to reevaluate the justification for this exclusion.

HD has not been associated epidemiologically with exposure to ionizing radiation. That finding (Boice, 1992) is supported by more recent reviews (UNSCEAR, 2000), and susceptibility of HD to radiation induction is categorized as very low or absent (Mettler and Upton, 1995).

HD is a rare condition (less than 1% of all new cancers diagnosed in the United States) that is morphologically distinct from non-Hodgkin's lymphoma, almost never has a leukemic component, and has clinical and histologic features that suggest a chronic infectious process. It is more common in males than females and more common among whites than blacks. It has a bimodal age distribution with peaks in young adulthood and after the age of 60 years. The overall incidence of HD has been decreasing, but it is increasing in young adults and decreasing more in adults over 40 years old. Survival rates have improved dramatically during the last 3 decades because of improved treatment methods. Various risk factors have been evaluated, including viral infections, particularly those associated with Epstein-Barr virus; occupational exposures to chemicals, particularly those used in woodworking industries and in the form of herbicides and pesticides; genetic factors; and primary immune deficiency (Schottenfeld and Winawer, 1996).

Conclusion   Epidemiologic studies fail to show any significant association between HD and exposure to ionizing radiation. The committee found no epidemiologic basis for designating Hodgkin's Disease as a RECA-compensable disease.

Colon Cancer

Downwinders and onsite participants are eligible for compensation under RECA for colon cancer, but uranium miners, millers, and ore transporters are not and the committee was asked to reevaluate the justification for this exclusion.

Colon or colorectal cancer is the third most common type of cancer and accounts for 10% of all cancer deaths among men and women in the United States (Jemal et al., 2004). Several risk factors have been identified, including genetic predisposition and other familial and hereditary factors, diet, and inflammatory bowel disease (Schottenfeld and Winawer, 1996).

Colon cancer is moderately susceptible to induction by radiation and has been associated with external exposure to low LET radiation at high doses and high dose rates in the atomic-bomb survivors and in some groups of patients treated with radiation for ankylosing spondylitis or for gynecologic conditions (NRC, 1990). Radiation doses to the colon from internally deposited uranium are, however, very low (UNSCEAR, 2001).

Neither uranium miners nor millers have been found to be at increased risk for colon cancer (Roscoe, 1997; Pinkerton et al., 2004). Roscoe found a deficit in intestinal-cancer mortality among Colorado Plateau uranium miners (SMR = 0.7, 95% CI = 0.4-1.2). Similarly, Pinkerton and colleagues found a deficit in colon-cancer mortality among uranium millers on the basis of 12 deaths (SMR = 0.63, 95% CI = 0.33-1.11). Although a statistically significant increase in deaths from colon cancer (24 deaths, SMR = 1.56, P = 0.025) was reported among workers employed in the United States as radium-dial painters before 1930, when large quantities of radium were ingested, the increase was found not to be related to initial radium intake (Stebbings et al., 1984). Studies by Clarke et al. (1996), Voelz et al. (1997), and Wing et al. (2004) have also failed to find any evidence of an increased risk of colon cancer among worker populations exposed to radiation from internally deposited radium or plutonium.

Conclusion   There is epidemiologic evidence of a dose-related increase in the relative risk for colon cancer in the Japanese atomic-bomb survivors and in some patient populations externally exposed to low LET radiations. However, there is no evidence of an increased risk of colon cancer in miners, millers, or ore transporters associated with their internal exposure to alpha-particle radiation. Thus, we found no basis for designating colon cancer in these occupational groups as compensable under RECA.

Testicular Cancer

Testicular cancer is another type that is not compensable in any of the populations designated under RECA or under other radiation compensation programs in the United States. Nonetheless, the committee was asked to reevaluate the justification for this exclusion. The susceptibility of testicular cancer to radiation induction is deemed very low or absent. It is generally regarded as being nonradiogenic, and risk estimates of the incidence of or death from this cancer type have typically not been asserted (Preston et al., 1994; UNSCEAR, 2001; Preston et al., 2003).

Testicular cancer, which is rare in the United States (about 1% of all cancers in males), is more common worldwide among whites than blacks or Hispanics. It is due almost exclusively to undifferentiated germ cells remaining in the testis. Its age distribution is bimodal; embryonal carcinoma peaks between the ages of 15 and 35 years, and declines after the age of 40 years, and a small rise after the age of 75 years is reported because of the increasing incidence of testicular lymphoma. Maldescent of a testis (cryptorchidism) is the major risk factor associated with testicular cancer but is unlikely to be the initiating event. Exposure of the mother to exogenous steroids (such as diethylstilbestrol) during pregnancy may play a minor role in the development of testicular cancer (Roth et al., 1992).

Conclusion   No epidemiologic evidence suggests that testicular cancer is induced by ionizing radiation. Thus, there is no basis for designating testicular cancer as compensable under RECA.

Prostate Cancer

Prostate cancer is not listed as a compensable disease for any of the RECA populations. Nonetheless, the committee was asked to elaborate on the justification for its exclusion. Prostate cancer is the most common type of cancer diagnosed and the second most common cause of cancer death after lung cancer (32% in men in the United States in 2004). The American Cancer Society estimates that prostate cancer will account for 13% (about 230,000) of all new cancer cases and 10% (about 30,000) of all cancer deaths in US males (Jemal, 2004). It occurs only rarely in men less than 40 years old, but the incidence doubles for each decade of life thereafter (Ross and Schottenfeld, 1996).

Under RECA, none of the designated populations is compensated for prostate cancer. On the basis of earlier epidemiologic studies, the susceptibility of prostate cancer to induction by ionizing radiation is very low or absent, especially in connection with the chronic or low exposure potentially experienced by the RECA populations. Studies of incidence and mortality among the atomic-bomb survivors have shown little indication of increased risks of prostate cancer during the 45 years of followup (Preston et al., 1994; Preston et al., 2003). There is no epidemiologic evidence that uranium miners are at increased risk for prostate cancer (Roscoe, 1997).

A study of a cohort of workers employed at the UK Atomic Energy Authority found an increased risk of prostate cancer among a small group of workers who had relatively high total external radiation doses and who also had been monitored for internal radionuclide contamination (Beral et al., 1985). A followup case-control study (136 cases in men with prostate cancer diagnosed between 1946 and 1986 and 404 matched controls) explored the relationship between cases and occupational exposures, particularly to radionuclides (Rooney et al., 1993). Although prostate cancer was statistically significantly increased in men known to be internally contaminated with or at risk of exposure to any of several radionuclides, no association was found with exposure to uranium. The most recent update of mortality (1946 through 1997) in this cohort (Atkinson et al., 2004) found no significantly increased risk of prostate cancer (200 deaths, SMR = 92, 95% CI = 79.9-102.2). That finding is consistent with those of the several followup studies of other cohorts of nuclear-industry workers in the United States, the UK, and Canada (Cardis et al. 1995; Zablotska et al., 2004; Howe, 2004; Howe et al., 2004).

Conclusion   There is no convincing epidemiologic evidence that prostate cancer is a radiogenic disease or that the RECA populations are at increased risk for it.

Uterine Cancer

Members of the public asked why downwinders are not eligible under RECA for compensation for uterine cancer but are compensated for ovarian cancer.

The main reason for the apparent discrepancy in compensability is that ovarian and uterine cancers differ in their susceptibility to induction by radiation. On the basis of earlier epidemiologic studies, ovarian tissue is moderately susceptible to radiation induction of cancer, whereas the susceptibility of uterine tissues is very low to absent (Mettler and Upton, 1995). Neither *BEIR V*, ICRP (1991), nor UNSCEAR (2000) has reported risk estimates for uterine cancer. The most recent report of mortality among the atomic-bomb survivors also failed to show an increased risk of uterine cancer (Preston et al., 2003); the incidence of benign uterine tumors (Yamada et al., 2004) showed a dose-related increase in risk that decreased with time since exposure.

Conclusion No convincing epidemiologic evidence suggests that uterine cancer is induced by ionizing radiation, so there is no basis for its compensation under RECA.

Brain Tumors

Nonmalignant brain tumors are not compensable under RECA in any RECA-eligible populations.

A well-documented increased incidence of meningioma has been found in Hiroshima atomic-bomb survivors. The incidence increased with dose and time. The incidence of meningioma among the Hiroshima survivors in 5-year intervals after 1975 was 5.3, 7.4, 10.1, and 14.9 cases 10^{-5} PY. The incidence classified by a distance from the hypocenter of 1.5-2.0 km, 1.0-1.5 km, and less than 1.0 km was 6.3, 7.6, and 20.0 cases 10^{-5} PY, respectively. The incidence classified by dose to the brain of 0-0.099 Sv, 0.1-0.99 Sv, and more than 1.0 Sv was 7.7, 9.2, and 18.2 cases 10^{-5} PY, respectively (Shintani et al., 1999).

A later study by Preston et al. (2002) of the same cohort of atomic-bomb survivors with respect to radiation dose found a statistically significant dose-related excess of nervous-system tumors (NSTs), (ERR/Sv = 1.2, 95% CI = 0.6-2.1), schwannoma separately (ERR/Sv = 4.5, 95% CI = 1.9-9.2), and all NSTs other than schwannomas (ERR/Sv = 0.6, 95% CI = 0.1-1.3). The risks of several of the other individual NSTs, including meningioma, were increased, but the increases were not statistically significant. On the basis of those findings, the authors concluded that there was an increased risk of NSTs even with radiation doses of under 1 Sv (100 rem).

A 2004 study (Yonehara et al., 2004) of the medical histories of some 80,160 Japanese atomic-bomb survivors found about a 6% increase in the risk of some type of tumor in the brain or spinal cord over a lifetime. However, for schwannomas, the risk rose to about 40%, although even with this increase these tumors are rare. Schwannomas are tumors of the nerve sheath and usually occur along nerves of the spine and along the auditory nerve in the brain; they are dangerous because of their location. Of the more than 80,000 survivors, the authors found 55 cases of schwannoma, which is about 20 more than would be expected to occur in a typical population without known radiation exposure. They also found 35 pituitary adenomas and 88 meningiomas. As in the 1999 Shintani study, there was a strong dose dependence for meningiomas (P = 0.004) but no significant correlation between incidence rate and age at exposure.

Conclusion RECA did not include compensation for benign tumors. Two benign brain tumors, intracranial meningiomas and intracranial schwannomas, have been found to be increased following high dose, high-dose rate exposures in the atomic-bomb survivors. Due to the low brain doses received by downwinders and onsite participants, there is little likelihood that an increased occurrence or incidence will be noted that could be attributed to nuclear-weapons testing.

Cancer Clusters

Several members of families with thyroid disease were brought to the attention of the committee as indicative of cancer clusters related to radiation from fallout. Familial clusters of disease are well known and new knowledge of genetics adds new tools for use in epidemiology inquiries into their pathologic basis. Individual immunologic differences and differences in intensity of exposure to agents in the environment modulate disease susceptibility, and recent advances in genetics make it increasingly possible to identify persons who are more likely than others to develop a disease, whether it is infectious or malignant. Inherited genes confer differences in cancer risk proclivity, and mutations that develop during life also influence individual risk. Ovarian, breast, and thyroid cancer are examples of cancers of which there is familial clustering, with multiple members of a family likely to be afflicted. In some cases, family members share the gene mutation; in others, the genetic basis is less well established. The apparent clustering of thyroid disease in some families in high-dose areas is consistent with the familial linkage that is a common feature of thyroid disease. The frequency of such clusters would be expected to increase in populations in which the incidence of thyroid nodules and thyroid cancer is increased, as in heavily exposed regions.

Knowledge regarding the clustering of cancer in specific people is hampered by many factors: multiple causative agents, long latent periods, inherited familial disease patterns, and the lack of a signature of a specific cause, especially when multiple factors are involved.

It is well known that cancer is common and that common things can aggregate. A simple example illustrates the frequency with which one might expect to sense an apparent cluster of cancers in a family, a neighborhood, or a community (Tversky and Kahneman, 1974; Rothman, 1987). Cancer causes about 25% of deaths in the United States (National Vital Statistics Report, 2002; 50:16:13). In a family of two, if there were no common cause, there would be a $75\% \times 75\% = 56\%$ chance of no cancers, a $18.75\% \times 18.75\% = 38\%$ chance of one cancer, a $25\% \times 25\% = 6\%$ chance of two cancers, and therefore a 44% chance of one or two cancers. In a family of three, there would be a $75\% \times 75\% \times 75\% = 42\%$ chance of no cancers, a 42% chance of one cancer, a 14% chance of two cancers, a 2% chance of three cancers, and therefore a 16% chance of two or more cancers. In a family of five, there would be a 37% chance of two or more cancers and a 10% chance of three or more cancers. In a neighborhood of 20 people, it can be shown that there would be a 91% chance of three or more cancers, a 59% chance of five or more cancers, and a 10% chance of eight or more cancers. In a small village of 100 people, it is virtually certain that there will be 20 or more cancers and an 85% chance that there will be 30 or more cancers. Because such statistical clustering may be perceived as having a common cause, it will be worrisome and more likely recalled. The increased availability of examples and the inference of a common cause may lead many people to assume that they were caused by exposure to ionizing radiation whereas in reality they may merely demonstrate that common things are common.

The issue regarding potential cancer clusters is complicated for several reasons:

• Many suspect cancer clusters are reported, and almost all are readily discounted on minor inquiry. Every year, state and local health departments are asked to respond to more than 1,000 inquiries on such matters, and they are able to respond to only a few inquiries.

• Environmental clusters are especially difficult because it is hard to establish the potential cause, and to select appropriate comparison groups for statistical analysis.

• Rare events often appear to come in groups. If the outcome is winning in gambling, that is agreeable and accepted as the way things should be. If the outcome is undesired, such as an adverse health effect, one seeks to identify its cause to avoid further unwanted consequences, and one seeks compensation if it is available. As the population ages, cancer incidence and mortality increase, and increased numbers of cancer clusters among old people are to be expected.

• Environmental contamination from any source attracts concern regarding possible adverse health effects. Cancer distribution in populations living near nuclear power plants has been investigated to evaluate possible increases in radiation-related disease, but epidemiologic studies have failed to find a positive correlation; this might be expected because the low radiation doses released in normal plant operations (Jablon et al., 1991).

Conclusion   Validated clusters of disease have involved particular exposures resulting in disease of a specific or closely related type (such as infectious disease, and industrial exposures). Many diseases of similar types occur sporadically and most clusters can not be attributed to specific causes. When they occur with increased frequency and in greater intensity, then a causal connection may be established. No evidence of unusual or unexpected clustering cancer or other diseases in exposed populations is known to the committee, although our attention was called to a possible association with multiple sclerosis. Compensation for persons with RECA-eligible disease should be compensated on the basis of their radiogenicity and the PC/AS value adopted by RECA.

Other Cancers

RECA compensated downwinders and onsite participants for specific cancers (see Table 2.1). No questions were addressed to the committee regarding the cancers not addressed specifically in this section. Those include cancers of the bile ducts, esophagus, gall bladder, non-Hodgkins' lymphoma, pancreas, pharynx, small intestine, stomach, and urinary bladder. Risk coefficients increased above those expected in a nonexposed population were the basis for their current RECA compensation, and no suggestions for changes were brought to our attention in the information-gathering meetings or in our review of new published findings. Data from the atomic-bomb survivors provides the major quantitative basis for RECA decisions concerning compensation for those conditions.

NONMALIGNANT DISEASES

Chronic Renal Disease Among Uranium Miners

It was brought to the committee's attention that under RECA, uranium miners are not eligible for compensation for chronic renal disease (CRD), a nonradiogenic disease, although uranium millers and ore transporters are eligible for such compensation (see Table 2.1). The committee asked to justify that exclusion.

Scientists have established that exposure to high doses of soluble uranium causes CRD and other damage to the kidneys in animals and humans (Hursh and Spoor, 1973). However, insoluble uranium, as generally found in uranium ore, has not been associated with CDR (Clayton and Clayton, 1981). As noted in Chapter 4, uranium millers, in contrast with miners, have the potential for exposure to soluble compounds of uranium during conversion of uranium ore to yellowcake, and renal dysfunction among actively employed millers has been identified, although the clinical significance of the finding is unclear (Thun et al., 1985). Millers may have a potential for the chemically toxic effects of uranium compounds to the kidneys. However, the recently updated studies of mortality among uranium miners (Darby et al., 1995; Roscoe, 1997) and millers (Pinkerton et al., 2004) have failed to show an increase in the risk of CRD in miners or to support earlier findings of a statistically significant risk of CRD in millers. Those studies are limited by small numbers and other methodologic problems in their ability to identify such risks.

Conclusion　　The committee concludes that there is no good epidemiologic evidence of increased risk of CRD among uranium miners and thus no epidemiologic basis for including it as a compensable disease in miners.

Chronic Lung Disease

Cor pulmonale, pneumoconiosis, pulmonary fibrosis, and silicosis are currently compensated under RECA for uranium miners, millers, and ore transporters. No issues relating to compensation for those diseases were raised at the four information-gathering meetings.

Autoimmune Disorders

Several people asked the committee for its recommendations directed toward a reappraisal of the justification for reimbursement for specific autoimmune diseases. The pathophysiologic mechanisms underlying autoimmune disorders (such as rheumatoid arthritis, lupus erythematosus, multiple sclerosis, type 1 diabetes, autoimmune thyroiditis, multiple sclerosis, and related diseases) are not well understood, but they are presumed to share some common features, each condition having variable presentations. With the possible exception of autoimmune thyroiditis, none has a clear relation to radiation exposure.

Autoimmune processes with increased circulating antithyroid antibodies can lead to clinically manifest hypothyroidism. A definitive diagnosis of autoimmune thyroiditis according to current WHO standards is complex. It depends on various combinations of specified concentrations of antithyroid peroxidase, antithyroid globulin, and thyroid-stimulating hormone in combination with thyroid abnormalities detected with ultrasonography, palpation, cytology (when available), and surgical pathology (when available). The different studies of irradiated populations rarely provide data adequate to support a definitive diagnosis. And a positive correlation with dose is rarely demonstrated.

Of the many published studies on irradiated populations, the only one in which many of those considerations were addressed involved 27 cases of autoimmune thyroiditis that were observed in 2,587 members of the Adult Health Study (atomic-bomb survivors). A positive correlation was found with doses up to about 0.7 Gy; the correlation inexplicably decreased at higher doses (much below cell-killing levels). None of the studies of other irradiated populations (Chornobyl, Marshall Islands, Southern Utah, and Hanford) has demonstrated a positive correlation between autoimmune thyroiditis and thyroid dose. Current cohort studies by the National Cancer Institute (NCI) of children after the Chornobyl accident may afford a more definitive answer when the prevalence data and dose correlations are analyzed.

Conclusion   There are no convincing epidemiologic data from which to calculate radiation risk estimates for autoimmune diseases and thereby to justify their inclusion as a new compensable category of disease under RECA.

Benign Tumors: Thyroid Nodules

Thyroid Nodules

None of the RECA populations is compensated for thyroid nodules. Thyroid nodules are common in the general population, and they increase in number with age. Benign thyroid nodules are more frequent in persons who live in regions with low dietary iodine. The diagnosis of thyroid nodules is based on palpation of a neck mass, ultrasonographic or nuclear-medicine screening, and an open or closed biopsy (typically a fine-needle aspiration) to determine the nature of the lesion. At operation and at autopsy, pathologists frequently find microscopic clusters of cells (micropapillary-microfollicular lesions), which are often referred to as occult cancer, or "pathologist's cancer," as opposed to cancers that are thought to affect health adversely. While conducting routine autopsies at the Mayo Clinic, Woolner LB (Surgical Forum, 1954) (Mortensen et al., 1955) found, as they took more sections through the thyroids of persons who died from nonthyroid diseases, a proportionate increase in the number of microscopic nests of thyroid-cancer cells. The literature does not ordinarily classify such lesions as clinically significant. Occult cancers are by definition found only when the glands are removed surgically or at autopsy.

Benign thyroid nodules have been found to be increased after high thyroid doses from radiation beams (typically x rays) and in patients after exposure to iodine-131 (^{131}I). In heavily exposed Marshall Islanders, thyroid nodules were increased by as much as 8 fold more than thyroid cancer especially after childhood irradiation (Robbins and Adams, 1989). The Republic of the Marshall Islands settlement has compensated exposed persons who had thyroid cancer and benign thyroid nodules (the differential compensation depended on whether surgery was performed and, if so, its extent). Among Hanford downwinders, thyroid nodules were increased in frequency, but the magnitude of the increase was not significant at the low ^{131}I doses received (Davis et al., 2004b).

Hypothyroidism

Data on the induction of hypothyroidism after exposure to high radiation doses come from the radiation-therapy literature. No reports of an increase in hypothyroid rates in children who received x ray therapy for other than thyroid conditions (tinea capitis, thymic enlargement, and lymph-node Rx) have been published. It is difficult to attribute diminished thyroid-function (hypothyroidism) to low doses of ^{131}I administered to patients in whom there was reason to suspect the presence of a thyroid disease at the time the dose was administered. After higher doses used in therapy for hyperthyroidism, the results are more clear cut. A large fraction of adults given high doses of ^{131}I in therapy for thyrotoxicosis become hypothyroid. After the first year, thyrotoxicosis patients treated with ^{131}I become hypothyroid at a constant rate of 2.3-4.4% per year (the increment depends on ^{131}I dosage) (Becker et al., 1971). A review of thyroid-function status after ^{131}I therapy for Graves' disease in 116 patients under 20 years old revealed a similar but steeper response. After doses of around 50 Gy, 30% of the children were hypothyroid in a year. The rate of increase slowed thereafter; 40-50% of the subjects were hypothyroid in 10 years (Read et al., 2004).

In patients with autoimmune diseases, such as Graves' disease, the body reacts against the gland; by what is now known to be an autoimmune process. It is not clear how much of the decrease in thyroid function after ^{131}I is due to the radiation and how much is due to the immunologic damage to the gland's functional capacity. It is likely that the same level of damage from ^{131}I to a normal thyroid gland would require a higher dose than would a Graves' disease patient's gland. A threshold dose of 2-4 Gy has been postulated for hypothyroidism induced after external beam radiation with x or gamma rays (Williams, 1991). Threshold doses of 10 Gy for hypothyroidism after ^{131}I and 2 Gy after external x- or gamma-ray radiation have also been suggested (NCRP, 1991). Both internal and external photon sources were noted to have induced hypothyroidism with thresholds of about 50 and 20 Gy, respectively. No increased incidence of hypothyroidism has been found among the NTS downwinder, Hanford or Chornobyl populations, and it is unlikely that thyroid doses from NTS fallout could exceed threshold levels in persons without preexisting thyroid abnormalities.

Conclusion   Hypothyroidism is increased after high radiation doses delivered externally or internally. Data on the incidence of hypothyroidism caused by ^{131}I in normal populations are lacking. A strong dose-related increase has been noted in patients who have autoimmune thyroid disease. No evidence of an increased incidence of hypothyroidism has been noted in any of the populations exposed to increased doses from [131]I. There is no convincing evidence that the incidence of hypothyroidism is likely to be increased by the doses received from NTS fallout by RECA-defined populations.

Type 2 Diabetes

Although Type 2 diabetes (non-insulin-dependent) is not known to be increased after radiation exposure, the disease is reported to be increased in Native American populations. A disproportionately high fraction of uranium miners were Native Americans. The complications of diabetes include CRD, which cannot be differentiated clinically from effects of high doses of uranium in the kidney. The doses of soluble uranium compounds that miners are likely to achieve are unlikely to reach or exceed thresholds at which CRD has been observed. However, the committee considered the possibility that there is a synergistic interaction between diabetes (and its propensity for CRD) and enhanced sensitivity to uranium compounds. The question is being addressed by research at the University of New Mexico, so further consideration of this issue awaits release of the study results.

Conclusion   There is no convincing evidence that the incidence of type 2 diabetes is likely to be increased by the radiation doses received by RECA-defined populations.

Cardiovascular Disease and Stroke

A small but statistically significant increase in cardiovascular-disease and stroke mortality (in 1950-1985) with increasing dose was identified among atomic-bomb survivors (Shimizu et al., 1992). The dose-response data indicate that the risk is likely to be negligible below 0.5 Sv. Analysis of data on mortality (1950-1997) from noncancer diseases except diseases of the blood and blood-forming organs has strengthened but not explained the association with respect to nonmalignant circulatory, respiratory, and digestive system diseases (Preston et al., 2003). The increase in risk is about 14%/Sv, which is about 10% less than the increase in the risk of radiogenic cancer and appears not to be influenced by age at exposure or attained age. A review of data from 26 studies of populations exposed to radiation doses of 0-5 Sv, showed that six of the studies had reasonable power to detect a cardiovascular effect if one existed. One study found supporting evidence, but five did not. McGale and Darby (2005) concluded that epidemiologic data have not provided clear evidence of increased risk of circulatory diseases after doses of 0-4 Sv, as had been suggested by the atomic-bomb survivor studies.

Conclusion   There is no convincing evidence that cardiovascular or cerebrovascular disease is likely to be increased by the radiation doses of the magnitude received by RECA-defined populations.

Cataracts

Radiation-induced cataracts are known to occur after high doses of ionizing radiation to the lens of the eyes. Most of the reported cases have been in adults who received relatively high doses in occupational exposures. Studies in the atomic-bomb survivors documented cataracts in children within the first 10 years after their exposure; some of the increase was attributed to neutrons. Radiation-protection standards have long recognized this fact and assigned high relative biological effectiveness factors (RBEs) to neutrons and charged particles (ICRP, 1991). Reports on the Chornobyl accident have also suggested an increased incidence in cataracts in adults (Junk et al., 1998).

Well-documented studies carried out in Scandinavia in infants less than 1 year old who received radiation therapy to the face (for treatment of hemangiomas) found a significant increase in cataracts many years later. Typically, there was a big difference in the dose to the two eyes and a strong correlation between cataract frequency and dose to the eyes closest to the beam. The increased frequency of cataracts was best represented by a linear dose-response relationship (Wilde and Sjostrand, 1997). The lowest dose (0.1 Gy) at which very mild effects were noted was higher than that expected from NTS fallout (CDC-NCI, 2001).

Conclusion   There is no convincing evidence that the incidence of cataracts is likely to be increased by the radiation doses received by RECA-defined populations.

In Utero Exposure to Radiation

Increased sensitivity of the fetus to the effects of many toxic agents, including ionizing radiation, is well documented. The subject has been reviewed recently and extensively in ICRP (2001a; 2001b) and UNSCEAR (1993, 2000) and in Mettler and Upton (1995, Chapter 8). Much of the detailed information is derived from animal studies buttressed by human data. In the first week of pregnancy, the fetus is growing most rapidly, and fetal sensitivity to radiation and other toxins is highest at this stage. The probability of damage depends on the amount and kind of fetal exposure (radiation, chemical, viral, hypoxia, and so on) and the time during pregnancy when the injury is received. The earliest effect noted in animal studies during the preimplantation period is loss of the shedding of the damaged embryo. Effects induced later in pregnancy depend on the fetal tissues; most sensitive are tissues that are experiencing the greatest growth rate then. Effects can include somatic or germline mutations, congenital malformations, and decreased organ cell mass (the functional impact expressed as decreased functional capacity observed later in life). Mental retardation has been noted in offspring of Japanese atomic-bomb survivors who were exposed during weeks 8-15 in utero, when neuronal migration rates are highest. On the basis of early estimates of dose (gamma and neutron), the risk is compatible with a linear nonthreshold, but a threshold dose of around 20 rem could not be excluded (UNSCEAR, 1993; ICRP, 2001a and 2001b).

Data on cancer after whole-body irradiation of the fetus come from studies of the Japanese atomic-bomb survivors, and from studies of results of using diagnostic medical x-ray during pregnancy (pelvimetry). Two cases of leukemia were observed among 807 atomic-bomb survivors exposed in utero—marginally higher than the number observed in the comparison group (P = 0.054). There was no evidence of a dose response correlation in the in utero period, as there were no cases of leukemia in the higher dose groups (Delongchamp et al., 1997).

X-ray pelvimetry was a common diagnostic procedure in the era before ultrasonography, and magnetic resonance imaging became available. It was used to detect and avoid delivery problems due to disproportions between the maternal pelvic anatomy and fetal anatomy, particularly head size and fetal position. Epidemiologic studies have documented an increased incidence of leukemia in the exposed offspring. Current estimates indicate that the doses received by the fetus were 10-20 mGy and that there was about a 1.4 fold increase in the risk of leukemia thereafter (MacMahon and Hutchinson, 1964); a British study reported an OR of 1.23 (95% CI = 1.04-1.48) (Mole, 1990). The lack of consistency between the findings in the atomic-bomb survivors and in x-ray pelvimetry patients raised concern that the selection of patients for pelvimetry may have been biased by inclusion of mothers who had problems that foretold complications. In such a case, the maternal condition would have been in some individuals responsible for the procedure—that is, an effect caused the procedure, rather than the procedure causing the effect—thereby compromising the validity of cause-effect conclusions. Additional pertinent discussion is presented in Boice and Miller (1999), Mettler and Upton (1995), and Brent (1999).

The radiation dose to the fetus from radionuclides in the environment depends on the element, on the amount ingested or inhaled by the mother, on whether the material passes via the placenta into the fetus and how long it remains in the different organs. The primary radionuclides of interest from fallout are iodine-131 (^{131}I), cesium-137 (^{137}Cs), and strontium-90 (^{90}Sr).

Radioactive Iodine

^{131}I taken into the mother's body concentrates in her thyroid, and what is in the body that does not concentrate there can cross the placenta and accumulate in the fetus's thyroid. The dose it receives depends on the mother's intake and the fetal stage of development. The fetal thyroid first appears as a discernable organ at 10-12 weeks after fertilization, so ^{131}I intake before then is not thought to be detrimental to its subsequent structure or function. The risk of thyroid damage is presumably highest around the transition from the first to second trimester, and ^{131}I continues to be a threat to the fetal thyroid throughout the rest of pregnancy. Additional fetal total-body dose is received from gamma rays from ^{131}I in the mother, but the fetal thyroid dose is 0.1-1% of what the fetal thyroid would have received if the ^{131}I had been there. Similarly, the dose to fetal tissues from ^{131}I in the mother is less than 1% of the dose received by the mother's thyroid.

In extreme circumstances, very high doses to the fetal thyroid can lead to thyroid ablation and severe problems (cretinism) in the newborn if not rapidly corrected. Smaller doses can lead to depressed thyroid function. The risk of thyroid complications is likely to be highest after in utero exposure, although the magnitude of the risk is not well established. Of the 12 people who were exposed in utero to fallout at the Marshall Islands, two were found to have adenomatous nodules (one exposed to 190 cGy at 10 weeks of gestation, and one to 870 cGy at 23 weeks), and one had a probable occult papillary cancer (after exposure to 110 cGy at 33 weeks) (Howard et al., 1995). Three cases with pathology in a population of 12 (25%) is a remarkably high incidence rate.

Radioactive Cesium

Cesium behaves like potassium in the body. It accumulates in circulating blood cells and in muscle. Intake into the body is primarily through the eating of meat from animals that grazed on contaminated lands, but it can also be through the eating of other foods when surface contamination is high. The estimated thyroid and red marrow doses to an adult from NTS fallout was estimated at 0.009 mGy for those tissues from ^{137}Cs and 0.002 mGy from ^{134}Cs, both of which radionuclides emit beta and gamma rays (NCI-CDC, 2003). Stannard (1988) reviewed research related to fallout radionuclides in the environment. For more detailed data, see http://www.atsdr.cdc.gov/toxprofiles/tp157-c3.pdf, accessed February 28, 2005.

Radioactive Strontium

Strontium behaves like calcium in the body and localizes primarily in bone. The organs with the highest dose are bone, bone marrow, and the lower large intestinal wall. More data based on direct measurements in bone are available on ^{90}Sr than on any of the other fallout nuclides (ICRP, 2001a; 2001b). The range of ratios of activity of ^{90}Sr in fetal bones to activity in maternal bones is 0.5-1.0 (Roedler, 1987). The dose to adult bone marrow from NTS fallout to the adult bone marrow is 0.02 mGy.

The Environmental Protection Agency (EPA) has published a set of internal-dose conversion factors for standard persons of various ages (newborn; 1, 5, 10, and 15 years old; and adult) in its *Federal Guidance Report No. 13* (EPA, 1999). For example, EPA has estimated that the dose equivalents after ingestion of 1 Bq of ^{90}Sr are 2.77×10^{-8} and 2.77×10^{-7} Sv, respectively, for the adult and infant (assuming an integration time of 50 years for an adult following the initial exposure). For ^{89}Sr, these values are 2.57×10^{-9} Sv and 3.59×10^{-8} Sv, respectively. Age-specific dose coefficients for inhalation and ingestion of any of the radioactive isotopes of strontium by the general public can be found in ICRP 71 (ICRP, 1995) and 72 (ICRP, 1996), respectively. Dose coefficients for inhalation and ingestion of strontium radionuclides can be found in EPA *Federal Guidance Report No. 11* (EPA, 1988). Dose coefficients for external exposure to radioisotopes of strontium in air, surface water, or soil contaminated to various depths can be found in EPA *Federal Guidance Report No. 12* (EPA, 1993). There is no evidence that ^{90}Sr doses from NTS fallout would rise to the point where increased incidence or mortality from bone cancer or leukemia would be detectable.

Beta or gamma emitters taken into the mother's body and from external radiation from terrestrial activity during pregnancy contribute to doses to the mother and fetus (whole-body dose, including dose to thyroid, bone marrow, and other organs). Fallout from the NTS weapons tests occurred during a long span of time relevant to an individual fetus's vulnerable period. It is unlikely that detrimental effects would be discernible from the low doses received during in utero exposure. However, due to heightened radiation sensitivity assigned to the in utero period, it should be added to the NTS testing period during which the mother was eligible for RECA-defined benefits.

Conclusion   Sensitivity to radiation exposure is higher in utero than at any other time of life. In the downwinders, the largest dose contribution to the fetus came from ^{131}I and from external exposure. The contributions from ^{137}Cs and ^{90}Sr taken into the body are significantly less than the other pathways. The committee concludes that it is important that the in utero period be included in determining residence eligibility and radiation dose of NTS releases for all subjects at risk during the fallout-testing interval.

PSYCHOLOGIC CONSEQUENCES OF RADIOLOGIC THREATS

No radiobiologic evidence indicates that interactions between radiation and molecules in the body cause the psychologic conditions or sequelae chronicled in humans after an event that involves a threat or actual release of radiation or radioactive materials. Thus, there is no credible evidence of a positive dose-effect correlation.

In recent decades, however, physicians and other scientists have become increasingly aware of the psychological consequences of a variety of real or perceived threats, including those involving ionizing radiation; they have documented that such responses can seriously affect the health of individuals and communities (IOM, 2003a; Schlenger and Jernigan, 2003). Various articles and reviews have also reported findings of several studies of the psychologic consequences observed among survivors of radiologic threats that occurred before 1990 (Bromet et al., 1990; IAC, 1991; Ricks et al., 1991); these studies confirmed earlier findings of increased emotional, behavioral, and psychologic stress that in some cases have persisted for many years in survivors of radiation accidents.

More recent information on the cascade of effects of emotional threats that involve radiation covers the nature of and risk factors in psychologic responses to such threats and their health consequences (Ginzberg, 1993). Psychologic effects have been described as the major public-health outcomes seen to date of the Three Mile Island (Kemeny, 1979) and Chornobyl reactor accidents (NEA/OECD, 2002). Separating effects that may have been related to the accidents themselves from those triggered by their social, economic, and institutional effects is difficult; adding to the complexity, in the case of the Chornobyl accident, was the dissolution of the former Soviet Union.

The psychologic outcomes observed among individuals and populations after sudden and unexpected radiation threats or events range from mild transient signs and symptoms of anxiety—in its more severe form, generalized anxiety disorder (GAD)—to major depression or minor depression (dysthymia) and post-traumatic stress disorder (PTSD). Those conditions, of course, are not peculiar to radiation threats or events.

In the aftermath of sudden and unexpected events, those diagnoses have been most prevalent in people with acute radiation injuries, physically unharmed children, mothers of young children, and the disadvantaged. They tend to persist for long periods after the event but gradually decline in intensity; they can be exacerbated by events such as anniversaries and heightened media interest that rekindle memories of the event. The effects can also be seen in physically unharmed persons at locations remote (low dose) from threats or events. However, their duration declines with increasing distance from the site of the threat (Collins and de Carvalho, 1993; Ursano et al., 1994; Becker et al., 2001; Adams et al., 2002; Bromet et al., 2002; Yamada and Izumi, 2002) and presumably with decreasing intensity of news reports concerning the incident. Lee (1996) introduced the term *chronic environmental stress disorder* to describe the psychologic effects observed in populations subject to additional stressors associated with an sudden catastrophe, such as the Chornobyl accident.

The populations at risk of radiation exposure from the US nuclear-weapons program with which we are concerned in this report did not, by most accounts, undergo sudden, unexpected catastrophic events, nor did the weapons program cause or threaten serious physical or emotional harm in the short term. Rather, as described to the committee by members of the public, the atmospheric tests often were exciting, apparently safe, sometimes dramatic events and spectacles that were announced in advance. The public and test-site workers observed the events, sometimes with encouragement to do so, with little or no guidance as to personal safety or protection. They and many physicians and scientists were largely unaware, especially in the early years of the nuclear-testing program, of the potential for long-term harm from fallout.

Later observations of unexpected, unaccounted-for health effects, accumulating knowledge and growing public awareness about the delayed health risks of radiation exposure, and gradual loss of trust in authority and professional figures could have prompted considerable stress in those populations and increased their risk of GAD, depression, and other psychiatric disorders. In contrast, uranium miners were working in environments already known to be harmful but without adequate information or protections to limit radon exposure. Their loss or lack of trust in the system as they or their colleagues developed lung cancer and other respiratory diseases that had previously been linked to their work exposures also could have contributed to stress-related anxieties and depression. The committee has been unable to identify any data that show that the psychologic effects of such chronic environmental factors have been evaluated either during the testing and mining periods or, more recently, among the RECA or other populations with similar radiation-exposure experiences.

Anxiety disorders of the types observed most frequently after radiation events also are among the more common psychiatric conditions among the US general population. Many are treatable but often go unrecognized. Prevalence rates from community-based surveys are 1.8-3.3% for depression within the preceding month and 4.9 -17.1% for depression at any time in life (Pignone et al., 2002); an earlier report put the lifetime prevalence of depression at about 20% (Mulrow et al., 1999). The lifetime prevalence of GAD is estimated to be 5% (Fricchone, 2004). It is unlikely that former or current psychiatric disorders in individuals in the RECA populations are directly attributable to the biologic interactions between radiation and DNA or other cellular targets.

Conclusion   There is no convincing epidemiologic evidence that the incidence of mental disorders is increased by the radiation doses associated with the radiation exposure of the downwinders or other RECA-eligible populations. And, there are no data whereby the psychologic effects of the exposure experiences can be evaluated in the RECA populations. Thus, there is no epidemiologic basis of compensation under RECA for psychologic disorders that developed during or since the end of the US weapons programs, but screening for depression may warrant consideration because RECA populations as a group might be at higher risk than the general population.

Based on currently available scientific evidence, **the committee recommends that no additional diseases be added to the list of diseases that should be considered for compensation under RECA.**

ADDITIONAL POPULATIONS OCCUPATIONALLY AT RISK FOR RADIATION EXPOSURE

Additional Occupational Groups Working in Underground Mines

The committee also considered additional occupational groups at risk for radiation exposure in uranium mining and milling operations, specifically, core drillers and geologists. The committee concludes that core drillers and geologists who worked in the underground mines should be considered in the same category as uranium miners. They worked side by side with the miners collecting samples to assay the ore bodies, and they would have been subject to the same exposures as the miners.

Core Drillers and Geologists Working on the Surface

Many core drillers and geologists were involved in exploratory work on the surface. Using drilling and other techniques, they sampled the subsurface soil to locate and define the extent of ore bodies. The committee could not locate sufficient sampling data on their work environment to evaluate the magnitude of possible exposures. Their work generated dust loading, but much of the material drilled through is overburden rather than uranium ore, so in general the dust would not be expected to have radionuclide concentrations as high as the ore itself. Exposure to radon and its decay products would be expected to be relatively low because the work was on the surface.

Such exposure to airborne dust could lead to a nonradiation hazard and give rise to some forms of nonmalignant respiratory disease, such as silicosis. The severity of those exposures depends highly on the type of soil that is being drilled and the resulting air concentrations. Crystalline silica in particular has a low threshold limit value time-weighted average as recommended by the American Conference of Governmental Industrial Hygienists (ACGIH, 2004), but the committee is not aware of any epidemiologic studies that indicated an increased incidence of nonmalignant respiratory disease among the workers in question.

Conclusion   The committee concludes that there is no convincing evidence that radiation exposure of core drillers and geologists performing exploratory work in uranium areas resulted in adverse health effects. The committee proposes that the National Institute for Occupational Safety and Health or another appropriate government agency conduct a hazard assessment of the conditions in which exploratory core drillers and geologists worked and determine whether there was a significant risk of exposure to hazards linked to RECA-compensable diseases. If so, the committee proposes that these workers be considered for inclusion under RECA.

ADDITIONAL POPULATIONS ENVIRONMENTALLY AT RISK FOR RADIATION EXPOSURE

Nuclear Testing: Downwinders and Onsite Participants

The committee reviewed the locations where nuclear-weapons tests were performed. The current RECA downwinder population is concentrated in the area around the NTS, and the 1997 NCI ^{131}I report (NCI, 1997) dealt with emissions from the NTS.

In RECA, Congress found that fallout from atmospheric nuclear tests exposed people to radiation that is presumed to have caused an excess of cancer and that this risk was borne by these people to serve the national security interests of the United States. The United States has conducted nuclear-weapons tests in areas other than the NTS, and populations exposed to fallout from these tests may also be considered as possible candidates for RECA compensation if Congress so chooses. The tests in question include the Trinity test near Alamogordo, New Mexico, and the Pacific tests. Onsite participants in the tests are already included under RECA, but RECA coverage may be extended to the downwinder populations in those areas.

Over the last several years, there has been a concern about the health effects associated with radioactive fallout that reached Guam during the testing of nuclear weapons in Micronesia. The Pacific Association for Radiation Survivors was formed. In 2002, a blue ribbon panel, authorized by the government of Guam, submitted the *Committee Action Report on Radioactive Contamination in Guam between 1946 and 1958*.

In March 2004, Robert Celestial provided written and oral testimony to the committee indicating that Guam did receive fallout from nuclear-weapons testing in the Pacific. He included statements from retired Navy Lt. Bert Schreiber, who testified that "the Geiger counters were off scale" in November 1952. In addition to this, various support ships deployed at Bikini Atoll during Operation Crossroads were sent to Guam and elsewhere for decontamination.

In April 2004, the congressional delegation from the Pacific Island Territory of Guam submitted a petition to Congress to amend RECA to include Guam in the jurisdiction of downwinders and onsite participants.

The committee initiated an independent assessment of the radiologic consequences related to the weapons tests in the Pacific to people living on Guam. The details of the assessment are presented in Appendix C.

Conclusions   As a result of its analysis, the committee concludes that Guam did receive measurable fallout from atmospheric testing of nuclear weapons in the Pacific. Residents of Guam during that period should be eligible for compensation under RECA in a way similar to that of persons considered to be downwinders.

The committee concludes that available evidence does not show that the general population of Guam was subjected to unwarranted radiation exposure resulting from the decontamination of naval vessels. Persons who have proof of their employment by a federal agency or its contractor in the process of decontaminating ships affected by fallout are already eligible for compensation as onsite participants under RECA.

Uranium Mining and Milling Materials Used for Construction or Other Purposes

The committee received testimony about the use of pre-December 31, 1971, uranium mine tailings and overburden in home construction. The experience with the use of uranium mill tailings in construction of homes and other buildings in uranium-mining areas indicates the potential hazard of this practice especially given that most people spend most of their time at home and many live in the same homes for decades. Others spend much time working in buildings that may contain tailings. Consequently, even a relatively small exposure rate could lead to an appreciable lifetime exposure.

This is a potentially important source of radiation exposure to the public. In the case of uranium mill tailings, the subject was addressed through the Uranium Mill Tailings Radiation Control Program, which evaluated and remediated "vicinity properties" and inactive mill-tailings piles and mills at 22 inactive mill sites across the United States. Earlier work had also been done by the federal government and the state of Colorado to remediate properties at Grand Junction, Colorado. That work was accompanied by some 4,000 measurements of indoor radon and radon decay products primarily in Grand Junction but also near other uranium mill locations, to determine whether particular buildings had radon concentrations high enough to require remediation. Uranium mine tailings would not be expected to have radionuclide concentrations as high as mill tailings, but the potential does exist for above-background concentrations of radium-226 (^{226}Ra) that would cause high radon concentrations in buildings.

The committee recognizes the hazard posed by the use of mine tailings in buildings. Historically, this has been addressed by remedial action programs, including scoping surveys to evaluate the extent of possible problems, and actions, if needed, to restore buildings to acceptable radon concentrations. The question for the committee is whether compensation should be available for people whose health has been affected by the use of tailings in building material.

Compensation is awarded by the US government under RECA because of its responsibility for the direct harm produced by nuclear-weapons development—mining, milling, ore transporting, and activities associated with aboveground testing. Responsibility for harm resulting from the use of mine and mill tailings by others for activities outside the scope of nuclear-weapons production is complex. First, such harm is an indirect rather than direct consequence of nuclear-weapons development. Second, it is plausible that shared responsibility exists for the harm produced, especially in cases in which the potential hazards associated with the tailings were known by both the citizen-procurer and the mill or mine under contract to the United States. The responsibility would rest with the user of the materials, the US government, and possibly the contractor who controlled the materials. Whether one party bore more responsibility than another is, in part, an empirical question about knowledge of harm, control of materials, contractual arrangements, and so on. Third, if all parties were ignorant of the potential hazards associated with use of the materials, the duty to compensate weakens; compensation would be become supererogatory, "beyond the call of duty."

The subject is ethically and empirically complex. **The committee recommends that the appropriate agency reviews the data on radiation exposure levels obtained inside dwellings constructed from mill and mine tailings. The committee also recommends that their findings regarding potential health consequences of such exposures be evaluated to determine whether the PC/AS values based on these exposures rise to or exceed the levels used in RECA compensation.**

Emissions from Uranium Mines or Mills

People living near uranium mines and mills may be exposed to airborne and waterborne effluents from them. Resulting radiation-associated risks are expected to be dominated by exposure to radon and its decay products. Radon emissions from underground uranium mines increased during the late 1950s when the mines began intensive ventilation. Radon is released from uranium mill tailings piles, as well as mine and mill ore storage piles. Windblown particulate emissions were also generated from mine and mill operations, tailings piles, ore storage piles, and overburden storage piles.

Liquid discharges included water from uranium mines where the ore bodies were in aquifers and seepage of process liquids from mills. Radiation doses from those discharges generally are not expected to exceed those from the airborne emissions.

The committee is not aware of any measurements of radon concentrations in offsite areas from uranium mining (as opposed to uranium milling) from 1942 to 1971. Radon concentrations were measured in the 1960s by the Public Health Service around uranium mills at Durango and Grand Junction, Colorado, and Monticello and Salt Lake City, Utah (Shearer and Sill, 1969). Of 44 stations in areas around but not over the tailings piles, only two had above-background annual average radon concentrations of at least 37 Bq/m^3 (1 pCi/L), specifically, 37 Bq/m^3 (1 pCi/L) and 96 Bq/m^3 (2.6 pCi/L). Using the 96-Bq/m^3 concentration, an equilibrium fraction of 0.4, and an indoor fraction of 0.7, as used by the EPA (Marcinowski et al., 1994), we calculate a bounding radon exposure of 0.38 WLM/year. A person (nonsmoker) born in 1927, exposed at 0.38 WLM/year for 30 years (1942–1971), and having lung cancer diagnosed in 1990 at the age of 63 years, would have a PC/AS of 0.12 (*BEIR VI* duration model) or 0.25 (*BEIR VI* concentration model).

The committee recommends that the appropriate agency reviews historical data on radon concentrations in off-site areas near tailings piles of uranium mills used to produce uranium for the US nuclear-weapons program. The agency should determine whether exposures to those concentrations in off-site areas could result in PC/AS values that meet or exceed the RECA compensation criteria. If so, the agency should take the necessary steps to have these populations included in RECA.

DEFINED INTERVALS FOR WHICH COMPENSATION IS GRANTED

Uranium Mining, Milling, and Ore Transporting

During the public information-gathering meetings, the committee was asked to consider recommending extension of the uranium mining and milling interval in Section 5(a) of RECA. RECA compensation applies to uranium miners, millers, and ore transporters who worked between January 1, 1942 and December 31, 1971. The decision to stop compensation in December 1971 was taken because the United States purchased no uranium for weapons programs after that date. The decision was based on the issue of responsibility and liability rather than radiation exposure and the resulting health effects.

With respect to radon and its decay products, no significant difference exists between exposures in 1971, which are covered under RECA, and those in 1972, which are not. Any decision to compensate uranium workers exposed after 1971, however, would be based on considerations of limits of responsibility. Such issues would be outside the scientific considerations discussed in this report.

Onsite Participants and Downwinders

The period covered by RECA for onsite participants and downwinders (January 21, 1951-October 31, 1958, and June 30-July 31, 1962) is the period of atmospheric testing of nuclear weapons at the NTS.

Radioactive material was also released from tests at the NTS other than the atmospheric tests. For instance, the underground Baneberry Test in 1970 inadvertently vented and was reported to have released 3.0×10^{15} Bq (80 kCi) of ^{131}I (NCI, 1997). Releases from such tests were generally not as large as those from the atmospheric tests, and the tests were included in the 1997 NCI ^{131}I study.

The committee recommends that the radiation doses and estimates of risks from the radioactive releases from all NTS nuclear weapons tests, including underground tests that resulted in atmospheric releases, be included in determining the PC/AS.

GROUPS AT RISK OF EXPOSURE OUTSIDE RECA'S TIME-SINCE-EXPOSURE INTERVALS

Several people testified to the committee that, in practice, the length of time spent in an affected area or since the first exposure is determined by date of birth and does not include the period in utero. This discrepancy means that some people are ineligible for RECA compensation because they do not meet the existing "time-since-exposure" criterion even though they were in utero and their pregnant mothers were in the area and at risk for exposure during the testing period.

In Chapter 6, the committee has recommended that a PC/AS-based process be used to determine the eligibility of compensation claims for people exposed to radiation in fallout. The PC/AS is determined from an estimate of the radiation dose that a person has received, and this dose must include any dose received in utero. In addition, determining PC/AS will take into account latency periods of each cancer type. Consequently, the committee's recommendation already considers in utero exposures in determining eligibility.

OTHER ISSUES OF PUBLIC CONCERN REGARDING ELIGIBILITY FOR COMPENSATION

As pointed in Chapter 1, we heard an argument from the Navajo Nation that miners can use affidavits under some circumstances to establish employment history but that millers and ore transporters cannot. People are concerned about their failed attempts at using affidavits in establishing proofs of eligibility. The

Department of Justice confirms that uranium-mining employment history may be substantiated by affidavit under certain circumstances, but the use of affidavits to establish employment as a miller or ore transporter is not allowed (*Federal Register*, Vol 69, No. 56, pgs. 13630-1). The committee could find no relevant difference to warrant that restriction on the use of affidavits and believes that it creates an unjustified inequity. This is not, strictly speaking, a scientific matter. The Congress may wish to consider re-examining the restriction and allow millers and ore transporters to submit affidavits as proof of employment.

Likewise, we heard argument from the Navajo Nation that an affidavit should be allowed as proof of presence or residence for downwinder claimants. The Department of Justice confirmed that an affidavit is not allowed as proof of presence for Native American downwinder claimants, although they are allowed to use them to establish employment (responses from the Department of Justice to the committee's questions, March 16, 2004). Members of the Native American community did not have the proofs of presence—such as utility bills, telephone bills, and mailing addresses—that were available to those living off reservations. The Department of Justice works as well as it is able within the current law to help such people. The committee finds a relevant difference between the Native American's ability to establish residence and that expected of the non-Native American population. To achieve equity of treatment regarding the processing of claims, the committee believes that Native Americans should be allowed to submit affidavits as proof of residency. The Congress may wish to consider re-evaluating this restriction.

8

Ethical Framework

Ethics, justice, and equity were recurring themes in the messages of many people who spoke to the committee at its information-gathering meetings in Arizona, Utah, and Washington, DC. Those themes held a central place in the information presented to the National Research Council representatives in Idaho, and in the letters sent by citizens who could not attend an information-gathering session. People wanted the committee to hear their distress over what they believed were past unethical actions or what they perceived as current inequitable legislation. The testimony persuaded the committee to include in this report a discussion of the ethical framework within which the Radiation Exposure Compensation Act (RECA) and Health Resources and Services Administration (HRSA) policies for the Radiation Exposure Screening and Education Program (RESEP) reside. That ethical framework includes

- The ethics of a compensation program and the concomitant duties to
  - Right a wrong.
  - Ameliorate or restore a loss.
  - Compensate for the effects of a loss.
- The ethics of medical and compensational screening.

We discuss each of those aspects of the ethical framework in the sections below to bring some clarity to the concerns expressed to us by the many people who shared them with us.

THE ETHICS OF A COMPENSATION PROGRAM: RECTIFICATORY AND DISTRIBUTIVE JUSTICE[1]

Justice is the part of ethics that is concerned with fairness. In his work *Nicomachean Ethics,* the Greek philosopher Aristotle introduces us to justice as a multifaceted phenomenon (Aristotle, Book V), focusing on distributive and rectificatory justice. In our own century, notable work on justice has expanded our understanding of this important ideal (Rawls, 1971; Nozick, 1974). Although scholars of ethics and practitioners of justice have studied the subject of justice since the early Greek philosophers, it is only in the last century that they have focused on rectificatory justice (Roberts, 2002).

Rectificatory justice plays the main role in providing an ethical framework for legislation like RECA that seeks to compensate people. Rectificatory justice concerns fairness in transactions between people who have experienced a wrong. The concepts of rectificatory justice help us to understand how to remedy wrongs. Distributive justice concerns fairness in the allocation of a good and helps to answer some questions about how one equitably distributes compensation across a population.

Rectificatory Justice

Three elements are at the heart of rectificatory justice: a wrong, a loss, and the effects of a loss. Duties or obligations are associated with each of those elements. In the case of a wrong, the duty is to right the wrong; with respect to the loss, the duty is to ameliorate or restore the loss; regarding the effects of the loss, the duty is to compensate for them (Roberts, 2002). Those elements, their concomitant duties, and their remedies are explored below.

The Duty To Right a Wrong

The committee heard much oral testimony and read pages of written testimony about disease, loss of life, and pain and suffering that were attributed to exposure associated with the US government's nuclear-weapons program. People called those matters to our attention on ethical grounds, claiming that the federal government had wronged them.

What is a wrong? Moral philosophers and legal scholars generally agree that, under the concept of justice, a wrong is related to an action that is taken against another's rights. How do we tell if an action invades another's rights?

---

[1]The ethical framework we provide for compensation legislation is similar to the scholarship in legal studies that applies the concept of corrective or rectificatory justice to theories of tort law. The main figures in this field include George Fletcher (1972), Richard A. Epstein (1973), Richard Posner (1981), and Jules Coleman (1982).

As one scholar puts it, "If I take something of yours over which you have the power to exclude and alienate, then I wrong you. On the other hand, if I take something of yours over which you have a power only to demand compensation, then (provided I compensate you) I have not wronged you" (Coleman, 1994).

Whether or not the federal government and its agents engaged in an unjustified wrong depends on a number of factors. Although stakeholders hold the government responsible, there may have been activities during some periods for which no moral responsibility exists or in which the government was justified in its actions. The need for strict secrecy, the lack of knowledge of radiation effects, and the unavailability of hazard-protection technology in part justified the federal government activities and in other circumstances excused the federal government from culpability. However, as events progressed, and secrecy became less important in achieving desired outcomes, as knowledge of radiation effects increased and hazard protections were available but ignored, the moral responsibility of the federal government and its agents increased with respect to not merely the infringement, but the violation of persons' rights[2] (ACHRE, 1995).

There were two common themes in people's characterization of the wrong. Those who spoke with or wrote to the committee believe that they were disproportionately subjected to unnecessary risks and costs to their health and life. They often asked, "Why did *we* have to experience these things?" They also believe that it was wrong not to seek their consent to bear the risks. They asked, "Why weren't we told so that we could have protected ourselves or been protected by our parents, teachers, employers, and others in authority?" One's life and choices in how to live it are, generally, those things one should be able to retain power over. Many of the people who contacted the committee perceived the actions of the federal government as an invasion of their rights to life and to the exercise of autonomy to make decisions in how to live their life.

If these wrongs did occur, how to we right them? What is the appropriate remedy for the wrong incurred? Ethicists agree that an apology is an appropriate remedy (Gill, 2002). The importance of the apology is that it acknowledges the wrong. An apology is a significant part of rectificatory justice. Without an apology, there is no formal admission of moral responsibility. Sometimes the lack of an apology is appropriate, particularly when there is no moral responsibility. When it is not, however—that is, when a wrong occurs—justice requires an apology. The reason is that an apology which is offered in an appropriate form addresses directly the injustice itself; it names the wrong, and it acknowledges responsibility. RECA included a formal of apology in its 1990 legislation. "The Congress apologizes on behalf of the Nation to the individuals described in

---

[2]This passage implies that there are distinctions to be made between moral culpability vs. responsibility and violation of rights vs infringement of rights which we do not develop here. For further discussion see Coleman, 1994.

subsection (a) and their families for the hardships they have endured" (Public Law 101-426, October 15, 1990).

It is helpful to distinguish *righting a wrong* from *restoring a loss* incurred as a result of the wrong. Some people who lived in areas downwind of or were participants at the nuclear test sites, worked in mines or mining operations, or transported the ore may have been wronged. They might not, however, have experienced any loss of health because of the unjust action; they might not have contracted a disease. Not having experienced a loss of health does not negate the wrong. Those people may not need remedies that will ameliorate a loss,[3] nor are they owed compensation for the effects of such a loss, but they are owed an apology for having been subjected to the injustice of failing to respect their personhood for no apparent relevant moral reasons. Although an apology is an appropriate response to the wrong itself, it should not be confused with amelioration, restoration, or compensation. An apology is not a form of compensation, and it does not restore a loss. Nonetheless, it is an important element of rectificatory justice.

The Duty To Ameliorate or Restore a Loss[4]

Rectificatory justice is concerned with the loss that may occur when a person experiences a wrong.[5] As we noted above, many people who contacted the committee asserted that the losses they experienced were due to wrongs committed by the federal government or its agents. Coleman has put it as follows: "The duty imposed by corrective justice—namely to render compensation—requires not just a loss, but a wrongful one, and not just causation, but responsibility (Coleman, 1994).

What constitutes a loss for the RECA population? Some testimony to the committee identified a person's loss from the time of possible exposure of the body to ionizing radiation under the belief that such exposure permanently altered cell structure. The committee considered the possibility that a loss might be construed as an unknown future harm that is currently present at the cellular level. However, the radiation signature is undefinable in that way. For that reason, we focused our examination in Chapters 4 and 7 on diagnosable diseases.

Until a disease presents itself, a person has only an increase in the probability that exposure to ionizing radiation will result in loss of health and life. Those who were exposed only to natural background radiation do not have that increase

---

[3]The loss could include psychological health. RECA does not cover psychological disorders, as defined by the *Diagnostic and Statistical Manual for Mental Disorders* (DSM), version IV, such as posttraumatic stress disorder, for reasons we discuss in later chapters.

[4]Loss and harm are often used interchangeably in law and in common parlance. One could stipulate definitions that would distinguish the measurable damages of losses from the damages that are harder to measure that we call harms. For simplicity, we do not stipulate distinct definitions.

[5]Rectificatory justice is also concerned with *gains* that may occur when a person experiences a wrong. For our purposes in this report we limit our discussion to loss.

in probability due to NTS fallout (see Chapters 5 and 7). Nevertheless, some persons assert that having been put at such risk is sufficient to establish a loss. For that to count as a loss under tort law the risk must cause an injury. The type of injury we associate with risk is a loss in security (Simon, 1992; Weinrib, 1987; Coleman, 1982). RECA compensation is not based on that view of loss.[6] Nevertheless, as we pointed out in Chapter 6, probability plays an important role in establishing eligibility for remedies once a disease is diagnosed.

The remedy for the loss sustained is restoration. "Restoring the exact same thing that was taken ensures that no tangible loss remains accounted for" (Roberts, 2002, page 15). Logically, restoring the loss due to disease means restoring the person to health. Clearly, if life is lost, restoration is not possible; but it is also true that, without returning a person to a premorbid state, restoration in the strict sense also may not be possible. Restoring someone to health is not the same as, for example, returning a stolen necklace to its rightful owner. For that reason, we also use the word *amelioration* when we describe the remedy associated with a wrongful loss. *Amelioration* allows us to include a range of medical practices, from screening to medical treatments, aimed at benefiting a person's health.

The medical screening and education program provided by RESEP can be understood as a remedy by forestalling (early diagnosis and care) or ameliorating a loss of health that may have resulted from the government's actions. Legislators claim to have created RESEP to provide an opportunity for an early intervention to protect against such a loss.

Other responses to the ethical concern of amelioration of the loss of health are possible or even morally preferable. For example, RECA populations are unevenly covered through existing additional compensatory legislation for the medical services associated with the eligible diseases. As we saw in Chapter 2, some RECA populations will receive medical services for their diseases from the time they file a successful claim if they also qualify for compensation under the Energy Employees Occupational Illness Compensation Program Act (EEOICPA) and the Radiation-Exposed Veterans Compensation Act (REVCA). Other populations have medical services covered under other programs or hospital practices, such as Medicare, Medicaid, private health insurance, and indigent care. But, none of those programs or practices is intended as a remedy for a loss of health due to a person's involvement in the development of nuclear weapons.

Restoration or amelioration of a loss is not identical with compensating a person for the effects of such a loss of health and life. The distinction does not rest on what can be restored and what cannot be (or is not) restored. It is a distinction between the loss and its effects. The important distinction between *restoring or ameliorating* a loss and *compensating* for the effects of a loss was

---

[6]Michael Simon considers that treating risk exposure itself as an injury for which compensation in tort law is possible but rejects this view as incoherent. He argues, instead, for a public law (like criminal law) that provides compensation for victims.

expressed in the following statement by Senator Orrin Hatch (R-Utah) on the eve of passage of PL 1515:

> I am extremely grateful to the interested and concerned constituents who helped in the drafting of the RECA amendments. Many times, their heartfelt stories helped lead to provisions in the legislation which can only help improve the program. For example, in one meeting on the bill held in St. George, Utah, a woman explained to my office that the compensation program, while well-intended, could never make families who had experienced radiation-caused illness whole again. She expressed her feeling that the greater good could come not from compensating individuals, but from instituting programs which will help families detect potential illness earlier, allowing them to be treated more successfully and cost-effectively. From that conversation was born the new prevention grant program, which I believe will prove to be extremely successful.

The woman points out the difference between the diseases for which access to successful, cost-effective remedies was not available and compensation paid out which did not affect such health-related remedies. Compensation is not intended to and will not directly restore health or life. However, if some measures are available to ameliorate the loss of health, rectificatory justice requires that they be considered in proportion to the loss incurred. Nonetheless, several people testified about their large medical bills that have exhausted their compensation payments, leaving unpaid balances. The remedy best associated with the loss of health is some explicit provision of medical services for the diseases associated with radiation exposure. RECA does not make that provision; other programs, such as EEIOCPA and REVCA, do. Which measures are taken may be a matter of utility, depending on the costs, risks, and benefits associated with them. We discuss this in more detail, including a recommendation, in Chapter 10.

The Duty To Compensate for the Effects of a Loss

A loss has its effects, ranging from the immediate pain and suffering due to the loss to longer-term effects in the quality of life that a person experiences. Justice asks us to compensate for those effects, to soften their impact if possible, or to make them easier to bear. What remedy is available to provide for that?

Compensation is generally regarded as the appropriate response in rectifying the effects of a loss. Compensation represents an attempt to rectify the imbalance resulting from the effects. Compensation is full insofar as it is equivalent to the value of what is lost, including the cost represented by the pain and suffering associated with the loss. Throughout RECA, compensation is described as partial. Generally, when compensation and an apology are accepted, although the loss has not been restored, the moral debt is met.

Not all compensation programs are based on the effects of a wrongful loss. In fact, having to experience the effects of a loss of health and life has been reason enough for the US government to compensate individual citizens and their families. Veteran compensation programs are generally based on that reasoning. In the act of soldiering, when risks are revealed as much as is practically possible and are minimized, no wrong is incurred when a soldier is injured in war, but compensating people for the effects of their loss is nevertheless appropriate. The decision may be made to compensate people for the effects of a loss even if their ownership rights or rights to enjoyment or some other rights were not violated. For example, a group may be compensated for the effects of some loss for the sake of social solidarity, or to send a message of compassion to the larger society even though there is no obligation under a wrongful loss to compensate. Compensation may be offered prospectively to people for their willingness to accept additional risks, such as special payments to military person serving in special capacities or to entities willing to accept environmental risks. We have various reasons for wanting to provide compensation for a loss, but they are not all based in a wrongful loss. According to its history (see Chapter 2), RECA does appear to be based on rectificatory justice.

Equity

Are there any additional ethical requirements in compensating for the effects of such a loss? RECA stakeholders who testified before the committee were concerned about how compensation was allocated. They pointed to perceived inequities in the eligibility criteria for compensation—differences in age at disease onset, differences in county of residence, differences in diseases compensated, differences in work history or job performed, and difference in access to equitable allocation of compensation based on ethnicity.

It is appropriate to challenge the criteria that establish eligibility if like cases are not treated alike. Many uncompensated downwinders found it difficult to distinguish between themselves and compensated downwinders living in nearby counties. To some uranium workers, such as core drillers, there seemed to be no difference between the diseases they contracted and those contracted by their fellow miners and millers. To Navajos, it seemed unfair that the requirements of proof were still impossible for them to produce even though their residence and work history were well known among the members of the Navajo Nation. Finally, researchers who testified to the committee believed that there was little evidence that would establish important differences between hazards to persons living within 300 miles of one of the atmospheric testing grounds and those living within 3,000 miles.

Justice seeks to treat people fairly.[7] That does not mean that all persons must be treated equally. Distributive justice allows for differences in how people treat

---

[7]Justice as fairness is one of the central ideas of John Rawls. Other conceptions of justice, for example, utilitarian, communitarian, and so on have been advanced by scholars. For a discussion of those theories as well as the distinction between formal and material principles of justice, see Beauchamp and Childress, 2001.

each other. People do not need to be treated equally to be treated equitably; in fact, treating people equally may be a violation of equity. To treat people equitably is to give them what is due to them, which is not necessarily exactly what everyone else gets. Equity is a moral concept of right proportionality. *Equity allows us to treat persons differently if there are relevant differences between them, but it enjoins us to not discriminate when there are no such differences.*

Equity requires consistency of treatment. Like cases should be treated alike. The investigations of this committee revealed widespread inconsistencies and differences in who receives compensation for harm incurred when there seem to be no relevant differences between people. As we noted in Chapters 5 and 6, the committee faced the ethical challenge of finding a scientific basis for compensation that enhances equity so that like cases are treated alike. What is it that makes establishing a PC/AS-based compensation methodology equitable? We develop an answer to this below.

Causation

Tort law seeks to preserve equity by requiring that the same set of conditions be used to establish negligence and eligibility for compensation. A key element of tort law is the establishment of a proximate cause or a "reasonably close causal connection between the conduct and the resulting injury" (Prosser, 1971). An important health detriment that follows exposure to ionizing radiation is the increased risk of developing any of several types of cancers, referred to as radiogenic cancers. Can we say that there is a reasonably close causal connection between the activities of uranium mining, milling, and ore transporting and lung cancer, and other nonradiogenic diseases? We are able to make that claim for miners' activities, on the basis of the amount of exposure and the observed disease in the miners, but we are less so with respect to the lower doses and risks posed by fallout.

Radiogenic cancers can be caused by other agents or factors. No symptom or marker has been identified that unambiguously identifies that an individual cancer was caused by radiation. Thus, when cancer is diagnosed in a person, it is not possible to determine with absolute certainty whether it was caused by radiation or by some other factor. That creates a problem in trying to establish eligibility for compensation for radiogenic cancer.

As we noted in Chapter 2, failures to win remediation in the court system led citizens to turn to their congressional representatives for redress through legislation. The legislative history of RECA portrays a weakening of the use of causation as a basis of compensation. That is at least in part because of the difficulties in establishing causation owing to the state of scientific and medical knowledge. On the eve of the passage of the RECA amendments of 2000, Senator Orrin Hatch (R-Utah) stated:

Through advances in science, we now know so much more about the effects of that radiation than we did in the late 1950s and 1960s. In fact, we know so much more today than we did in 1990 when Congress passed the original compensation program, the Radiation Exposure Compensation Act. Our current state of scientific knowledge allows us to pinpoint with more accuracy which diseases are reasonably believed to be related to radiation exposure, and that is what necessitated the legislation we are considering today.

In cases where there is sound scientific evidence of harm, it is reasonable to award compensation. The science is still evolving, and the committee was especially challenged to find a scientific method with which to recommend a more equitable approach to establishing eligibility for compensation. We believe that the use of probability of causation/assigned share provides this equitably approach to compensation.

Compensating on Probabilities

Nevertheless, significant ethical questions remain in compensating on probabilities. Those questions become more challenging since knowledge of a disease's cause in a specific person is almost never known with certainty. At least three questions complicate compensation decisions. First, when there is little way to know whether a given person has sustained a loss to health even if we assume that many people who have sustained it, what is the best way to compensate? Second, as Upton and Wilson (2000) have asked, "How does society compensate someone who has a disease which has a number of possible causes and for which the assignment of causation can only be expressed on a probabilistic basis?" Third, how is fairness achieved in a compensation program for the effects of a loss when those effects can differ widely from one individual to another?

We may agree in general that the exposures from nuclear testing and fallout did contribute to some additional cancers, but because each individual's excess risk could be small in some RECA and other populations, we have little ability to say which cancers were caused and which cancers were not caused by radiation. As we saw in Chapter 5, the PC method seeks to rectify this concern by accounting for the possibility that one's disease could have been caused by factors other than radiation exposure from the activities associated with nuclear-weapons development. It also provides ways of taking into account age, place and time of exposure, milk consumption, and estimated radiation dose.

We saw in Table 2.4 and in our discussions in Chapters 5 and 6 that other US radiation exposure compensation programs (REVCA and EEOICPA) use PC and compensate people by using either a complex formula or a single flat rate. To remind the reader, in 1982, British Nuclear Fuels Ltd. (BNFL) and related trade unions initiated a compensation program based on PC and cancer

TABLE 8.1   Ethical Concerns and Remedies in RECA and RESEP

Ethical Concern	Remedy	RECA and RESEP
Righting the wrong or injustice itself	Offering an apology	RECA includes an apology
Ameliorating or restoring the loss of health	Providing medical services	RECA provides no medical services directly; medical services are provided only indirectly to some RECA populations and through other programs, such as REVCA (atomic veterans) and EEOICPA; RESEP supports detection of disease
Compensating for the effects of the loss	Compensation	RECA provides monetary payments at a flat rate

mortality. In 1987, the program was extended to cancer incidence, for which people were compensated but at different rates; compensation is awarded on a sliding scale based on selected PC values. Upton and Wilson (2000) suggest that a decision first be made on a sum for compensation for the disease in question and people then be paid a fraction of the sum equal to the sum times the PC, with a de minimis cutoff at some selected figure. Some compensation schemes try to account for the differing effects of a loss. Discussions of the duty to compensate become more challenging if there is very limited information to use in distinguishing the particular people who have suffered losses. Should whole communities, rather than individuals, be compensated? Do the difficulties associated with individualized causation require that we abandon compensation payments to individual persons? Programs that use PC/AS allow the AS to establish eligibility for an individual, not for a community, because of widely varied doses in communities and because of the presumption that individuals sustain the harm. Nevertheless, there is no simple logic that will tell us how compensation schemes should work. It is beyond the scope of what science can tell us.

In sum, attending to the ethical foundation of a compensation program is critical. The ethical basis rests in rectificatory justice. It is important to provide the appropriate remedies to the various aspects of such justice. Table 8.1 displays those remedies and the extent to which RECA and RESEP accommodate them.

THE ETHICS OF MEDICAL AND COMPENSATIONAL SCREENING

The RESEP legislation proposes screening for 19 malignancies in downwinders and seven malignant and nonmalignant conditions in uranium miners, millers, and ore transporters (see Chapter 2). In its interim report (NRC, 2003a), the present committee addressed the ethical issues associated with such screening. Just as the committee expands its discussion of screening in Chapters 9 and 10, it expands its discussion of the ethics of screening here.

We necessarily divide screening into medical screening and compensational screening. Medical screening has the underlying purpose of improving health outcomes (see Chapter 9). Most uses of the term *screening* refer to medical screening. But the possibility of receiving compensation for a detected disease lies in the background of any screening program that RESEP grantees promote, so it is helpful to define another type of screening, compensational screening. Compensational screening has the underlying purpose of identifying persons who may be eligible for compensation under RECA, but not the improvement of health outcomes (see Chapter 10).

The populations we visited and the grantees we heard from presented themselves or are recruited for screening both because of its medical benefits and because of the possibility that a compensable disease may be detected. Screening for either of those reasons could be supportive of the other.

We pointed out in our interim report (NRC, 2003a) and in Chapter 9 that routine medical screening is generally not recommended for RECA diseases except for cancers of the breast and colon and for chronic renal disease; screening for other cancers is medically problematic. The criteria used to determine whether a cancer screening program is useful are the general principles of screening adopted by most medical practitioners. Therapeutic benefits, high prevalence, perhaps high sensitivity,[8] and adequate specificity tests are needed, and these are discussed in detail in Chapters 9 and 10.

The same general screening principles affirm the ethical preferability of screening for some non-RECA eligible diseases. In Chapter 9, the committee examines those diseases to determine whether there are sufficient grounds for recommending screening. In a few instances, although the diseases do not have sufficient support for RECA eligibility, the committee nevertheless finds reason to comment on screening for them as a good medical practice. The duty to act beneficently by providing a good rather than the duty to right an injustice by ameliorating a wrongful loss is what justifies this recommendation to screen. Depression is such a disease. Screening is not recommended as a remedy to restore a loss because there is no proven connection between the disease and radiation. It is not justified as a measure of rectificatory justice but HRSA may want to consider screening as a beneficent act. Were such diseases to become RECA-eligible, their screening protocols would probably remain unchanged, but the rationale for screening would change, and the costs of screening could be covered as part of the remedy to ameliorate or restore a loss.

---

[8]A somewhat insensitive but specific test may be reasonable for screening, depending on the circumstances.

The committee found that it not only is medically unsound to screen for many RECA diseases but also may be morally irresponsible or unethical. The practices and principles of medical screening are infused with ethical considerations. Ethics is not tacked onto the practices. The utilitarian approach is widely used in the decision to screen. That tradition affirms the use of a calculus to achieve an acceptable balance between the risk of harm and the probability of benefits of a particular action (directly), program, or policy. The moral objective of the utilitarian approach is to determine which action or a rule (indirectly) brings about the greatest good or prevents the most harm. The principle of utility may be applied to both individual actions and population wide programs or policies. Ethical duties (actions or rules) are derivatives of the calculus. That means that how one ought to act or what moral rule one should follow is determined by the value that the action or the use of the rule produces (Bentham, 1961; Mill, 1998).

The decision to screen is complex, as discussed in Chapters 9 and 10. The utility calculus is itself complex and based on several assumptions. Insofar as we can rely on the truth of the empirical information used in the method, we have developed a valuable tool for medical screening decisions.

Another equally respected tradition in ethics is the deontologic tradition, derived from the Greek word for *duty* (*deon*), which focuses on the action or rule itself, and not on its consequences, to determine one's moral duties. That tradition reflects the primary value that we place on autonomy (freedom) and respect for persons. Actions that reflect such respect for a person's ability to choose for himself or herself are morally permissible, not because engaging in them brings about some extrinsic value (the greatest good or prevention harm) but simply because of their intrinsic value or rightness.

Discussing that other tradition is important when we articulate the ethics of screening. A utility calculus may not have all the answers. A potential RECA stakeholder may ask to be screened for one of the diseases for which the scientific and medical communities do not recommend screening. To ensure that a person's autonomy is respected, a medical practitioner must take time to inform the stakeholder of the potential harm, the unlikely benefits, and the costs of screening. Valuing autonomy is not necessarily in conflict with valuing utility. Nonetheless, the stakeholder may persist in his or her request for a screening test. The stakeholder's desire to engage in what is judged to be a futile screening protocol does not oblige a medical practitioner to offer the screening protocol. Respecting the stakeholder's autonomy might instead entail the negative duty of not preventing him or her from pursuing this course of action. That raises further questions about autonomy and its place in a hierarchy of values, including prevention of harm to the patient (Beauchamp and Childress, 2001).

Moral philosophers have given the action of overriding a person's autonomy for her or his own good a name—paternalism.[9] Some instances of paternalism are clearly morally permissible, or even morally required or obligatory, especially when the ability to make informed choices is curtailed by age or inability to comprehend. That type of paternalism is referred to as *soft paternalism. Hard paternalism*, however, is generally more difficult to defend when informed consent is reliable, because it requires that the stakeholder's autonomy be overridden in favor of the medical practitioner's judgment of potential harm.[10]

The ethics of screening is further complicated by compensational screening. The stakeholder who wishes to be screened (or the grantee who creates screening programs with establishment of eligibility as the primary purpose) is perceived to have miscalculated the real harm of screening with no offsetting or overriding benefits. Again, that might not be true. Knowing full well the harm of screening, stakeholders may wish to exercise their freedom to make a risky choice.

A stakeholder's decision must be truly his or her decision and not the result of having been overreached by enthusiastic and well-meaning medical practitioners or family members. Although persuasion and some forms of influence are morally acceptable, coercion is not. The prospect or hope of receiving compensation exerts its own pressure on the stakeholder. To add to that pressure by recruiting people into screening programs that are unlikely to have medical benefit is unethical. Families that stand to gain from compensation might do well to avoid overzealous screening and wait for postmortem examinations. At the same time, a stakeholder's decision must be based on information that he or she is capable of understanding (Dworkin, 1976).[11] How information is valued complicates matters. In some populations covered by RECA, people come from cultures that devalue modern medicine or supplement it with alternative medical practices. And in some cultures, imparting negative information to a patient may fly directly in the face of cherished values. For example, although their findings apply primarily to traditional Navajos, Carrese and Thodes (1995) conclude that discussing negative information conflicts with the Navajo concept of *hozho* and was viewed as harmful by the Navajo informants in their study. A participant's

---

[9]Because the gender of the medical practitioner that prevents ill-advised screening may be either female or male, it might be more appropriate to call this maternalism or parentalism; the common parlance in medical ethics is *paternalism*, however, and the analysis that makes the important distinction between "hard" and "soft" is historically related to *paternalism.*

[10]Not all instances of paternalism represent the distinction between utility and autonomy. In many cases, the judgments of utility simply differ. The patient may see more benefit in the screening protocol, on the basis of different values and assumptions in the utility calculus, than does the medical practitioner.

[11]Gerald Dworkin calls these features of autonomy *authenticity* and *procedural independence.*

beliefs about the effectiveness of allopathic medicine or alternative medical practices or views about the value of disclosure of risks are not by themselves sufficient evidence that a participant lacks the characteristics needed to exercise independent, nonstandard judgment.

To enhance the autonomy of a stakeholder who pursues compensational screening, we advise counseling that provides evidence-based information about screening while respecting a stakeholder's choice. At a minimum, no stakeholder should be screened for purposes of compensation without first meeting other administrative eligibility conditions—such as residence, occupation, or other criteria, such as PC/AS, if established by Congress. We continue this discussion in detail in Chapters 9 and 10.

CONCLUSION

In response to the many ethical concerns voiced by the stakeholders and experts who testified before the committee, we provide an ethical framework before proceeding to address the committee's charge. We clarify and discuss the ethical concerns in this chapter to which we have referred in earlier and upcoming chapters of our report. These concerns are included in two main subjects: the ethics of a compensation program and the ethics of medical and compensational screening.

9

Medical Screening

In this chapter and the next, the committee turns to issues of screening in the context of the Radiation Exposure Screening and Education Program (RESEP), which is administered by the Health Resources and Services Administration (HRSA) in the Department of Health and Human Services. The broader framework is the role that RESEP plays relative to the Radiation Exposure Compensation Act (RECA) and its amendments, as administered by the Department of Justice (DOJ).

HISTORICAL CONTEXT OF SCREENING IN RESEP

The original 1990 RECA legislation does not mention screening for the diseases it covers. The omission may have arisen because RECA's primary purpose is compensation. The inaccessibility to services for determining eligibility for compensation led to a change in the original RECA statute, but Sec. 2, "Findings" of the RECA amendments of 2000, still does not explicitly mention screening. However, screening could be construed as implied in Congress' sixth finding (Public Law 106-245 [S. 1515] July 10, 2000 Radiation Exposure Compensation Act Amendments of 2000 106 PL 245; 114 Stat. 501) that

> it should be the responsibility of Federal Government in partnership with State and local governments and appropriate healthcare organizations, to initiate and *support programs designed for the early detection*, prevention and education on radiogenic diseases in approved States to aid the thousands of individuals adversely affected by the mining of uranium and the testing of nuclear weapons for the Nation's weapons arsenal. (emphasis added)

Sec. 4 of the RECA amendments gives an explicit requirement for screening. The HRSA administrator must make competitive grants available "for the purpose of *carrying out programs to screen individuals* described under section 4(a)(1)(A)(i) or 5(a)(1)(A) of the Radiation Exposure Compensation Act (42 U.S.C. 2210 note) for cancer as a preventative health measure" (PL 106-245 Sec. 417C(b)(1)) (emphasis added). In other words, RECA, as amended, does require cancer screening as a preventive health intervention for the purpose of early detection (CFR, Vol. 67, No. 83, page 21256).

To carry out that mandate, HRSA created RESEP and announced the first fiscal-year competitive application cycle on April 30, 2000, in the *Federal Register*. In its announcement, HRSA described the legislation as providing the authority for competitive grants for "individual cancer screening" (CFR, Vol. 67, No. 83, page 21257). Although the amended legislation might be read as linking screening to cancer, this CFR section provides for the "availability of $3.0 million to eligible entities for the purpose of carrying out programs to screen eligible individuals for cancer and other radiogenic diseases" (ibid) (emphasis added). Because RECA legislation covers both nonmalignant and malignant radiogenic diseases, RESEP is consistent with the overall intent of the amended RECA in not limiting screening to cancer.

HRSA charges its Bureau of Primary Health Care with administering the RESEP grant program. On May 2, 2002, and June 5, 2003, HRSA published a program information notice providing guidance to potential grantees about the RESEP competitive process. This guidance includes eligibility requirements, program expectations, review criteria, and award factors for the grants.

RESEP grantees are expected to pursue multiple core activities: education, screening and early detection, referrals for medical treatment, eligibility assistance, quality assurance, staffing, data collection, finance, performance reports, and outreach. We examine only the screening and early detection component here.

Screening and Early Detection in RESEP

HRSA treats screening and early detection as a single activity. The notice did distinguish between nonmalignant and malignant radiogenic diseases. RESEP identifies basic screening protocols and differentiates basic screening and an array of steps collectively considered referrals for medical treatment (advanced testing, diagnosis, evaluation, and treatment), as follows:

Screening and early detection:
• Basic screening protocols.
Referrals for medical treatment:
• Advanced testing.
• Diagnosis.

- Evaluation.
- Treatment.

Table 9.1 lists the HRSA requirements for a screening and early detection program with the basic screening protocols for both nonmalignant and malignant diseases as the agency categorized them. Not all the tests listed (in the table), however, detect clinical diseases and processes that have been associated scientifically with exposure to radiation.

Referrals for Medical Treatment

If basic screening protocols reveal abnormalities in patients, then the providers must refer the patients to a hospital, a special clinic or an imaging center, or doctor's office for advanced testing. Table 9.2 elucidates what HRSA means by *advanced testing*

THE NATURE OF SCREENING

For our purposes, screening is of at least two different types: (1) screening for *medical* reasons, with the underlying purpose of improving health outcomes; and (2) screening to identify persons who may be eligible for *compensation* under RECA. This committee distinguishes these two concepts with respect to the underlying purposes of RESEP, using *medical screening* for the former and *compensational screening* for the latter. This distinction is not traditional but is required because RESEP specifies screening for some diseases not traditionally recommended for medical screening to improve health outcomes. Because a term is required to describe this RESEP concept, we coin the term compensational screening.

This chapter is concerned largely with epidemiologic and statistical factors related to medical screening, the limitations of medical screening, and the implications for screening in RESEP. We comment as well on future research, and we introduce issues that extend beyond the purely technical questions of screening itself. Chapter 10 discusses these findings in the context of RECA and RESEP, which are oriented more toward compensation than toward medical interventions.

Screening Definitions

Screening has numerous meanings. One general meaning is sifting or filtering objects to separate what is wanted or desirable from what is not, as when a miner during the Alaskan gold rush screened for gold. In medical contexts, screening is commonly directed at groups of individuals who are asymptomatic, using relatively speedy, inexpensive tests and procedures. Such screening is

TABLE 9.1   Screening Protocols in RESEP

Screening Protocols	Compensable Nonmalignant Radiogenic Diseases	Compensable Malignant Radiogenic Diseases
Medical and occupational history	Must include date of exposure, place, duration of employment, and tobacco use	Must include date of exposure, place, duration of employment; special attention to symptoms of and risk factors for primary cancers or other diseases covered by RECA; should also include tobacco, alcohol, and caffeine use
Physical examination	Emphasis on pulmonary, cardiovascular, and renal systems	Complete examination to include all cancers covered by RECA
Chest radiography	Standard posterior-anterior view chest radiograph for presence of radiologic fibrosis, silicosis, or pneumoconiosis	As indicated by physician
Pulmonary-function testing	As needed, can include spirometry, lung volumes, arterial blood gases, and DLCO[a]	See physical examination
Routine testing	Other routine laboratory work and electrocardiography as required	Routine laboratory work as indicated by physician
Other Program Requirements Included in Screening and Early Detection		
Followup	Patient contact via telephone; report results to patient, primary care physician, or both; periodic re-evaluation	See case management below
Case management	Not required	Extensive followup to ensure that care was received; documented monitoring of patient progress; all operative, consultative, procedural, and pathology reports and physician, hospital, and health-care facility discharge summaries maintained

[a]DLCO, the diffusing capacity of the lung for carbon monoxide (CO), i.e., measurement of carbon monoxide transfer from inspired gas to pulmonary capillary blood (American Thoracic Society, 1995).

TABLE 9.2   Advanced Testing Protocols in RESEP

Advanced Testing for Compensable Nonmalignant Radiogenic Diseases	Advanced Testing for Compensable Malignant Radiogenic Diseases
• Measurement of lung volumes and diffusion capacity (if not performed earlier) • Resting and exercise arterial blood gases if not medically contraindicated • High resolution computed tomography and computed tomography-scans; chest tomography, bronchoscopy, ventilation and perfusion lung scanning, pulmonary angiography, and thoracentesis, • Pleural biopsy, magnetic resonance imaging, and positron emission tomography scans if required • 24-hour urine studies and supplementary blood tests if not previously performed • Renal ultrasonography, radionuclide scanning, magnetic resonance imaging, and renal biopsy if required	• Endoscopy • Tissue biopsies and fine-needle aspiration • Imaging studies, including computed tomography scans, magnetic resonance imaging, mammography and other breast imaging techniques, radionuclide imaging, ultrasonography, regular x-rays and contrast studies

generally intended to distinguish people with some undiagnosed and even unsuspected ailment or pathologic condition from those who are well (at least with respect to that ailment or condition). Screening may also be used to detect risk factors for disease that may give individuals a higher than average probability of illness; it may also be directed at individuals who are known or suspected to have risk factors.

Typically, medical screening is directed at large numbers of apparently well individuals (for example, the public at large) with the aim of finding those who have undiagnosed problems, genetic traits, or other characteristics that may benefit from medical intervention. The underlying assumption is that early detection can improve outcomes through early treatment or changes in lifestyle. Screening is used to facilitate appropriate care early in a disease, before serious signs, symptoms, or complications develop. An Institute of Medicine (IOM) committee stated that such screening for cancer "refers to the early detection of cancer or premalignant disease in persons without signs or symptoms suggestive of the target condition (the type of cancer that the test seeks to detect)" (IOM, 2003b, p. 156). Positive results from screening tests are usually not conclusive, so confirmatory tests are needed before a diagnosis is established; if such a diagnosis is made, then followup, referral, and treatment would typically be expected to ensue.

Other terms are sometimes used in conjunction with or as synonyms for *medical screening*. Among them are *case-finding* and *early detection*, although these terms have not been used with total consistency.

The distinction rests on whether such testing is done in the context of an established provider-patient relationship. We use *patient screening* to refer to testing within such a relationship and *case-finding* if the person screened and the testing program or provider have no established relationship. Thus, this distinction refers to the context of the screening program. Much of the discussion in this chapter and the rest of this report applies to both contexts. When what we say refers properly to both contexts, we simply use the term *screening*.

PRINCIPLES OF SCREENING

Basic Tenets

Several principles should guide an analysis of screening and case-finding, whether for medical purposes, for compensation, or both.

First, screening asymptomatic individuals presumes that efficacious and effective therapies exist for treating the disease either in its early (preclinical) stages or at later stages upon clinical detection (that is, upon diagnosis in patients with signs or symptoms). If that is not the case, then either screening is needless or treatment should await clinical detection.

Second, screening always carries some risk of harm. These risks arise from the tests themselves, from diagnostic labeling (whether accurate or not), and from false-positive results that lead unnecessarily to further tests and possibly therapeutic interventions. These risks or harms must be balanced against the assumed benefits of screening. This balancing is especially important because medical tests almost always carry a risk of physical, social, or psychological harm; in the RESEP context, these harms must be weighed against the possibility of compensation. At low probabilities of disease, screening engenders substantial risk of false-positive results and their consequences. Screening for more than one disease amplifies that problem. Screening for one disease using multiple tests is also problematic.

Third, particularly for compensational purposes, medical screening tests are appropriate only after the individuals are fully informed of the risks that these tests pose (see Chapter 8).

Fourth, screening may provide useful information about related diseases. One example is screening for diabetes among people who have high blood pressure or high blood lipid concentrations but not symptoms or signs of diabetes (Harris et al., 2003); positive results of diabetes screening (especially impaired glucose tolerance or impaired fasting glucose) may call for changes in the management of the other conditions (perhaps more than for the possible diabetes itself).

Fifth, if publicly supported screening programs are to be effective, ethical, and equitable, some means must be available for screened populations to gain access to appropriate followup, diagnosis, and therapy.

Sixth, once a person is screened, he or she should, in general, be informed of any clinical information that the test revealed. Screening, even for compensational purposes, may provide information that should be linked to actions that might improve the medical situation for the person screened. Once screening for compensational purposes has been done, the patient has already been exposed to the risks of the test and of a possibly false-positive result. Having undertaken this risk, the screened individual should have the opportunity to receive whatever medical benefit such information might provide.

Stated another way, in general once a person is screened, he or she should be informed of any clinical information that the test revealed. In most medical circumstances, withholding incidentally discovered information would be unethical (see Chapter 8). This principle, however, must also reflect some sensitivity to the screened individual's culture, background, and preferences for receiving information (especially bad information). Individuals of different cultural backgrounds evince different preferences for information (Blackhall et al., 1995). For example, traditional Navajo culture takes a worldview that language shapes reality and that revealing negative information (truth telling) may "cause" those bad outcomes to occur. Moreover, an important element of traditional Navajo culture is to think and speak in a positive way and to avoid thinking or speaking in a negative way (Carrese and Thodes, 1995; Gostin, 1995).

Seventh, screening programs must use reliable and valid tests and procedures. Of particular importance are sensitivity, specificity, and positive and negative predictive values. Increasingly, policy makers, clinicians, and others concerned with screening issues are called on to deal with odds ratios, likelihood ratios, and receiver operating characteristic (ROC) curves.

Eighth, in specific populations, judgments about the feasibility of screening must consider access to health care and health insurance.

To summarize, for general screening to be useful on a population basis, certain circumstances must exist (Frame and Carlson, 1975; Eddy, 2004):

- The condition must be present in the population and have an important effect on the quality and length of life.
- The incidence of the condition must be sufficient to justify the risks of the screening.
- Acceptable methods of treatment or prevention in the preclinical or early stage of disease must be available.
- The condition must have an asymptomatic period during which detection and treatment would substantially improve clinical outcomes, such as reducing morbidity or mortality.
- A screening protocol or test that is sufficiently accurate and acceptable to patients and that has reasonable costs must be available to detect the condition in the asymptomatic period.
- The benefits of screening must exceed its harms.

- The benefits of a screening program must justify its costs (including its induced costs) and its use of resources.
- The health care system must be able to provide referral services that address followup (further evaluation and continuing care) of patients identified by screening.
- Patients who choose to participate must be fully informed of the risks posed by screening and the benefits of screening and freely consent to be screened.

Applications to RESEP

The committee believes that in the RECA-RESEP legislation screening aims principally not at diagnosing and treating patients but rather at facilitating compensation for the harms caused by uranium mining and related activities and by the aboveground nuclear tests of the 1950s (discussed in Chapters 2 and 7). Although the legislation speaks to screening for the purpose of improving medical outcomes, early diagnosis of the majority of diseases identified for RESEP screening (such as lung cancer, pancreatic cancer, and ovarian cancer) improves health outcomes little or not at all. Furthermore, the current RESEP screening protocols include tests, such as pulse oximetry and spirometry, that have not been shown to improve health outcomes in asymptomatic populations.

Most tests used for screening are also those used to establish a medical diagnosis. When a patient's signs or symptoms are the reasons, or part of the reason, for using a given test, the procedure is more properly called medical diagnosis than screening. The hallmark of a screening program is identification of preclinical (that is, undiagnosed) disease. Patients or populations that are screened do not have clinical evidence of disease or, if they do, then neither the patients nor their health care providers have yet appreciated that evidence. The target for which the screening is undertaken can be a propensity or risk factor (presumably modifiable) for developing a disease, a disease itself, or a complication of a disease that the patient is known to have. For convenience and readability, we use the term *disease* to refer to all three entities.

In many screening scenarios, the likelihood of the disease that the screening program seeks to identify is relatively low (low pretest probabilities). Moreover, most tests and screening protocols are imperfect in that their specificity is less than 100% (meaning that their ability to rule out disease is not infallible). Because of those two factors, screening can produce substantial numbers of false-positive results; depending on the aftermath of screening when false-positive results occur, the potential benefits of screening may be decreased (or may even not outweigh the risks of error). Because screening tests commonly have sensitivity below 100%, false-negative results will also occur, but these errors may be far less problematic in the screening setting (see discussion later in this chapter). Nonetheless, a negative screening test does not guarantee that the patient does not have the disease, risk factor, or condition being sought.

As discussed later in this chapter, a tradeoff often exists between false-positive and false-negative results. Typically, for a given test, altering the test or its interpretation to decrease one of these errors will increase the other. Thus, establishing the criterion for or definition of a positive test result requires well-reasoned balancing of the consequences of each kind of potential error and the frequency of the disease in the population to be screened.

CONTEMPORARY SCREENING PROTOCOLS

General Issues

Many groups have explored traditional medical screening, especially screening for various forms of cancer. Among the authoritative sources are the US Preventive Services Task Force (USPSTF) (Harris et al., 2001; http://www. ahrq.gov/clinic/uspstfix.htm) and (until mid-2004) the Canadian Task Force for Preventive Health Care. Appendix E presents current information on cancer screening tests of interest to this committee; some are for RECA-designated diagnoses and others for nonradiogenic cancers; as a more complete context for considering medical screening in primary care, we also present there a fuller account of the types of screening tests the USPSTF recommends or strongly recommends for general populations other than RECA populations. As seen in Appendix E, relatively few screening tests for cancer have been recommended for widespread use. The limited number or nature of such recommendations turns on several issues.

In many instances, no effective treatment for preclinical disease exists or a treatment appears more effective than it truly is. A common issue that arises here is *lead time bias*, which refers to an apparent improvement in survival if survival is measured from the detection of disease and not from the onset of disease. Some authors argue that such early identification can be beneficial even if overall survival is unchanged, because a patient and his or her family might be able to plan better for the patient's limited lifetime. A common counterargument is that such early labeling of a patient as having cancer (or another serious ailment) may diminish quality of life during the lead time. The benefit or the loss is surely very patient-specific. Thus, screening should be predicated on extensive counseling and informed consent for both practical and ethical reasons.

The effectiveness of a treatment can be overestimated because of *length bias*, which refers to the fact that disease that develops more slowly (such as a slow-growing or relatively benign cancer) may be more likely to be detected by a screening test, whereas a rapidly progressive disease is more likely to present clinically. These patterns mean that a substantial proportion of disease that is found on screening (as contrasted with disease that presents clinically with signs or symptoms) will be, on average, less aggressive. Sometimes, screening might detect disease that might never become clinically evident. This problem is well recognized when screening for cancers of the prostate, lung, thyroid, and brain; in fact, the term "incidentaloma" is sometimes used to describe the detection of such disease. The patients detected by screening have an excellent prognosis, of course. Such patients inflate survival rates for patients with disease identified by screening but overall unnecessarily consume society's resources and generate unreasonable anxiety and potential complications for the patient without substantial benefit.

Another common reason that cancer-screening programs receive unfavorable recommendations rests on the occurrence of false-positive results (or, said another way, on the imperfect specificity of almost all medical tests). Because clinicians must inform patients about each positive result and because they often recommend additional diagnostic studies, such false-positive results carry substantial risks. They include the risk of physical harm during the additional studies (the workup, which not infrequently can involve invasive diagnostic testing). Other issues include psychologic harm (from anxiety and from labeling, insofar as patients consider themselves as "damaged goods" and societal harm and discrimination (in employability and insurability). The same psychologic and societal harms may affect family members as well. In addition, identifying disease that benefits little from treatment (lung or pancreatic cancer, for example) can lead to futile surgery, which can engender major costs, pain, disability (from reduced lung volume in the case of pulmonary neoplasms), and surgical mortality.

Although most screening tests are themselves of low risk and noninvasive, some screening tests expose individuals to ionizing radiation. Such exposure itself can be oncogenic (cancer-causing). Recently, investigators have calculated that the oncogenic risks posed by some radiographic studies outweigh potential benefits from screening (Brenner and Elliston, 2004).

The third common reason for unfavorable recommendations about medical screening has rested (openly or covertly) on economic arguments. Our world of limited resources is increasingly constrained with respect to health care and faces many competing demands. Thus, committees that produce clinical practice guidelines, insurers that fund medical care, and even providers that treat individual patients may consider the relatively high cost of some proposed screening programs in light of other uses to which the resources might be applied (the opportunity costs of health care services forgone). Whether these issues are dealt with formally or informally is a matter for the groups in question, but all decisionmakers need to consider how best to deal with such tradeoffs in resource-constrained circumstances.

The economic cost incurred as a result of false-positive tests is important. One recent article documents that a large proportion of persons screened for cancer (43%, higher among men than among women) had at least one false-positive result, and more than four of five of these patients had followup care (Lafata et al., 2004). Medical costs in the year after the screening were statistically significantly higher among those with false-positive tests than among those with negative tests (on average, about $1,000 higher).

The economic costs of a screening program thus extend well beyond the screening test itself. They include the required infrastructure for providing the test and the followup for abnormal results, the counseling that must be undertaken before the screening test is performed and when informing patients about abnormal results, and the choices they then must make. (Note that we use the term *patients* here for both patient screening and for case-finding because, even in the latter context, once clinicians inform individuals about an abnormal result, they become patients and require a provider-patient relationship.) Economic costs of the screening program must also include the costs of the therapeutic or preventive intervention to be undertaken in patients whose positive results are confirmed. In settings in which the benefit of treating a patient with a disease that has been detected early is small, therapeutic costs may not be economically reasonable for society to undertake.

Prevalence and Benefit

In assessing the therapeutic benefit of a screening program, one must consider the perspective of the population that is screened, not the subpopulation that is identified as having disease and then treated because of the program. In the population to be screened, disease prevalence can markedly limit benefit.

Consider a screening program for a single cancer for which early detection provides a huge therapeutic benefit of 10 years of added life expectancy, as illustrated in Table 9.3. Assume that a screening test has a high sensitivity of 90% and that the prevalence of disease in the population to be screened is 2 cases per 1,000 individuals. Further assume (theoretically) that the screening test has perfect specificity (100%—a positive result is pathognomonic of disease). In a population of 10,000 patients screened, 20 will have the cancer and 18 will be identified and treated early. These 18 patients will gain 180 years overall (10 years of life per patient detected). The 20 patients who have cancer will, on average, gain only 9 years of life (180/20). The 10,000 patients who

TABLE 9.3 Effect of Perspective on Benefit of Good Treatment

Group	Years of Life Saved	Number of Individuals	Gain per Individual
Patients identified	180	18	10 years
Individuals with disease	180	20	9 years
Screened individuals	180	10,000	0.018 years = 6.6 days

are screened will gain, on average, only 0.018 years, or 6.6 days of life, per patient screened.

All persons screened are subjected to the risks posed by testing and costs of this testing. Thus, decisions about implementing screening programs must include a comparison of the (perhaps large) benefits to the few persons with disease to the much smaller benefits that would accrue, on average, to all persons screened and to the far more numerous risks and costs that would accrue to patients free of disease. Of course, the details would depend on the magnitude of those benefits, risks, and costs.

Issues Related to Cancer Screening

For all these reasons, relatively few cancer screening programs have been endorsed in toto; of those endorsed, many are typically constrained to particular age groups or categories defined by other risk factors or variables. Annual screening for cervical cancer was long a standard of care (see Appendix E), but annual testing in low-risk women who have had several prior negative tests is no longer thought to be necessary in many settings. Mammography has been shown to detect breast cancer early enough for outcomes to be improved, but the frequency of such screening and the age at which it should begin (or end) have engendered much discussion. Colonoscopy has been shown to be effective in finding colon cancer early enough to improve prognosis; it may even be a preventive measure if premalignant lesions (adenomas) are removed before they evolve into cancer. However, some debate remains about whether colonoscopy will remain the first choice for colorectal cancer screening—especially if costs are taken into account (Pignone et al., 2002).

A particularly thorny example that has been carefully examined in the literature involves routinely screening men for prostate cancer with the serum prostate-specific antigen (PSA) test (Harris and Lohr, 2002). This protocol is controversial for several reasons. First, more prostate cancers are slow-growing than fast-growing (aggressive), and many more men die *with* prostate cancer than *of* prostate cancer. Second, the PSA test has a substantial false-positive rate; most men who get a positive (abnormal) PSA test will get further workup (such as biopsy), but because biopsies may not reliably distinguish between slow- and fast-growing cancers, most men will receive treatment. Finally, standard treatments (such as surgery and radiation) have substantial morbidity (such as impotence and incontinence). Recent studies have raised the issue of false-negative results, the occurrence of histologically proven cancer in men with low serum PSA (Thompson et al., 2004). For these reasons, some groups have recommended PSA screening only after extensive discussion with each patient. This view reflects the ethical principles introduced in Chapter 8 and the tenets of shared decision-making taken up later in this chapter. Similar issues arise in medical screening for most RECA-compensable diseases (although the negative consequences on patients' quality of life have not been as carefully studied) because most patients with those diseases have not been shown to derive substantial health benefit from screening.

Routine screening of asymptomatic individuals for lung cancer, pancreatic cancer, ovarian cancer, gastric cancer, and many other conditions in the RECA has not been recommended (see Appendix E). More directly related to the RESEP program is an IOM committee's recommendation against routine screening for thyroid cancer in populations downwind of the Nevada Test Site (NTS) (IOM-NRC, 1999).

EPIDEMIOLOGIC, STATISTICAL, AND CLINICAL ISSUES OF SCREENING TESTS

The complexities of interpreting the risks posed by and the results of screening tests (whether in general or for RESEP) are not widely understood. To address the gap, we present a brief primer to assist policy-makers, clinicians, and potential screenees in evaluating the desirability and feasibility of the types of medical screening or case-finding implemented by or contemplated for RESEP. This material, only hinted at in the committee's interim report, forms a critical basis of our later recommendations about screening. We hope that this material will be useful to anyone considering obtaining any screening test.

Probabilities, Odds, Prevalence, and Incidence

To interpret the meaning of any test result, one must know what the likelihood of disease was in an individual before the result if that test was known. One can estimate a pretest likelihood in many ways: from a clinical study, from a database, from a prediction rule or a logistic regression equation, or on the basis of clinical judgment. When it is taken from an epidemiologic study of a population of presumably similar individuals, it is often called prevalence. It represents an estimate in a population at a given time (see Chapter 3).

If the individual to be screened was known to be free of the disease at some time in the past (that is, if the individual has had a negative screening test N years before), then the relevant pretest likelihood is the incidence of (newly diagnosable but undiagnosed) disease in those N years. If disease incidence in a population of similar individuals were Y cases per year, then one would expect the pretest probability of disease to be its interval incidence or $1 - e^{-Y \times N}$, if we ignore those individuals who present clinically or who have died (of any cause) in the interim. If both Y and N are small, then the interval incidence can be approximated by $Y \times N$.

As explained in Chapter 4, probabilities of disease can also be expressed in terms of odds (omega, Ω), which are more convenient for some calculations. Odds are the number of persons with an event or feature divided by the number without the event or feature—that is, the ratio of positive to negative outcomes that are either observed or predicted. Probability (a proportion) and odds (a ratio) can be easily converted to each other through simple formulas: if p is a probability and if Ω is the corresponding odds, then $\Omega = \dfrac{p}{1-p}$ or $p = \dfrac{\Omega}{\Omega+1}$.

What Is a Positive Screening Test?

Throughout this chapter and this report, we use the terms *positive* and *abnormal* to describe a test result. We consider here how those terms might be defined.

Many laboratory tests produce quantitative results, such as a hematocrit of 33%, a PSA of 5 nanograms per milliliter (ng/ml), or an oxygen saturation of 80%. Such results are often categorized as either normal or abnormal on the basis of a defined normal range or some criterion of positivity. Even imaging studies can have an associated threshold criterion of detection; for example, one might choose to report as abnormal a thyroid ultrasonographic examination or a helical chest computed-tomography scan with a nodule greater than 3 mm in diameter but choose not to report smaller lesions, perhaps because they are unlikely to be meaningful.

Some laboratories define the normal range for each test they perform according to results in a population free of known disease. Many such test results are normally distributed; the so-called normal range is the mean plus or minus two standard deviations. Such definitions of normality, however, ignore the distribution of test results in patients *with* a given disease. In most circumstances, the distributions of results in patients with disease and in healthy patients overlap. In these cases, any criterion of positivity means that some patients free of disease will have an abnormal result (a false positive); conversely, some patients with disease will have a normal result (a false negative).

Moreover, many tests provide information on more than a single disease. That means that a single rational, mathematically proper criterion of optimization (a tradeoff of false-positives and false-negatives results) may not be possible to design. In such circumstances, laboratories are most likely to report a normal range.

In Figure 9.1 and later figures, we display a probability density distribution (vertical axis) of test results (horizontal axis) for different populations. In Figure 9.1, the populations represent patients with no disease, patients with disease 1, and patients with disease 2. We display patients with no disease along a negative vertical axis to provide visual separation from the populations with disease.[1]

In all three distributions, the greater the distance from the horizontal (zero) axis, the higher the probability. The two vertical lines delineate the normal range,

---

[1]For illustrative purposes, we drew each distribution as normal or Gaussian.

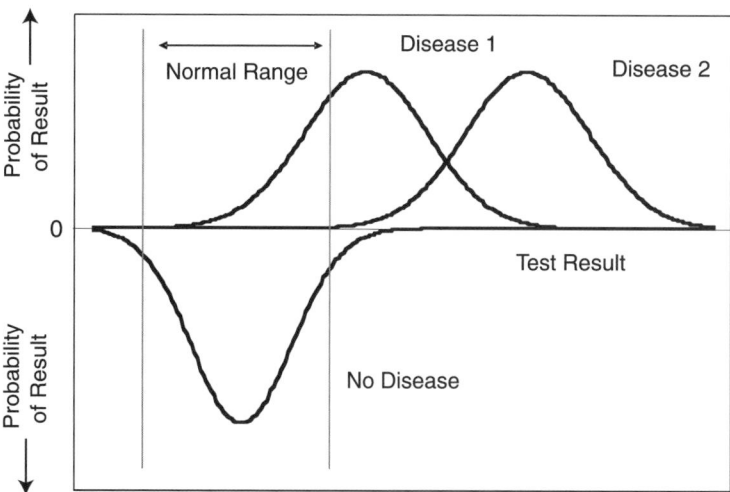

FIGURE 9.1   Probability distributions of laboratory normal range and two diseases.

here set as the mean plus or minus two standard deviations in the normal population, encompassing 95% of patients without disease. The number of individuals with disease 1 and disease 2 who have an abnormal test result (which is a result greater than or to the right of the upper cutoff) will depend on the overlap of the distributions of results in disease 1 and disease 2 and the normal population. Such laboratory-based definitions of normality may create different degrees of overlap for different diseases.

When one reports quantitative test results of this type as either positive or negative, one is discarding what could be useful information. For example, a PSA of 5 ng/ml carries different information about the likelihood of cancer of the prostate (apart from the issue of slow- or fast-growing disease) than does a PSA of 8 ng/ml. However, having a simple criterion of positivity allows clinicians to interpret the meaning of the result more readily in the context of a particular patient who may have a particular pretest likelihood of disease, which often will be based on clinical presentation, history, demographics, and even physical examination or other known test results.

With a criterion of positivity, the performance of a test can be described on the basis of two characteristics: sensitivity and specificity (Figure 9.2). Sensitivity describes a test's performance in patients known to have disease; it is the likelihood of a positive result given disease. It is also known as the true-positive rate; the false-negative rate is its complement. Specificity describes a test's performance in patients known not to have disease; it is the likelihood of a negative result given the absence of disease. It is also known as the true-negative rate; the false-positive rate is its complement.

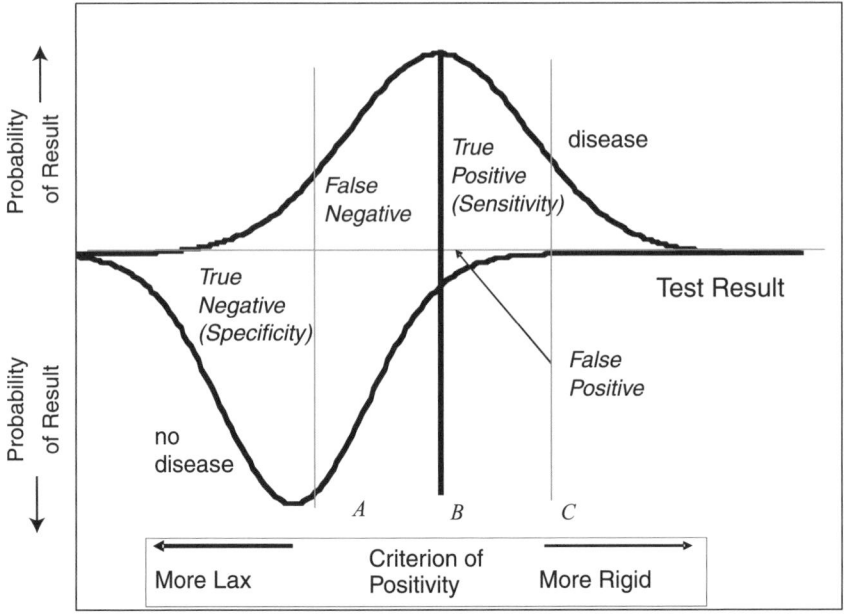

FIGURE 9.2   Probability distributions of diseases and test results.

Figure 9.2 displays test results on the horizontal axis, and the probability of a given test result on the vertical axis for patients with disease (going up) and patients with no disease (going down). The three vertical lines display three different potential cutoffs or criteria of positivity, A, B, and C. The heavy line B is a middle ground that represents a compromise between sensitivity and specificity.

The line to the left (A) represents a more lenient—"lax"—definition of positivity; this definition increases sensitivity (including a greater proportion of patients with disease to the right of the criterion) while decreasing specificity (including a greater proportion of patients without disease to the right of the criterion). The line to the right (C) represents a more strict—"rigid"—definition of positivity; this definition decreases sensitivity (including a smaller proportion of patients with disease to the right of the criterion) while increasing specificity (including a smaller proportion of patients without disease to the right of the criterion). Those performance characteristics can be transformed into a variety of interpretation aids, of which the receiver operator characteristic (ROC) curve is particularly useful.

The Receiver Operator Characteristic Curve for a Test

A critical question in interpreting screening tests is how and where to establish a criterion of positivity—that is, the cutoff (cutpoint) for that particular test as being either positive (abnormal) or negative (normal). That is especially relevant in a screening situation in which the test is applied to many similar patients and the purpose of the test is to categorize patients into two subsets: those with positive results, who will be referred for evaluation or treatment (or both), and those with negative results, who will be reassured.

The optimal cutpoint depends critically on the extent to which the distributions of results in healthy individuals and in those with disease overlap. It also depends on several other factors, including the prevalence of disease in the population to be screened, the importance of early identification of individuals with disease (so that they might receive the benefit of early treatment or, in the case or RECA, compensation), and on the burden of falsely classifying healthy individuals as persons requiring further evaluation or treatment. Viewed from a different perspective, the benefit of early treatment is just an alternative way of expressing the burden of having treatment delayed.

These comparisons typically presume that a test is to be performed. The determination of an optimal cutpoint is applied to interpreting a test. If we broaden the decision problem to include whether to perform the test at all (Pauker and Kassirer, 1980), the situation becomes more complex. Although arguably the optimization decision should include that perspective, clinicians usually do not take that perspective when establishing a cutoff. More typically, clinicians select an optimal cutoff, thereby establishing the test's operating characteristics, and only then do patients and clinicians choose whether to perform the test.

Typically, distributions of screening tests overlap. Often, the relationship between the false-positive rate and the true-positive rate is monotonic: increasing one increases the other. That relationship defines the ROC curve (Figure 9.3).

In Figure 9.3, the horizontal axis displays the false-positive rate for a given criterion of positivity; the vertical axis displays the true-positive rate corresponding to the same criterion of positivity. A perfect test (i.e., one for which the false-positive and false-negative rates are both zero) would have an ROC curve that rises along the vertical axis from the origin until the point corresponding to a true-positive rate of 1.0 and a false-positive rate of zero (upper left corner of the graph) and then passes to the right along the upper border of the graph to the upper right hand corner. The degree to which an ROC curve departs from this ideal is a measure of the general reliability of the test. This can be represented as the area between the test's ROC curve and a 45 degree diagonal line corresponding to a test whose true-positive rate equals its false-positive rate for all possible cutoff criteria. Such a test would carry no information.

The three hypothetical criteria shown in Figure 9.2 correspond to the three points (A, B, and C) on the ROC curve in Figure 9.3. Criteria that are more lax (criteria toward the left in Figure 9.2) fall toward the upper right end of the ROC curve; criteria that are more strict (criteria toward the right in Figure 9.2) fall toward the lower left end of the ROC curve.

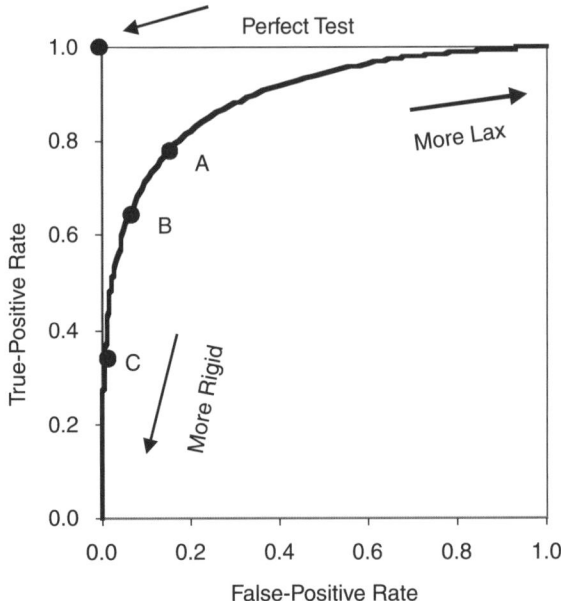

FIGURE 9.3   Receiver operating characteristic (ROC) curve.

For example, consider criterion C, which is rather strict. The area under the no-disease curve to the right of that line (the false-positive rate) in Figure 9.2 is roughly 1% and is plotted on the horizontal axis in Figure 9.3 as corresponding to point C. The area under the disease curve to the right of that line (the true-positive rate) in Figure 9.2 is roughly 35% and corresponds to the vertical axis value of point C in Figure 9.3.

The mathematics of optimization are complex (and discussed further in Appendix D). One can visually identify the optimal operating point (corresponding to the criterion of positivity or cut point that minimizes the combined burden of false-negative and false-positive results in a given population) as the point where the *slope* of the ROC curve (or its tangent) is numerically equivalent to

$$\frac{(1 - Prevalence) \times (BurdenFalsePositive)}{(Prevalence) \times (BurdenFalseNegative)}$$

The burden of a false-positive result is defined as the degree to which an individual without disease but with a falsely positive test result is less well off (on some outcome scale) than a similar individual without disease but with a truly negative test result. Similarly, the burden of a false negative is defined as the degree to which an individual with disease but with a falsely negative test result is less well off (on some outcome scale) than a similar individual with disease but with a truly positive test result.

In another formulation of the criterion for the optimal cutoff, the *reciprocal of the slope* of the ROC curve (or its tangent) should be numerically equivalent to the burden ratio. In this case, the reciprocal of the slope equals $\Omega_{\text{disease}} \times$ BurdenRatio, where Ω_{disease} is the odds of disease and BurdenRatio is Burden$_{\text{FalseNegative}}$/Burden$_{\text{FalsePositive}}$. Some authors call the burden of a false-positive result the "cost of a false positive" (C) and the burden of a false-negative result the "benefit of a true positive" (B). This approach yields an optimal cutoff at the point on the ROC curve where the reciprocal of the slope equals $\Omega_{\text{disease}} \times \text{B/C}$, the benefit-cost ratio. The reason is that a false negative—a test that should have been positive—causes the loss of the benefit of a true positive (the ability to treat the disease early or in the case of compensational screening in RECA populations, the ability to successfully seek compensation).

At high benefits of early detection, high prevalence of disease, or low burdens of false-positive results, the ratio will be small and the best cutoff point will lie along the flat area to the right of the ROC curve, where both the true-positive and the false-positive rates are high. This point corresponds, in other words, to the spot at which the criterion of positivity is lax or toward the left in Figure 9.2. Conversely, at low benefits of early detection, low prevalence of disease, or high burdens of false-positive results, the ratio will be large, and the best cutoff point will lie along the steep area to the left of the ROC curve, where both the true-positive and the false-positive rates are low; this point corresponds, in other words, to the spot at which the criterion of positivity is rigid or toward the right in Figure 9.2.

Thus, as one considers medical screening tests for the RESEP populations, one should consider the prevalence of disease in each case. Doing so implies a program that is administratively far more complex than RESEP (or HRSA grantees) now follows. Nonetheless, for the specialized purposes of RESEP and the populations it is supposed to serve, establishing different criteria for test positivity in different populations or even in different geographic areas—for which prevalence rates of compensable disease might vary—might be scientifically more appropriate and ethically more defensible.

For example, in a geographic area with little fallout, an unusually high proportion of screenees would turn out to be false positives; in fact, if no case of compensable disease were detected, all the positives would be false positives. The costs of those false positives might pose a major challenge that might be mitigated by choosing a more rigorous criterion of positivity. Although in principle an analogous argument might be made about false negatives, the only way that a high proportion (or all) of the negatives could be false negative is for the prevalence of the RECA-compensable disease for which screening is undertaken to be very high, certainly higher than 50%. That circumstance would be quite unlikely.

Gold Standards

When we consider the performance of a diagnostic test in establishing the presence or absence of a disease, the underlying assumption is that, separate from the test in question, a "gold standard" (sometimes called a reference standard) or definition exists that can classify the population to be tested into individuals with disease and individuals without disease. Sometimes this simple concept is not unambiguously met, because no clear external diagnostic gold standard exists (at least for many patients). Rather, the gold standard for some tests can be the clinical diagnosis, but that diagnosis itself sometimes depends on the results of the test being evaluated, and this leads to a kind of circular reasoning.

When the gold standard for diagnosis changes, the assessments of the performance of a diagnostic test may also change. The effects differ depending on whether the gold standard for diagnosis is made more strict or more lax.

If a newer or revised gold standard is stricter than a former one, then patients whom the newer standard classifies as having disease will tend to have more severe disease than would previously have been the case. Moreover, a diagnostic test being measured against the new gold standard is more likely to be positive in this smaller set of diseased patients. Thus, the test's sensitivity will increase. For example, if the stricter gold standard excludes very small cancers, such as cancer in situ or very localized disease, then a diagnostic test that depends on tumor mass will be more likely to be positive in this small subset of patients.

In addition, a stricter gold standard will tend to classify patients who have mild or minimal disease as not having disease. Hence, a diagnostic test being measured against the new gold standard is more likely to be (falsely) positive in the larger set of patients without disease, because the enlarged set of patients said to be free of the disease will actually include more patients who have mild or minimal disease. Thus, the test's specificity will decrease. Finally, making the gold standard stricter also decreases the measured (or estimated) prevalence of disease.

The opposite effects will occur if the revised gold standard that defines the presence of disease is made more lax. The sensitivity of a test measured against this revised gold standard will decrease because the group of patients classified as having disease will include more mild or minimal cases. Necessarily, the specificity of a test measured against this revised gold standard will increase because the group of patients classified as not having disease will include fewer mild or minimal cases. Making the gold standard more lax also increases the prevalence of disease.

These are complex and important relationships. Early in this chapter, we used the terms *strict* and *lax* to describe criteria for defining a positive test. In this section, we use the same terms to describe the criteria for defining the presence or absence of disease. Many gold standards are not fixed (that is, they also can vary); this fact complicates the analysis of relative test performance.

Consider the example shown in Figure 9.4A-D. Figure 9.4A shows the likelihood of a given test result in three populations of patients: those with severe disease, those with mild disease, and those with no disease. In Figure 9.4B, these three populations have been merged into two groups—patients with disease and patients with no disease—by defining disease to be only severe disease. The no-disease category now includes some patients with mild disease, and patients in this category are more likely to have a positive test result than are patients in the

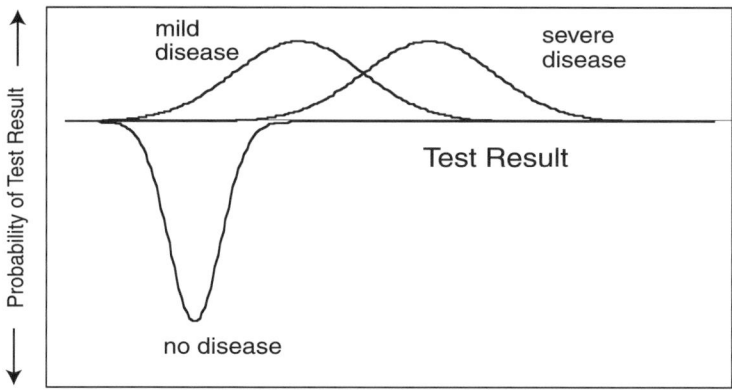

FIGURE 9.4A   Probability of test result in three populations defined by presence and severity of disease.

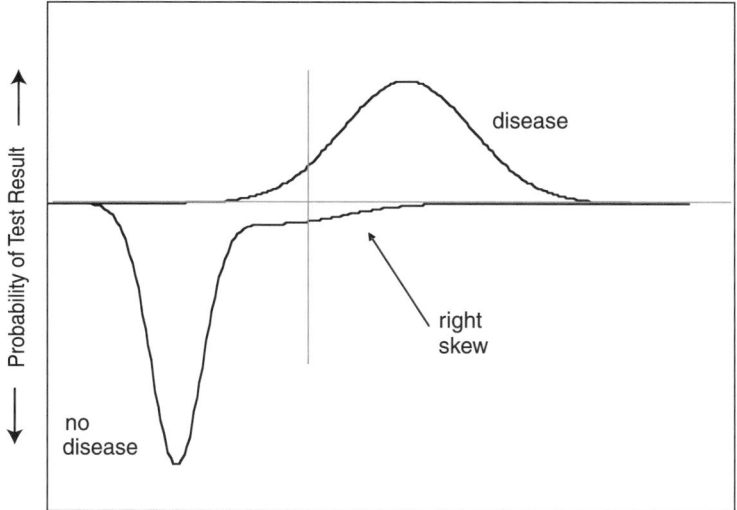

FIGURE 9.4B   Probability of disease in two populations: mild disease is classified as no disease.

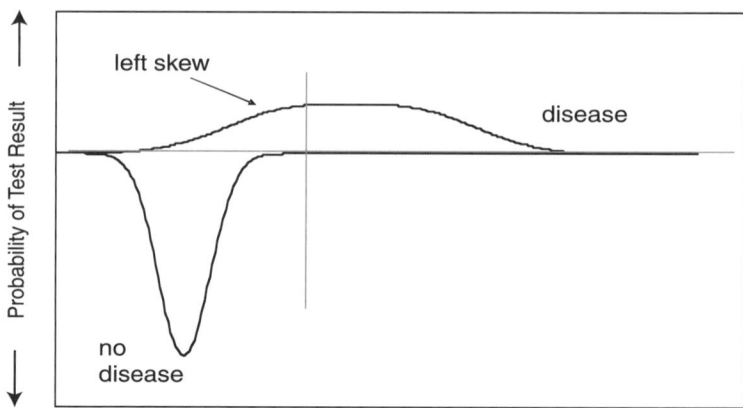

FIGURE 9.4C   Probability of disease in two populations: mild disease is classified as disease.

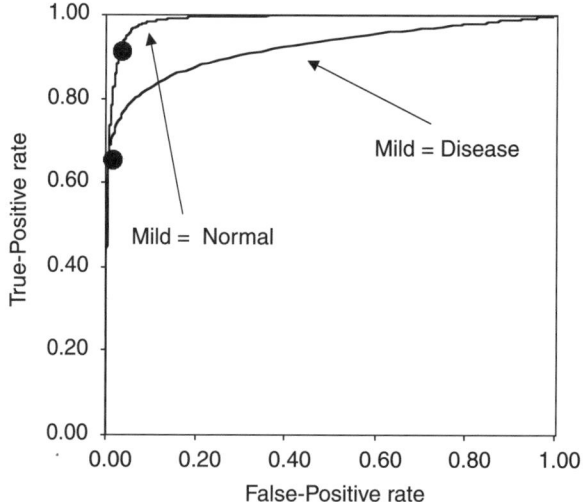

FIGURE 9.4D   Two ROC curves: mild disease classified as disease, and mild disease classified as no disease.

original no-disease population. With this classification, test specificity will decrease (for a given criterion of positivity) compared to just looking at the test in patients who have no disease at all. In contrast, test sensitivity will increase compared to a mix of patients with mild and severe disease because all patients now classified as having disease will have severe disease and therefore be more likely to have a higher test result.

The opposite circumstance arises in Figure 9.4C, in which patients with disease include patients with both mild and severe disease. In this case, test specificity will be higher because the population of patients with no disease no longer includes patients with mild disease. Similarly, with this new classification, patients with "disease" now include patients with less severe disease, and patients in this category are less likely to have a test result above the same criterion of positivity. For that reason, test sensitivity will be lower. These effects are summarized in the corresponding ROC curves (Figure 9.4D). The dot on each curve corresponds to the same criterion of positivity.

If the test under consideration is itself the gold standard for diagnosis, then it is by definition perfect. That is, a positive result establishes the diagnosis and a negative result excludes it, both with certainty. The specificity and the sensitivity of the test are both 100%. (These points are taken up further in Chapter 10 in the discussion of the RECA gold standard.)

The Meaning of a Positive Test

Sensitivity, specificity, and ROC curves apply to two populations: individuals without disease and individuals with disease. In contrast, the task that the clinician interpreting a screening test faces deals with an individual who is not known to be a member of either population. Rather, that individual is known to have some chance (the pretest probability) of being in one population or the other.

Faced with a positive test result, the clinician and the patient must consider what it means. The effect of a positive test is to increase the probability of disease from the pretest likelihood (often the population-based likelihood or prevalence) to a revised, posttest probability. Calculating a revised probability of disease depends on three factors: the pretest probability or prevalence of disease, the test's sensitivity, and the test's specificity.

Consider, as shown in Figure 9.5, 1,000 patients screened, of whom 1% (10) have disease. If test sensitivity is 80%, 8 of the 10 will have a positive test result. Of the 990 patients without disease, if the test specificity is 70%, 30% or 297 patients will also have a positive test result. Thus, 305 patients will have positive test results; of these, 8 or 2.6% will actually have disease (the revised probability of disease). Thus, a positive test result increases (in this case, more than doubles) the probability that the patient has the disease from 1% (the state before testing) to 2.6%; a negative result decreases it (in this case, reduces by two-thirds) from the original 1% to 0.3%.

Clinicians, patients, and policymakers can legitimately ask: Is 2.6% high enough to justify the risks and costs of further workup? Is 0.3% low enough to forego further workup and reassure the patient that no disease has been detected? Answers to these critical questions depend on a host of other, nonstatistical matters, such as the benefits, risks, and costs of treatment, and the risks and costs of further diagnostic evaluation (Pauker and Kassirer, 1980).

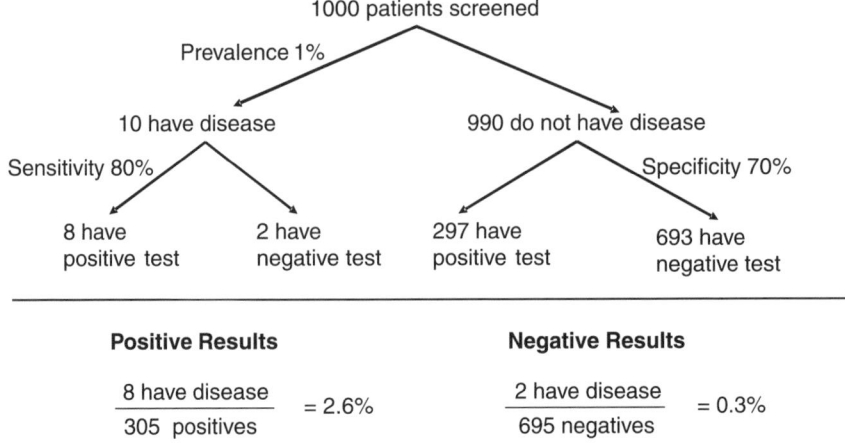

FIGURE 9.5   The meaning of a positive test.

These three factors are combined in a process known as Bayesian analysis. In the simple case of a test that is either positive or negative and a disease that is either present or absent, the revised or posterior (posttest) probability (also known as the positive predictive value) is calculated as

$$\frac{Prevalence \times Sensitivity}{(Prevalence \times Sensitivity) + ([1 - Prevalence] \times [1 - Specificity])}$$

For example, as in Figure 9.5, if the prevalence were 1%, the sensitivity 80%, and the specificity 70%, then the revised or posterior probability of disease after a positive test would increase to

$$\frac{1\% \times 80\%}{(1\% \times 80\%) + (99\% \times 30\%)}, \text{ or } 2.6\%.$$

In the screening situation (when the prior or pretest probability of disease is low), the major drivers of the posttest probability are the test's specificity and the disease prevalence. If one assumes that the test's sensitivity is reasonably high, one can approximate the posttest probability as

$$\frac{Prevalence}{Prevalence + False\ Positive\ Rate}, \text{ in this case } \frac{1\%}{1\% + 30\%} \text{ or } 3.2\%.$$

This simple approximation allows the clinician or the patient who was or will be screened to judge quickly the meaning of a positive test result.

The above reasoning applies to single screening tests that are looking for a single disease. The issue is far thornier when one considers the simultaneous application of several screening tests that are looking for several diseases.

The Conundrum of the False-Positive Cascade

RESEP proposes screening for 20 malignant conditions in downwinders and onsite participants, seven malignant and nonmalignant conditions in uranium millers and ore transporters, and six other conditions in miners. The specificity of each of the proposed screening tests for these separate conditions is likely to be well below 95%. For instance, the specificity of CA-125, a tumor marker that is often present in the peripheral blood of patients with ovarian cancer is limited; CA-125 is high in 88% of patients who have endometriosis, a far more common entity, as well as in right heart failure, cirrhosis and tuberculous ascites. Similarly, the fecal occult blood test (a screening test for colon cancer) and breast self-examination, clinical breast examination by a clinician, and mammography (screening tests for breast cancer) have low specificity.

Since Galen and Gambino (1975) described the effect of panels of clinical chemical tests, physicians have understood that obtaining a battery of tests is highly likely to produce false-positive results—"red herrings" that demand further study that can lead to various good and bad consequences. The general public may not understand that principle as clearly, however. Consider a typical laboratory test for which the normal range is defined in a healthy population as the test's mean plus or minus two standard deviations—meant to encompass 95% of the healthy population. That definition corresponds to a test specificity of 95% and a false-positive rate of 5%; this point is shown in Table 9.4, first row, columns 2 and 3 (in bold face).

Complicating this point is the fact that some laboratory tests do not define a normal range (an upper *and* a lower limit of normal). Rather, they specify either an upper *or* a lower limit. In that case, the likelihood of a false positive may not be 5%. Nevertheless, if the chance of a false positive in each test of a panel is not zero, a cascade of false positives will still develop as the number of tests in the panel increases.

Now consider a pair of two such independent tests applied to a healthy individual (Table 9.4, second row). Using our main example, the likelihood that both tests will be normal (what might be called joint specificity) is 95% times 95%, or 90.25%; the joint false-positive rate is 9.75% (100% - 90.25%). If three tests were performed on a healthy individual, the joint specificity would be 85.7% (95% times 95% times 95%) and the joint false-positive rate 14.3% (100% - 85.7%). For 12 such independent tests, the joint specificity would be 54% and the likelihood of *at least one* (falsely) positive result 46% (100% - 54%). Of course, if the tests are not independent, then this simple multiplication must be adjusted for the degree of positive or negative correlation among the tests.

TABLE 9.4 The False-Positive Cascade: Effect of Number of Tests and Specificity on Probability of Positive Test Results

	Level of Specificity[a] Equals											
	95%			90%			85%			80%		
	Probability That None, One, or Two Results Are Positive in a Healthy Population											
Number of Tests	N = 0	1	2	0	1	2	0	1	2	0	1	2
1	**0.9500**	0.0500		0.9000	0.1000		0.8500	0.1500		0.8000	0.2000	
2	**0.9025**	0.0950	0.0025	0.8100	0.1800	0.0100	0.7225	0.2550	0.0225	0.6400	0.3200	0.0400
3	**0.8574**	0.1354	0.0071	0.7290	0.2430	0.0270	0.6141	0.3251	0.0574	0.5120	0.3840	0.0960
4	0.8145	0.1715	0.0135	0.6561	0.2916	0.0486	0.5220	0.3685	0.0975	0.4096	0.4096	0.1536
5	0.7738	0.2036	0.0214	0.5905	0.3281	0.0729	0.4437	0.3915	0.1382	0.3277	0.4096	0.2048
6	0.7351	0.2321	0.0305	0.5314	0.3543	0.0984	0.3771	0.3993	0.1762	0.2621	0.3932	0.2458
7	0.6983	0.2573	0.0406	0.4783	0.3720	0.1240	0.3206	0.3960	0.2097	0.2097	0.3670	0.2753
8	0.6634	0.2793	0.0515	0.4305	0.3826	0.1488	0.2725	0.3847	0.2376	0.1678	0.3355	0.2936
9	0.6302	0.2985	0.0629	0.3874	0.3874	0.1722	0.2316	0.3679	0.2597	0.1342	0.3020	0.3020
10	0.5987	0.3151	0.0746	0.3487	0.3874	0.1937	0.1969	0.3474	0.2759	0.1074	0.2684	0.3020
11	0.5688	0.3293	0.0867	0.3138	0.3835	0.2131	0.1673	0.3248	0.2866	0.0859	0.2362	0.2953
12	**0.5404**	0.3413	0.0988	0.2824	0.3766	0.2301	0.1422	0.3012	0.2924	0.0687	0.2062	0.2835
13	0.5133	0.3512	0.1109	0.2542	0.3672	0.2448	0.1209	0.2774	0.2937	0.0550	0.1787	0.2680
14	0.4877	0.3593	0.1229	0.2288	0.3559	0.2570	0.1028	0.2539	0.2912	0.0440	0.1539	0.2501
15	0.4633	0.3658	0.1348	0.2059	0.3432	0.2669	0.0874	0.2312	0.2856	0.0352	0.1319	0.2309
16	0.4401	0.3706	0.1463	0.1853	0.3294	0.2745	0.0743	0.2097	0.2775	0.0281	0.1126	0.2111
17	0.4181	0.3741	0.1575	0.1668	0.3150	0.2800	0.0631	0.1893	0.2673	0.0225	0.0957	0.1914
18	0.3972	0.3763	0.1683	0.1501	0.3002	0.2835	0.0536	0.1704	0.2556	0.0180	0.0811	0.1723
19	0.3774	0.3774	0.1787	0.1351	0.2852	0.2852	0.0456	0.1529	0.2428	0.0144	0.0685	0.1540
20	0.3585	0.3774	0.1887	0.1216	0.2702	0.2852	0.0388	0.1368	0.2293	**0.0115**	0.0576	0.1369

[a]The joint specificity is the probability of having no positive test results when a set of N tests is applied to a healthy population.
NOTE: Items in **boldface** correspond to examples in the text.

The classic lesson of the false-positive cascade is taught to medical students: a panel of screening tests is likely to include at least one false-positive result. Each test's specificity and the number of tests performed affect these values, of course, and one can calculate them from the binomial distribution. If 20 tests were performed on a downwinder (as in some of the current RESEP programs) and if each test has a specificity of 80% (a value not atypical of the RESEP screening tests; Table 9.4, row 20, columns 11-13), then the likelihood that all 20 will be negative in an individual who does not have any of diseases being sought is only 1.15% or roughly 1 in 90. As shown, 5.8% of individuals truly without disease who are screened will have a false-positive result on one test, and roughly 14 percent will have false-positive results in two tests. Not shown in Table 9.4: roughly 20% will have false-positive results on three or four tests, roughly 18% on five tests, 11% on six, 5% on seven, and 2% on eight. An example of this phenomenon is reported in the results of the Prostate, Lung, Colorectal and Ovarian (PLCO) Screening Trial (Lafata et al., 2004), in which the joint specificity of four screening tests was 49.11% among men and 64.0% among women.

In general, joint specificity equals

$$1 - \prod_i Specificity_i,$$

or, if all the specificities are the same $1 - Specificity^N$. Even modest declines in specificity (moving across Table 9.4 from left to right) cascade into substantial increases in the likelihood of one or more false-positive results. In large measure, this phenomenon explains the great loss of enthusiasm for multiphasic screening, an approach popular in the 1970s (Olsen et al., 1976).

Multiple Tests for a Single Disease

In this section, we consider the case in which more than one test is used to look for the same disease; in this scenario each test has its own sensitivity (for example, sensitivity1 and sensitivity2) and its own specificity (for example, specificity1 and specificity2). Each of these operating parameters is less than unity because no test is perfect. The implication of such a screening approach, involving two or more steps, depends on how a positive screen is defined (Figure 9.6).

Definitions of a positive screen may center on whether one applies an "AND" or an "OR"[2] criterion and then judges test performance in terms of sensitivity, specificity, and positive predictive value. We discuss these points below, using the percentages in Table 9.5 to illustrate the argument.

If denoting the screening protocol as positive and proceeding to the next step (informing the patient, referring the patient, and beginning a confirmatory work-

---

[2]In this discussion, we use "OR" in the inclusive sense, that is "A OR B" means "A or B or both."

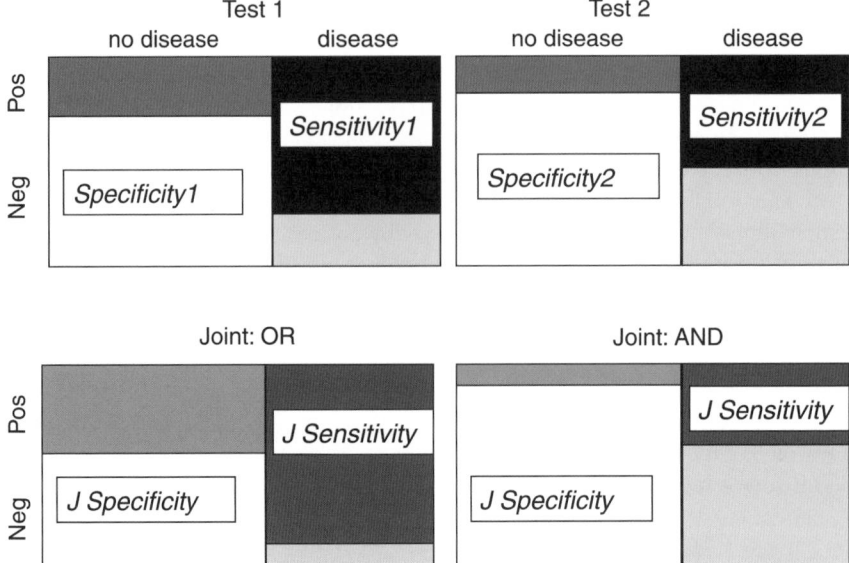

FIGURE 9.6   Multiple tests: AND versus OR.

up) together require that both tests be positive—a criterion often called the AND criterion—then the joint sensitivity of the testing pair is less than either individual sensitivity, and the joint specificity is higher than either individual specificity. If the tests are statistically independent, then the joint sensitivity will be Sensitivity1 × Sensitivity2, and the joint specificity will be $1 - [1 - Specificity1) \times (1 - Specificity2)]$.

Of course, many tests are not statistically independent of one another; instead, they may depend on the stage of disease. In such cases, the joint specificity of two or more tests interpreted by the AND criterion may not be as high as this expression might predict. As a special case, consider the circumstance in which the second test is a repeat of the first, perhaps even on the same tissue or blood sample. In that case, the two results would be very dependent, and one would not

TABLE 9.5   Interpretation of Two Diagnostic Tests

| Parameter | Individual | | Joint | |
	Test 1	Test 2	AND	OR
Sensitivity	70%	80%	56%	94%
Specificity	90%	95%	99.5%	85.5%
Positive Predictive Value				
Disease Prevalence 1%	6.6%	13.8%	52.8%	6.1%
Disease Prevalence 20%	58.6%	76.2%	95.7%	56.5%

expect the joint specificity to increase at all, save perhaps to account for truly stochastic laboratory variation. In general, if the two tests are correlated, then their joint specificity must be adjusted to account for the degree of positive or negative correlation.

For example, if the two sensitivities are 70% and 80% and the two specificities are 90% and 95%, then the joint sensitivity will be 56% and the joint specificity 99.5%. In a screening setting with low prevalence of disease, a higher specificity will probably produce an improved posterior (posttest) probability—that is, a higher positive predictive value. Such improved specificity with the AND criterion of combination might be useful if the prevalence of disease were low or if the burden of false-positive results were high, that is, when excluding the presence of a disease may have a high priority. If those tests were applied in a population in which the prevalence of disease is 1%, the positive predictive value if test 1 alone were positive would be 6.6%, the positive predictive value if test 2 alone were positive would be 13.8%, but the positive predictive value if both were positive would be 53%. In contrast, if the tests were to be applied in a population in which the prevalence of disease is 20%, the positive predictive value if test 1 alone were positive would be 59%, the positive predictive value if test 2 alone were positive would be 76%, but the positive predictive value if both were positive would be 96%.

Conversely, if denoting a screening protocol as positive and proceeding to the next step requires only that *either* test be positive—a criterion often called the OR criterion—then the joint sensitivity of the testing pair is higher than either individual sensitivity and the joint specificity is lower than either individual specificity. If the tests are independent, then the joint sensitivity will be $1 - [1 - Sensitivity1) \times (1 - Sensitivity2)]$, and the joint specificity will be $Specificity1 \times Specificity2$.

Using the same example as above, the joint sensitivity will be 94% and the joint specificity will be 85.5%. In a screening setting with low prevalence, a lower specificity will probably produce a lower positive predictive value. Again, in a population in which the prevalence of disease is 1%, the positive predictive value after either test was positive would be only 6.1%. In contrast, in a population in which the prevalence of disease is 20%, the positive predictive value after either test was positive would be 57%. The improved sensitivity of the OR criterion of combination might be useful if the burden of false-positive results were low and if finding a disease had high priority. As seen in Table 9.5, the difference between the AND criterion and the OR criterion is proportionally greater in the low-prevalence setting that is typical of screening.

Multiple Diseases and Conditional Probabilities

For simplicity, we have been dealing up to this point with a single disease and a cutoff criterion that classifies all test results into one of four categories: true positive, false positive, true negative, and false negative. That allows us to think in terms of the summary measures sensitivity and specificity. Unfortunately, the real world does not always support this simplification.

Consider a circumstance with three diagnostic possibilities: disease 1, disease 2, and no disease. Assume that we perform a test to establish the presence of disease 1 but that the likelihood of a positive test is different among patients with disease 2 than among patients with no disease. In such circumstances, the "specificity" of the test cannot be defined, because it depends on the relative likelihood of disease 2 and of no disease among patients who do not have disease 1. These factors make the mathematics a bit more complicated because one needs to think in terms of conditional probabilities (for example, the conditional probability of a positive test given the presence of disease 2). One can, however, derive unexpectedly useful diagnostic information by doing so. A negative test (presumably for disease 1) can actually change the relative likelihoods of disease 2 and no disease (Gorry et al., 1978).

Imaging Studies

The situation becomes substantially more complex if one considers imaging studies, such as chest x rays, computed tomography scanning (CT scans), magnetic resonance imaging (MRI), or nuclear scanning. For each of those studies, one does not have a single parameter or axis of results, such as an ejection fraction, a PSA concentration, or an oxygen saturation. Rather, the person interpreting the image can identify any of a large number of abnormalities; these might be a pulmonary nodule, a diffuse pattern of fibrosis, a pulmonary infiltrate, or an enlarged cardiac silhouette.

For most imaging studies, interpreters do not limit their reports to a single criterion of positivity, such as a pulmonary nodule of at least 5 mm in diameter or a renal mass. Instead, they typically report one or more of a substantial number of incidental findings if for no other reason than to protect themselves from an accusation of missing a cancer, one of the more common reasons for malpractice actions. For adrenal, pituitary, and thyroid images, these tiny lesions have sometimes been called incidentalomas.

Thus, estimating the specificity (true negative rate) of an imaging study is difficult. The literature can provide information about specific findings, for instance, how often a solitary pulmonary nodule larger than 5 mm in diameter is present in the absence of cancer. It does not tell us how often a radiograph of a healthy individual is reported as completely normal. In part, these decisions depend on how hard the imager looks. Because many incidental findings will generate further workup and anxiety among individuals screened with imaging techniques, the false-positive cascade and its burden will increase considerably if a screening protocol includes several imaging studies.

Repeated Studies

Our discussion of screening has implicitly considered a screening test or battery of tests at a single time. Many RECA diseases, however, evolve, either developing a new manifestation or reaching the threshold of detectability. That threshold may vary from test to test for a specific disease (see the discussion above of lead time bias and the inherent biases of using more sensitive screening tests).

Thus, we can ask, if an initial screening test for a disease is negative, should the test be repeated in the future? If so, when? The many groups that have recommended only a limited number of screening tests have not uniformly analyzed the complex question of retesting. They often address it informally or in terms of expert opinion.

For the initial screening test, the appropriate measure of pretest probability is the prevalence of disease in the population. If a screening test is to be repeated, then the pretest probability of disease is usually the interval incidence of disease, a number almost always lower than the prevalence. The interval incidence of disease refers to individuals known not to have the disease at the beginning of a period (because of a prior negative evaluation at that time) who develop the disease by the end of that period. If the initial screening test is negative, one may well be selecting a subset of the population with a lower-than-average propensity for developing the disease.

In such repeat screening scenarios, the pretest likelihood is often far lower for the second and later screens than for the first. Arguably, the evidence for repeating a test must be even stronger than it was for the initial screen. Some experts pose counterarguments, however. On the one hand, for instance, one might say that disease identified on a repeated screen is more likely to be rapidly developing, although whether more aggressive disease is more or less amenable to the benefit of early treatment is not clear. On the other hand, one might argue that disease identified on the repeated screen may be more likely to have existed in an earlier stage and thus be more amenable to early treatment. Answers must lie in scientific evidence and in studies that compare outcomes of various frequencies of repetition of screening. Few such studies have been reported.

Conclusions of Other Groups

As mentioned above, various advisory bodies have considered screening options for many diseases, and relatively few have been recommended for routine screening. We have presented a fairly extensive listing in Appendix E to convey where the current view of medical screening for primary care; materials there are taken from recommendations of the second or third *Guide to Clinical Preventive Services* of the USPSTF or, where appropriate, recommendations of the equivalent Canadian task force. We have included all "recommendation and rationale" statements concerning neoplastic diseases, whether they are specifically RECA conditions or not; we have also developed a listing of other "A" and "B" recommendations (respectively, strongly recommend and recommend) from the USPSTF for a variety of other conditions. The intent is to convey the extent of coverage of clinical preventive services as they apply to screening, to underscore the point that very few diseases are at present amenable to screening (in asymptomatic populations) with reliable and valid tests in which the benefits of screening likely outweigh the risks or harms of such screening.

We note, for example, that screening for cancers of the cervix, breast, and colon is recommended; screening for cancer of the prostate receives mixed reviews, and extensive discussion with the patient is suggested before it is undertaken. Screening for cancers of the lung, pancreas, ovaries, and thyroid is not recommended, in part because prevalence is low (ovary and thyroid) and in part because the therapeutic benefit is small (lung, pancreas, thyroid).

To modify those recommendations in RECA populations, one would need to believe that the prevalence of those diseases that offer some benefit of early detection and treatment is substantially higher than it is in other populations. As explained in other chapters, that proposition probably does not hold. First, we are now many years after exposure, so that the *excess* relative risk of the cancers has declined. Second, some downwinders had had relatively low exposures to the radiation particles that would be expected to induce most of the cancers. Of course, for many cancers and geographies the data about exposure include substantial uncertainties. Third, as the exposed population ages and comorbidities develop, the medical benefit of early detection in improving life expectancy wanes.

Thus, for downwinders and onsite participants, little rationale can be advanced for expanding medical screening recommendations in the RECA populations beyond contemporary recommendations for the general population (as in Appendix E) that are based on an assumption of achieving improved health outcomes through early detection and on an acceptable balance of benefits and harms.

For miners, millers, and transporters who were exposed to dust and silica, the same issues arise as in mining for other materials, for example, exposure to silica. In general, silicosis and coal workers' pneumoconiosis (CWP) are progressive diseases managed by minimizing exposure.[3] Few data suggest that early identification and treatment can modify prognosis, save once the fibrotic lung disease is established, save to decrease further exposure to silica (Sharma et al., 1991; Bates et al., 1992; Banks et al., 1993; Banks, 2005). Small trials of systemic or bronchoalveolar lavage with steroids have shown modest statistical but not clinical improvement (Sharma et al., 1991). Thus, early identification of silicosis in

---

[3] *Black lung* is a common term for coal worker's pneumoconiosis caused by excessive exposure to coal mine dust. Silicosis is related to the excessive quartz dust exposure (http://www.thoracic.org/news/atsnews/news0197.html#ats9, accessed for this purpose 10/2/2004).

asymptomatic individuals who are no longer actively employed as miners would be of limited medical benefit.

Silicosis is, however, associated with an increased prevalence of mycobacterial disease. Performing a purified protein derivative (PPD) or tuberculin test to screen former miners, millers, and ore transporters for tuberculosis is a rational approach. Such skin testing need not be predicated on demonstrating silicosis on a chest x ray. If the skin test were positive, following up with a chest x ray, and perhaps with other appropriate tests to confirm the presence of active disease (if the x ray were suspicious) would be the current standard of care. Offering prophylactic antituberculosis therapy to patients with positive skin reactions who do not have active disease might be valuable as well. The demonstration of silicosis on chest x ray might improve compliance with early treatment for latent mycobacterial infections. Some data suggest that lung cancer to have a somewhat increased frequency in patients with silicosis, but the data are inconsistent; moreover, early detection does not appear to confer much, if any benefit.

Several groups, including the National Institutes of Health and the National Kidney Foundation, suggest screening for early evidence of chronic renal disease in populations at risk for renal failure, such as patients with diabetes and hypertension. Recently, those recommendations have been extended to include a far broader population (Levey et al., 2003; National Kidney Foundation, 2002). If chronic renal dysfunction is identified early, various interventions such as angiotensin-converting enzyme [ACE] inhibitor drugs, diet modification, and lipid-lowering therapies, can slow the progression of renal disease with substantial benefit. Recent data suggest that the "MDRD" glomerular-filtration-rate (GFR) calculator, developed as a part of the modification of diet in renal disease study (Levey et al., 2003; http://www.nephron.com/mdrd/default.html, accessed 3/13/2005), is probably the most accurate means of identifying such patients. Because of the putative relationship between exposure to soluble uranium salts and renal failure (perhaps more acute than chronic) and because chronic renal disease is a RECA-covered disease for some uranium workers, screening millers and ore transporters for renal dysfunction is reasonable.

The committee finds insufficient supporting evidence to support additional medical screening or medical case-finding in the RECA populations beyond the level and type of screening advised for general populations and persons with occupational exposures similar to those of in miners in general. Specifically **the committee recommends that the Health Resources and Services Administration base RESEP medical screening efforts in asymptomatic individuals on robust scientific evidence that such screening improves health outcomes and that its benefits outweigh its risks.**

The committee believes that the current legislation and regulations that promulgate screening for the RECA-compensable diseases *for the purpose of preventing disease and improving health* arise because the distinction between medical screening and compensational screening, as explicated in this chapter, has not been clear. With that distinction in mind, **the committee recommends that HRSA not extend its medical screening beyond the generally accepted screening protocols that apply to the US population at large.** However, the committee further recommends that uranium miners, millers and ore transporters also be screened for diseases generally recommended for screening in other mining populations and that uranium millers and ore transporters be screened for chronic renal disease.

In particular, the committee observes that some screening protocols that HRSA grantees now use exceed these guidelines. In our judgment, HRSA should cease funding any medical screening efforts that do not conform to the recommendation. **Correspondingly, the committee recommends that, once an individual has been shown to be administratively eligible for compensation under RECA (including employment, residence, or a calculated PC/AS at or above some established cutoff criterion), the individual be offered medical screening recommended in generally accepted protocols that apply to the population at large (see Appendix E-2).**

FUTURE RESEARCH

The committee has referred earlier to authoritative screening recommendations from the USPSTF and other groups (see Appendix E). The dozens of systematic reviews prepared for the USPSTF and the related USPSTF statements of recommendations and rationales typically have extensive comments on future research issues, including the reviews of cancer screening technologies. These can be found on the USPSTF website (www.ahrq.gov/clinic/prevenix.htm) and in the peer-reviewed literature.

We briefly raised the question of the costs of screening in general and of the RESEP program in particular. Not much is known about the cost-effectiveness of screening, in part because of the unproven effectiveness of many screening modalities (such that examining cost-effectiveness has little meaning). Very little information is available on the cost-effectiveness of alternative screening strategies; the chief exceptions are screening for colorectal cancer and type 2 diabetes. Insofar as HRSA and RESEP engage in appropriate medical screening in the future, data from strong cost-effectiveness analyses would help HRSA to specify RESEP screening strategies that would optimize use of federal resources. We note in particular the need for more head-to-head comparisons of screening technologies and for collection of cost data for cost-effectiveness analyses. The committee encourages Congress to expand financial support for rigorous studies of diagnostic and screening tests, including cost-effectiveness analyses.

As documented elsewhere in this report, the six current HRSA grantees operate screening programs with quite different orientations. Some do screening similar to a broad-based annual physical examination, with attention to both

RECA-compensable disorders and to non-RECA ailments typical of the age and sex of patients that they see; their orientation might thus be characterized as more medical than compensational. Others (notably the Indian Health Service RESEP program in Shiprock, New Mexico) apparently direct their focus nearly exclusively to RECA conditions and do not expand their screening efforts beyond that; their orientation might be characterized as more compensational than medical. The committee believes that such variation poses problems of inequity that HRSA should investigate in more detail; one aim of such a study is to determine whether some individuals are receiving more services than others for equivalent expenditures of public dollars in RESEP. On the other hand, the best practices among current grantees are not clear. It would not be unreasonable for HRSA to undertake a trial to establish which protocols are most effective. Of course, participation in such a trial should also be predicated on obtaining informed consent for the trial from each participant.

The RECA populations being screened by HRSA grantees with RESEP funding are an interesting group to study insofar as they comprise individuals with presumably greater than average risks of disease. The sensitivity, specificity, and positive and negative predictive values (or reliability and validity) of a variety of screening tests may well differ in those populations from the values in populations of either average or low risk. Epidemiologic and clinical studies might well be undertaken to clarify the yield from various screening tests; such work would shed light on the utility of the individual tests and combinations of tests under those circumstances.

HEALTH-CARE ISSUES BEYOND SCREENING

Evolution of Technology

Apart from the issues we have raised above about medical and compensation screening, we note that some testing, particularly diagnostic testing, must and will continue. For uranium miners, for example, the increased risk of lung cancer and restrictive lung disease can be addressed through traditional services, such as pulmonary function tests and chest x-rays. Those methods have been applied to studies of underground miners (for example, of coal workers' pneumoconiosis in the United States and uranium miners in many countries) and can be said to be appropriate current practice as carried out by the current HRSA grantees.

However, as new imaging and sputum-cytology methods emerge, the HRSA grantees should be encouraged to consider new and better tests (especially those with test properties better than those of traditional screening or diagnostic tests) and abandon tests that appear to be less effective. Such evolution should, however, be based on evidence. The committee would generalize that point by saying that both HRSA and the medical professions more broadly must be sensitive to the evolution of useful technologies, adopting those of improved performance and discontinuing use of those approaching obsolescence.

Screening Recommendations from Other Groups

As reflected in this chapter and Appendix E-1, the committee used the USPSTF guidelines to help develop recommendations on medical or compensational screening of RECA populations. As the clinical and research communities develop new or improved screening methods and put them into practice, the USPSTF and other professional or clinical bodies will review such guidelines and, as appropriate, revise them.

The Institute of Medicine instituted a modern-day understanding of clinical practice guidelines as the behest of the then Agency for Health Care Policy and Research (IOM, 1990a; IOM, 1992; Lohr, 1998, 1999; Lohr et al., 1998). This evolution in clinical-practice guidelines since that time for both primary and specialty or referral care, may have implications for both RECA populations and the RESEP program. Updates to clinical practice guidelines of all sorts can be found through the National Guidelines Clearinghouse supported by the Agency for Healthcare Research and Quality (AHRQ) (www.guideline.gov) and, specifically for the USPSTF, through the AHRQ Preventive Services Web site (www. ahrq.gov/clinic/prevenix.htm) and the Put Prevention into Practice program (www.ahrq.gov/clinic/ppipix.htm).

The committee recommends that HRSA regularly monitor and follow screening guidelines developed by the US Preventive Services Task Force and published by the Agency for Healthcare Research and Quality

Psychologic Issues in Exposed Populations

Of concern to the committee are psychologic disorders potentially present among downwinders who at various times have not been fully informed about the risks, although small, of exposure to fallout from the NTS tests. Major depression is a particular concern because of the already high prevalence of depression in the general population, its serious potential consequences (such as suicide), and its responsiveness to treatment if it is identified. Generalized anxiety disorder, post-traumatic stress disorder, and a more chronic concern among these populations about their view of living in a "contaminated environment" (Kuletz, 1998, p336), has been called "chronic environmental stress disorder" (IOM, 1999), and it may be associated with NTS activity and its aftermath over the past 50 years. In the end, the committee could not amass sufficient scientific data or develop a plausible chain of reasoning to include these conditions as compensable under RECA.

In the setting of a nuclear accident or other catastrophic event, rapidly providing the public with accurate information can minimize the mental health effects (IOM, 1999, p106). For a variety of reasons, that was not always the case for downwind RECA populations. Some people who testified at the committee's information-gathering meetings described experiences and impressions that may place them at risk for some of these disorders. We heard a good deal about chronic emotional pain and sorrow related to the cancer deaths of many family members and friends. We also heard many of those who testified express anger and frustration at the government for (supposedly) misleading them and their predecessors about the dangers of uranium mining, the effects of fallout, and the meaning of the "dust" found on many crops, homes, and other locations. They presumed that this dust was a consequence of the nuclear-weapons tests at the NTS. Finally, many spoke of their strong sense of "not being heard" by either HRSA or DOJ.

Although downwinders were not exposed to a catastrophic threat (such as a nuclear accident), the paucity of accurate information provided to them for many years may have contributed to their psychologic burden. However, even here the committee did not find any scientific studies to validate the contributory effect of this paucity of information. In addition, as noted in Chapter 7, the committee has been unable to identify any data that evaluate the psychologic effects of chronic environmental stressors during either the testing and mining periods or, more recently, among RECA downwinders or other populations with similar radiation-exposure experiences. Hence, the committee does not believe that compensation, as outlined in Chapter 8, is appropriate for downwinders with psychologic disorders.

Nevertheless, the possibility exists that depression and generalized anxiety disorder are more prevalent in RECA populations than in the general population and might have been ameliorated if more complete information had been provided to these populations in the past. Of course, such information must be provided in a culturally sensitive an appropriate manner. The committee notes three important factors: (1) rapid, simple, low-cost screening tests for depression are available and have been validated (Pignone et al., 2002); (2) depression can have serious consequences that can be ameliorated by timely treatment; and (3) the USPSTF recommends screening for depression in the general adult population within the context of the delivery of health care services by institutions or providers with appropriate quality assessment programs in place (http://www.ahrq.gov/clinic/3rduspstf/depressrr.htm).

In sum, screening for depression constitutes good medical practice by providers with good quality-of-care procedures; the link between quality of care and ethics (Lohr, 1995) strengthens this conclusion and is implicit in reports from the Institute of Medicine spanning more than a decade (IOM, 1990b; IOM, 2001). Given the above, HRSA may want to consider screening for depression in their medical screening protocols. The low-cost, potentially high-benefit strategy of screening for depression, with appropriate referrals and follow through, may ameliorate some psychologic burdens in the RECA population.[4,5]

CONCLUSION

This chapter has reviewed in depth the conceptual and statistical underpinnings of screening for medical purposes. We have discussed the limitations and risks of screening for diseases for which early detection offers little benefit. We have also shown how screening in low-risk populations and screening for multiple diseases will produce a substantial burden of false-positive results. Finally, we have offered several recommendations to reduce potential harms and mistakes from the use of multiple screening tests among individuals for whom the medical benefits cannot outweigh the harms and the likelihood of compensation may be low. Some recommendations are directed at agents that would need to amend RECA (or RESEP) legislation; others are directed at steps that HRSA can consider without statutory changes. The next chapter picks up on the new issue in RESEP: screening for compensable disease.

---

[4]For the reasons provided in detail in this chapter concerning the potential risks of depression developing in exposed populations (about which we heard testimony from Drs. Robert Ursano and Evelyn Bromet), the seriousness of the depression in terms of morbidity and mortality from suicide, and the treatability of depression, committee members Stephen G. Pauker and Catherine Borbas find it inconsistent with good medical or public health practice merely to state that "HRSA may want to consider screening for depression" in RECA populations. It is their opinion that such screening should be "recommended," as it is for adults in the general population (see USPSTF). Further, it is noteworthy that few, if any, current HRSA grantees have listed screening for depression in their current protocols, emphasizing the need for this specific recommendation.

[5]Committee member Kathleen N. Lohr wishes to support a recommendation that HRSA expand its screening activities to include mental and emotional disorders (particularly major depression, generalized anxiety disorder, and post-traumatic stress disorder), following through as needed with appropriate referrals to medical and psychiatric care relevant to the diagnoses in question. The report still contains much evidence, both from the published literature and more anecdotally from the numerous presentations at the committee's information-gathering meetings, that major and/or minor depression may well have a prevalence in these RECA populations higher than that for the general adult population. Screening for depression in adults, with the qualifications noted in the report as to the medical infrastructure needed for high-quality care, is a formal recommendation of the US Preventive Services Task Force. Numerous easy-to-use screening methods exist, and the practice of such screening is spreading (Santora and Carey, 2005).

10

Screening for Compensation

As discussed in earlier chapters, amendments in 2000 to the Radiation Exposure Compensation Act (RECA) of 1990 created funds for education, prevention, and early detection of radiogenic cancers and other diseases. Of particular relevance here is the charge to the Health Resources and Services Administration (HRSA) to administer the Radiation Exposure Screening and Education Program (RESEP) to facilitate the RECA compensation program. Chapter 9 clarified that screening activities have two purposes and targets: medical screening to improve health outcomes and compensational screening specifically for RECA. In its interim letter report (NRC, 2003a), the committee addressed medical screening and concluded that no scientific evidence supported the view that medical screening beyond that ordinarily recommended for unexposed populations would improve health outcomes.

In Chapter 9, we examined epidemiologic, statistical, and clinical issues related chiefly to medical screening and explained the numerous drawbacks to programs that use multiple screening tests even for medical purposes. We noted the relative paucity of robust evidence that supports screening for many conditions (including many that RECA covers) and cited the authoritative sources of guidelines for medical screening of populations in general or those known to have specific medical risks. We observed that screening programs can be harmful because of false-positive test results (for example, the physical risks that their workup entails and the psychologic and social consequences of labeling individuals as potentially affected). We explained the problems of false-positive rates with multiple tests in various scenarios. Finally, we discussed the ramifications of these issues specifically for RESEP, giving particular attention to activities intended to promote health outcomes rather than simply document possible disease for compensation purposes.

In this chapter, we address compensational screening. This is the use of medical tests and administrative qualifications to identify individuals who might be candidates for compensation under RECA. This concept is required because RESEP specifies screening for some diseases not traditionally recommended for medical screening to improve health outcomes. Because a term is required to describe this RESEP concept, we coin the term compensational screening. RECA provides compensation for certain radiogenic and nonradiogenic diseases. Some individuals may wish to be evaluated to establish eligibility for compensation, and RESEP provides a mechanism for such evaluation.

RESEP also provides means of organizing direct or referral care for at least some screened individuals who are found to have abnormalities that need further investigation. It is not, however, specifically mandated to provide care to such persons (who might then be regarded as patients). As noted in Chapter 11, some HRSA grantees with RESEP awards apparently leverage their grant awards to provide at least some medical services to some patients under various arrangements for those with or without health insurance. Nonetheless, even though RESEP is identified as a medical screening program, it is properly seen more as a compensational screening program. Moreover, RESEP was not established to be a provider of medical services (of last resort or otherwise), although one interpretation of the services rendered is that they are meant to enhance prevention.

If populations in the RESEP screening program are medically underserved or reluctant to contact the medical system, diverting resources or providing additional resources to facilitate such contact might be more beneficial than devoting resources to enhance screening (medical or compensational); such screening is unlikely to improve outcomes. The questions of expanded contact, education, and outreach are taken up in Chapter 11.

Regardless of the underlying motivation, screening should be preceded by informed decision making in coordination with a clinician. The reason is that such evaluation has potential benefits and potential harms, and the balance between them may not be clear, especially in RECA-RESEP circumstances. For professional and ethical reasons, an explicit plan for appropriate follow-up services for compensable and noncompensable diseases is needed. Informed decision making is as important, if not more so, for compensational than for medical screening because of the low likelihoods of either monetary payments or health benefits with the former.

COMPENSATIONAL SCREENING ISSUES:
THE CORE CONCERN FOR RECA AND RESEP

In compensational screening, individuals being screened may have only a modest likelihood of having a compensable disease. Although the same may be said of some medical screening tests if the prevalence of the disease in the screened population is low, the motivating effect of potential compensation may well attract individuals with a low likelihood of a compensable disease who, all other things being equal, probably would not elect to be screened for medical reasons alone. Such persons may have only a low likelihood of receiving compensation, but they face the same risks of harm from the screening that individuals with higher likelihoods of disease would face. Hence, they should balance the likelihood of monetary (but not medical) benefit against the health risks of screening.

Reasonable individuals may well disagree about whether being screened solely for the purposes of seeking compensation is a good choice. Sometimes such decisions may reflect a lack of complete understanding of the data that describe the uncertainties engendered by screening; such data may turn on probabilities that themselves are quite low. For example, many individuals do not intuitively understand the implications of uncertainty or small numbers (such as probabilities under 1%), and this is a matter of both literacy and numeracy (discussed in Chapter 11). Many untrained individuals cannot adequately distinguish among, say, 1/100 (1%), 1/1,000 (0.1%), and 1/10,000 (0.01%); all are sometimes seen simply as small numbers. Many adults not only do not understand the meaning of small numbers, but they are also challenged in interpreting probabilities and statistics and applying such data to decisions in their daily lives (Schwartz et al., 1997).

Making rational choices about topics as complex as medical screening (wherein longer-term health benefits must be balanced against shorter-term risks) will certainly be near the limits of the ability of some persons to make well-considered decisions. The even more complex tradeoffs in compensational screening (for example, balancing medical risks against the sometimes small likelihood of monetary gain) could well push many people beyond their ability to make informed, rational judgments unless they are carefully supported throughout the process.

Thus, the ethics of screening for rare diseases and conditions among either general populations or populations targeted for screening on grounds other than disease itself (such as compensation) must be taken into account (an issue introduced in Chapter 8). Compensational screening needs to account for both the autonomy and the authenticity of the person; the decision must reflect the person's desires and must not be coerced.

As now structured, RECA is designed to compensate individuals in three main groups:

- Those who were exposed to radiation in the course of their employment.
- Those who were exposed during onsite testing of nuclear armaments.
- Those who were inadvertently exposed to fallout because they were downwind of nuclear tests.

RECA established criteria for exposure to radiation (based on timing, place of residence, and occupational activities). It also required that claimants demonstrate at least one of more than 2 dozen diseases that were deemed compensable. The list of compensable disorders was based on the best available knowledge in 1990 and was not totally rooted in science; it was modified in 2000 and again in 2002. The most recent regulations from the Department of Justice (DOJ) appeared in mid-2004. Chapter 2 provided details on this legislation.

RESEP was designed to educate and identify individuals who would potentially be compensated under RECA. To receive compensation—assuming that the administrative requirements of documenting the timing of residence and occupational activities were also met—RECA requires an exposed individual to meet specific published criteria for (that is, proof of) the presence of disease. Such proof can rest on a variety of imaging studies, functional studies (such as pulmonary function tests), physician reports, or autopsy reports (see DOJ, 2004). However, these RECA-acceptable proofs would not always establish the presence of disease with certainty.

We consider here several additional epidemiologic, statistical, or clinical issues in screening in the context of the main RECA and RESEP goals. Some of the issues relate to specific diseases and the probability of causation or assigned share (PC/AS—see Chapters 5 and 6); others relate to the information needed to prove the presence of disease.

Base Rates of Disease and Probability of Causation

One problematic issue in requiring exposed individuals to demonstrate the presence of disease is that radiation exposure is only one of several causes of most of these diseases. The RECA standard for compensation for any given individual does not depend on the cause of the disease or even on the probability of such cause (in that individual). Instead, it depends merely on proof that the individual has the disease and is administratively qualified to receive compensation.

Two conditions illustrate the complicated causal pathways that make definitive decisions about causation so difficult for RECA. One is a lifestyle issue: smoking behavior. The other is a highly prevalent, serious condition in some RECA populations: type 2 diabetes.

Lung cancer is the only RECA disease that is compensable in all five RECA populations (miners, millers, ore transporters, onsite participants, and downwinders). As detailed in other chapters, the contribution of radiation exposure to the etiology of lung cancer, now 35 to 60 years after exposure, is likely to be small, even among uranium workers, for whom the case is best made. The contribution is likely to be considerably lower among downwinders and onsite participants.

Tobacco abuse is the most prominent cause of lung cancer. It can be considered a choice that each individual makes. In some ways, then, the less healthy an individual's lifestyle is, the more likely it is that the individual will have a lung cancer and the more likely that he (or she) will receive compensation. That outcome must be considered a perverse effect of RECA, that is, an effect that would seem contrary to the presumed rationality of RECA.

In a different vein, type 2 diabetes is the most common cause of chronic renal disease in this country. It is highly prevalent among the Native American populations who were frequently uranium workers. Although the renal or other sequelae of diabetes are not a direct consequence of lifestyle choices, they do depend heavily on both self-management and medical care that may or may not be easily accessible for these populations. Chronic renal disease may occur with relatively high frequency in these populations for numerous reasons unrelated to radiation exposure. Hence, decisions about the probability of causation of this compensable disease may be difficult to make.

What Is Disease (RECA Gold Standard)?

RECA provides specific criteria for establishing the presence of disease, as defined by the legislation and published in the *Federal Register* (DOJ, 2004). Although the criteria have a basis in good medical practice (such as standards issued by the National Cancer Institute), they and the underlying test results in essence define the presence of disease as far as RECA compensation is concerned. Thus, the results of these tests and processes in fact constitute the gold standards for diagnosis in terms of RECA compensation, although not necessarily in terms of medical certainty. In fact, these standards state that if a disease can be established to a reasonable degree of medical certainty (at or above 50%, as discussed below), then the disease is present as far as RECA is concerned.

If one of these RECA gold standards were used as a (the only) screening test, both the sensitivity and the specificity of the compensational screening test would, by definition, be 100%. The problem of the cascade of false positives, discussed in Chapter 9 as an important issue in medical screening, would disappear, at least with respect to compensational screening. However, the patient would still face the cascade of medical false-positives results and the medical risks they engender, perhaps without medical benefit. Identifying a patient as eligible for compensation does not itself guarantee either that the patient has the disease or that the patient's health will be improved by such identification.

Tiered Requirements for Compensation

The requirements for proof of the presence of disease have both criteria that can be satisfied by meeting any *one* of several options and criteria that can be met only by satisfying *all* of several requirements. We explored those problems in detail in the discussion of "OR" and the "AND" combinations in Chapter 9. OR criteria in effect increase sensitivity measured against the true medical presence of disease as a gold standard, whereas AND criteria decrease sensitivity. The issue of specificity does not arise here because these criteria are the gold standard for compensation.

For purposes of decreasing the likelihood of (perhaps erroneous) compensation, using AND criteria will be preferred. For purposes of increasing the likelihood of (perhaps warranted) compensation, using OR criteria will be preferred. The latter would provide claimants some element of the benefit of the doubt with respect to the presence of a specified disorder.

Medical Certainty in the Context of RECA

One question that needs clarification is how certain is certain enough to permit compensation, especially when the tests that bear on a diagnosis are imperfect. One interesting phrase appears in the statutory sections about establishing the diagnosis of many compensable diseases: "from which the appropriate authorities . . . can make a diagnosis to a reasonable degree of medical certainty."

At least in courts considering such torts as medical malpractice, the concept of "reasonable degree of medical certainty" has often, but not consistently, been held to mean "more likely than not," or at least 51% (Lewin, 1998; Craig, 1999). Some legal experts question the "appropriateness of the phrase [*reasonable medical certainty*] as a legal standard" and even regard it as "oxymoronic" (Lewin, 1998; pp. 1-2). Nonetheless, the phrase has been in use since a 1916 opinion issued by the Illinois Supreme Court, and it has gradually become a fairly standard, although not sharply defined, term. It has been applied in worker compensation cases (Lewin, 1998). Textbooks about medical malpractice (Gutheil, 1998, p. 6) also point to the "more likely than not" (greater than 50%) criterion.

At times, the concept is rendered as "as likely as not." The distinction between "as likely as not" (at least 50%) and "more likely than not" (greater than 50%) may be largely semantic, in that in only very few circumstances would a calculated probability be exactly 50.000%. A probability of 50.001% would satisfy either meaning, whereas a probability of 49.999% would satisfy neither meaning—unless, of course, it was rounded up to 50%.

In the RECA-RESEP context, the problem is compounded because medical probabilities are never precise; indeed, probabilities consistently within a few percentage points of the true (but unknown) value might be regarded as unusually precise. If, as in the policies and procedures for the Energy Employees Occupational Illness Compensation Program Act (EEOICPA), the approach underlying RECA is to provide an exposed individual the benefit of the doubt, then the criterion should be "as likely as not," with reasonable rounding permitted.

Other Sources of Documentation of Disease for Compensation

Two other mechanisms for screening for compensational purposes might be considered: both appear to reduce risks to individuals, and one provides a way to obtain information that could not otherwise be obtained. We first discuss using autopsy data on individuals who die before any screening can take place; the autopsy yields definitive (diagnostic) information. We then consider whole body magnetic resonance imaging.

The RECA Autopsy

For RECA, autopsy reports are acceptable evidence for most diagnoses of compensable disease. That fact suggests several strategies for maximizing the potential for establishing the presence of compensable disease, assuming a post-mortem examination is ethically acceptable to the potentially affected individual and his or her family.

If a potentially compensable patient dies, obtaining a routine autopsy might be a reasonable strategy for maximizing the opportunity for the patient's family to receive compensation. Many families may be reluctant to obtain a postmortem examination; thus, a reasonable step might be for the clinicians caring for the patient to consider discussing this option with the patient and his or her health care proxy or next of kin. This step might be particularly relevant if the patient is chronically or terminally ill and especially if the patient is known to meet the geographic and temporal administrative criteria for compensation. If such an autopsy were approved for the purpose of looking for compensable disease, the pathologist should be so informed and should look specifically for any RECA-compensable diseases. Of course, posthumous establishment of RECA eligibility, whether by autopsy or by a family's application for posthumous compensation, raises the issue of whether such posthumous eligibility should qualify the individual's family or estate for retroactive reimbursement for the medical care already received.

In the Radiation-Exposed Veterans Compensation Act (REVCA), "under certain conditions, veterans who are retroactively awarded service-connection may qualify for reimbursement of certain medical expenses back to the date of the original claim filing; a family member could file on behalf of the deceased veteran. These cases would be referred to as "not previously authorized" claims and generally must meet three conditions: (1) treatment was for a service-connected condition or for a condition held to be aggravating an adjudicated service-connected disability; (2) a medical emergency; and (3) Department of Veterans Affairs (VA) or other federal facilities not feasibly available" (information taken from correspondence with Department of Veterans Affairs, March 31, 2005).

Unlike REVCA, under the Energy Employees Occupational Illness Compensation Program Act (EEOICPA), "medical payments are made only to a living energy employee if the claim is approved; no reimbursement is made for prior expenses, and no medical payments are made to survivors if the energy employee is deceased" (correspondence with Department of Labor, April 6, 2005)

The committee did not have the opportunity to focus on this conundrum and merely highlights it for careful legislative consideration.

Whole Body Magnetic Resonance Imaging

Although many kinds of imaging studies provide acceptable evidence of specific diseases, many are invasive or involve exposure to radiation. For example, a recent study suggests that the exposure from a routine computed-tomography (CT) body scan carries more risk (from radiation) than the small diagnostic benefit it provides in identifying diseases for which early intervention improves health outcomes (Brenner and Elliston, 2004). Magnetic resonance imaging (MRI), which is not invasive and does not subject patients to ionizing radiation, yields evidence of many diseases that would be acceptable for RECA purposes. That is, "positive" MRI results are acceptable means by which to establish the diagnosis of most RECA-compensable diseases. However, as is the case with most high-sensitivity imaging studies, a whole-body MRI would be associated with an unacceptable number of potentially false-positive test results (for example, incidentalomas), and these would subject the patient to workup and labeling and the risk of physical, social, and emotional consequences.

Nonetheless, if patients were to choose to set aside the issue of costs and the risks of medical, social, and psychologic consequences of false-positive results, they might request to have a "RECA whole-body MRI." An MRI would not expose the person screened to ionizing radiation. Moreover, it could be considered a gold standard for establishing the presence of RECA-compensable disease insofar as it produces information relevant to the body parts that any RECA-compensable disease might affect, albeit at enormous cost in terms of false-positive results. The feasibility and cost of obtaining an MRI of the entire body must also be questioned.

In short, MRIs might provide both the initial screening information and the diagnostic data relevant to RECA conditions and cause, on balance, less potential harm than the current "screening protocols" impose. If properly explained to the screenee, a negative "RECA whole body MRI" might help reassure patients that they do not have one of the RECA-compensable diseases. Set against this point is that, given the propensity for such an MRI (as an imaging test that identifies multiple diseases) to produce false-positive results, such reassurance for some patients may come at the cost of increased anxiety in many other patients.

Because such a plan would expose the screened individual to the risks of medical false-positive results, workup, and potentially futile treatment, careful, extensive (but culturally sensitive) informed consent is critically important. Regardless of the desirability or feasibility of pursuing such strategies for compensation-oriented screening, the committee reiterates its view of expanding medical screening to compensational screening. Specifically, in considering expansion of compensational screening, **the committee recommends that HRSA base decisions about screening primarily for compensation on recommendations drawn from credible scientific evidence that the proposed test provides reliable information about the presence or absence of specified RECA-compensable diseases.**

As suggested in Chapter 9, even if screening is performed primarily for purposes for identifying RECA-compensable disease, if such tests reveal any information about the potential presence (or absence) of other diseases, that information should, in general, always be shared with the screenee. In conveying such information, however, clinicians must be sensitive to the cultural traditions and other preferences for individuals to be told bad news or have negative information discussed in their presence (Gostin, 1995).

Considering the Medical Risks Posed by Screening in the Context of Compensation

Another complexity arises in the context of compensational screening. Because each of the screening tests except a postmortem examination places the patient at risk for a false-positive finding or even at risk for a true-positive result that identifies a disease that does not benefit from early intervention, such compensational screening creates some degree of medical risk. The risk may be physical or psychologic (or both). Labeling the patient or even the patient's family may produce adverse social effects.

Thus, clinicians should obtain truly informed consent about the consequences (medical, psychologic, and social) of compensational screening before undertaking such a program (see Beauchamp and Childress, 2001). Such consent is enhanced through techniques of shared, informed decision-making, perhaps especially for cancer diagnoses (Barry, 2002; Whitney et al., 2003; van Roosmalen et al., 2004; Sheridan et al., 2004).

During such informed consent and shared decision-making activities, the individual to be screened will be weighing the likely small chance of substantial monetary gain without medical gain against a more substantial chance of physical, psychologic, or social harm. Society allows people to gamble with their own health, but such decisions should be well counseled and fully informed. One might consider that undertaking such compensational screening is playing a lottery of sorts, in which one buys his or her ticket of entry not with money but by undertaking risk. In some ways, because such a gamble involves a risk of physical harm (from unnecessary tests and treatments), it could be seen as akin to playing a sort of Russian roulette with a monetary payoff attached to some outcomes.

Untrained people are not likely to understand the meaning of small numbers, in terms of both risk and payoff. Indeed, as elaborated in Chapter 11, many people have low numeracy skills in general. In this context, ensuring that individuals are informed of the likelihood that they will receive RECA compensation, the likelihood and consequences of false-positive results, and the likely medical benefit of early detection, is essential.

One part of appropriate counseling about compensational screening might be to use a decision-support tool that helps patients to understand their willingness to risk health consequences in exchange for the opportunity for monetary gain. Willingness to risk is, in effect, the obverse of one's willingness to pay for healthy life years—a concept and measurement method with decades of theory and application (Gold et al., 1996, p28). Developing and standardizing such a decision-support tool might be the type of activity that HRSA could usefully pursue for use by its grantees (or contractors), as outlined in Chapter 11.

Although such numeracy issues are beyond the scope of this report, the committee expresses its support for and concern about thorough discussion with potential screenees of the advantages and drawbacks of compensational screening. Specifically, **the committee recommends that any screening carried out under RESEP auspices be preceded by detailed counseling and informed consent that reflects an understanding of and sensitivity to the culture of the potential screenee.** When screening is primarily or exclusively for compensation (and when the likelihood of medical benefit is low or absent), a special obligation exists to be as certain as possible that the potential screenee is aware of all the risks and benefits of screening. Although compensational screening is unlikely to provide *medical* benefit, a successful applicant for compensation might receive such compensation sooner, perhaps while he or she still had the health to enjoy its benefits. Such early compensation might in many ways be analogous to the early treatment that motivates medical screening. Those issues should be made explicit when counseling a potential screenee.

Therefore, **the committee also recommends that counselors, when dealing with screening for compensation, ascertain that individuals proposed to be screened fully understand the associated risks, benefits, and likelihood of potential outcomes of screening.** Although ascertaining such comprehension is complex, a minimal standard might be to have the patient verbalize those risks and benefits in his or her own words. When the level of understanding seems inadequate, counselors would be expected to engage in a more detailed explanation couched in terms and using examples tailored to the individual's literacy and numeracy capabilities and taking into account cultural issues. Moreover, if a screening test produces a positive result, even one that suggests the possibility of a disease that is not compensable under RECA, the clinician or counselor should communicate that result to the screenee, in a culturally sensitive manner, and make an appropriate referral for followup.

The drawbacks of compensational screening, especially in the event of low prevalence rates and multiple tests, led the committee to consider the desirability of what is, in effect, a two-stage process. Specifically, **the committee recom-**

mends that RESEP screening be undertaken only if individuals satisfy administrative criteria for compensation before screening. By satisfying administrative criteria, we mean that patients need to be certain that they are already administratively eligible before any screening takes place. Meeting these criteria can include documenting employment or residence history (as appropriate), and, for cancers, obtaining an individually calculated PC/AS for the cancer being proposed for screening.

Tying Screening to Probability of Causation and Assigned Share

Criteria for Probability of Causation (Assigned Share)

In addition to considering the medical certainty in establishing the presence of a RECA-compensable disease, one must take into account the administrative requirements for eligibility for compensation. In Chapters 5 and 6, the committee proposed that Congress amend RECA by requiring the use of a calculated PC/AS for each exposed individual; in determining compensation, that value and its uncertainty (credibility interval) would be compared to the criteria that Congress establishes for a cutoff. As discussed by several experts (Lagakos and Mosteller, 1986; Robins and Greenland, 1991), such calculations may introduce a bias against individuals harmed by their exposure to ionizing radiation because such exposure can be a contributing cause of disease without being the predominant cause of disease. As a case in point: if an individual so exposed to fallout were to develop a thyroid cancer but, based on her residence history and age at exposure, her calculated PC/AS were only 35%, then exposure to ionizing radiation from fallout could have been an important factor in her cancer. Nonetheless, if the rules for compensation were to be set at a PC/AS of at least 50% (as likely as not), then the contribution of the fallout would go unrecognized and uncompensated. Of course, this shortcoming might be managed by setting the cutoff criterion for PC/AS at some value below 50%, by employing credibility limits around PC/AS estimates, or both.

As discussed in Chapter 5, we relied on contemporary scientific evidence and reasoning in recommending that RECA be amended to use a PC/AS approach. We did not, however, find a scientific basis for establishing a particular threshold value of PC/AS or a specific allowable uncertainty in an individually calculated PC/AS that should be used for determining administrative eligibility for compensation. Calculating an individual PC/AS and comparing its distribution with such threshold criteria are administrative activities that are predicated on the presumed presence of a RECA-compensable disease (see also Chapter 11).

Making such a calculation, however, does not require that screening or diagnostic tests already have been performed in that individual. The individual will need to identify the disease or diseases for which he or she wishes to be screened for (or obtain) the PC/AS calculation to be performed. Thus, calculating a PC/AS and making a judgment about administrative eligibility do not require that any patient first be screened and (thereby) subjected to the costs and risks of such screening tests.

Clearly, a RECA compensational screening program could look for the presence of any or all of the RECA-compensable diseases. Which tests or combinations of tests to perform should involve the patient's informed consent about risks and the often-small likelihood of benefits (primarily from compensation) of screening. The patient's choice of the diseases for which to seek compensation might reflect several factors: the preassessment of the maximum reasonably expected PC/AS in a population or geographic region (as called for in the explication of the committee's first recommendation); the likelihood that such a disease is present in a particular patient; the calculated PC/AS value for that patient; and the patient's desire to receive information about ailments about which he or she may be anxious.

Using the same threshold criterion for PC/AS as for establishing the diagnosis of a RECA-compensable disease may be tempting. Doing so requires interpreting the criterion; for example, it might be taken to mean "with reasonable medical certainty"—either "more likely than not" or "as likely as not." Because calculating PC/AS and making a diagnosis are distinct issues, however, the criteria (thresholds or cutpoints) may well be different. Currently, RECA does not use the PC/AS approach (although it had been proposed before RECA for other compensation programs), so it is silent on what the criterion or cutpoint should be.

Apart from a point estimate for a threshold is the question of uncertainty around the point estimate. For example, although EEOICPA uses a "more likely than not" criterion, it gives the exposed individual an enormous benefit of the doubt by requiring that the person have only at least a 1% chance that the true PC/AS exceeds 50%. That is, EEOICPA, like some other compensation programs, compensates on the basis of an upper credibility limit, analogous to the upper bound of a confidence interval, set in the EEOICPA case at 99%. Currently, when applicable in REVCA, radiation doses and information in the 1985 NIH radioepidemiological tables from the National Institutes of Health are used to calculate a PC using the version of the Interactive Radioepidemiological Program (IREP) software from the National Institute of Occupational Safety and Health. The PC is used in parallel with consideration of other factors, to make the determination of eligibility for compensation (correspondence with Department of Veterans Affairs, March 31, 2005).

As explained in Chapter 5, Congress might consider two different criteria in determining who might be due compensation. The first is the value of a specific threshold or cutoff of PC/AS above which compensation can be awarded (such as 50% for EEOICPA), but perhaps 30% or 10% or any other value (below 50%). The second is the use of an upper credibility limit such that the value of some specified upper percentile of the PC/AS uncertainty distribution is compared to the designated threshold. Compensation would be awarded when the value of the upper credibility limit exceeded the threshold.

With some knowledge or expectation about the distribution of PC/AS values and their associated uncertainties, Congress should be able to determine how the number of individuals compensated and the cost of the program (both RECA and RESEP) might vary with different criteria. The committee did not have the time or resources to examine the cost issues in detail.

From the perspective of populations now receiving compensation or possibly eligible for compensation, it is instructive to note that many individuals would have calculated PC/AS values well below a 50% threshold. Whether the upper percentiles of their uncertainty distributions might fall above the 50% threshold is less clear.

Should the Calculated PC/AS Affect How Screening Tests Are Interpreted?

As discussed in Chapter 9 and Appendix D, many medical tests (both screening and diagnostic) can have sensitivities and specificities that vary depending on the criterion of positivity chosen in a particular circumstance. By analyzing a test's receiver operating characteristic (ROC) curve with knowledge of the prevalence of disease in the population to be tested and the burdens of false-positive and false-negative classifications that the test results may impose, one can determine the optimal operating point on the ROC curve and its corresponding cutoff criterion.

That principle is often applied to screening. If the population to be screened has a low disease prevalence or if the burden of missing a disease is low, one uses a strict criterion of positivity. If the population to be screened has a high disease prevalence or if the burden of missing disease is high, one uses a less strict (more lax) criterion of positivity. Similarly, in the context of RECA and RESEP (after administrative eligibility has been established as suggested earlier), the criterion of positivity for the screening tests performed to identify compensable disease could be adjusted according to the prevalence of disease in the population. The burden of a false-negative result will be largely the denial of compensation (and potential emotional effects of such denial) because, as stated previously, most of these screening tests will not produce any improvement in health outcomes. The burden of a false-positive test, however, will be the physical, psychologic, and social risks described in Chapter 9.

A more difficult question—which may not even be practical to address—is whether the criterion of positivity of a medical screening test should be adjusted to reflect the individually calculated value of PC/AS or its uncertainty. If one is compensating for radiogenic disease, the calculated value of PC/AS might represent the pretest likelihood (for the screening test) of a specific disease caused by radiation in a given individual (such as thyroid cancer). If that is the case, an individual with a higher PC/AS might logically be screened with a more sensitive test—for example, thyroid ultrasonography with a cutoff for nodule size set at 2 mm rather than 5 mm.

One might take the perspective that a disease should be worthy of compensation only if it has clinical effects. In that case, a tiny or inconsequential cancer would not be compensable. However, currently RECA apparently does not adopt this perspective. RECA compensates for the presence of pathologically or clinically established disease, even if that disease has no bearing on the patient's prognosis. For example, the family of a patient with an asymptomatic pulmonary nodule, shown to be cancer, will be compensated even if the patient is struck and killed by a car, so that the cancer had no effect on his survival and produced no morbidity.

In addition to the challenging mathematics of how one should adjust the pretest probability of "disease" on the basis of PC/AS (although adjustments in the odds domain seem attractive at first blush), at least two practical challenges arise. First, in a screening program, how does one manage the logistics of adjusting test interpretation on an individual basis? Second, would using a more sensitive criterion change the apparent base rate of disease in the population, and would that affect the excess relative risk and the calculated PC/AS? Notwithstanding those challenges, whether and how to think about adjustments to a PC/AS approach warrant further examination.

FUTURE RESEARCH

Many patients dealing with the tradeoffs among medical risks, health benefits, and potential monetary compensation will require decision-support tools. Broadly speaking, such tools embody ways for patients and clinicians to share their decision-making logic. Other intellectual and emotional supports are also important.

We are not aware of proven models of shared decision-making protocols that are adequately designed to help people to manage those difficult tradeoffs in the RECA-RESEP context. Some tools have been developed for behavioral and decision-making research (for example, for prostate cancer screening decisions). The entire field of shared decision making and the tools to support it are important topics for future research, both in the domain of collaborative communication and in the domain of explicit mathematical modeling.

Thus, **the committee recommends that the Department of Health and Human Services support development of explicit decision models and approaches to shared decision making and related tools that enhance the ability of patients to participate in decisions that affect their care and prognosis. We recommend, in particular, that HRSA take responsibility for similar activities in the domain of compensational screening.**

In addition, the committee's proposed strategy for using individually calcu-lated PC/AS values for RECA eligibility determinations will require tools to make such calculations more convenient. The committee also calls for further exploration of the benefits, limitations, potential biases, and necessary extensions of the PC/AS approach.

HEALTH-CARE ISSUES BEYOND SCREENING

Followup Diagnostic and Therapeutic Services

One question that arises for of any medical screening situation and presum-ably for RECA and RESEP even for compensational screening concerns the obligations to patients inherent in the programs themselves and in the agency roles of health care providers. When applicants or patients are found, through RESEP screening, to be at special risk for a particular RECA-compensable con-dition, the expectation (and apparently generally the practice) is to refer them for appropriate diagnostic followup and therapy. When applicants or patients are found, through similar screening, to be at special risk for a noncompensable disease, those patients should be advised to get appropriate medical management, independent of its possible RECA connection.

For that process to be effective, health care practitioners have to be informed and stay abreast of knowledge in several domains. These include the special risks known to be associated with mining and related activities and health care issues that may be related to radiation doses received by covered individuals at special risk (for example, children in selected areas who drank large amounts or milk or consumed foodstuffs contaminated with iodine-131). In addition, if Congress amends RECA to adopt a PC/AS strategy for populations, geographic areas, and diseases not now covered, clinicians need to be knowledgeable about the method and the implications for their patients. The committee urges HRSA and appropri-ate professional societies and other groups to assist in educational activities for health care professionals related to these topics. We return to these matters in Chapter 11.

Health Insurance

In Chapter 8, we noted a distinction between the duty to restore a loss to health and a duty to compensate for the effects of the loss. These are distinct duties; compensation for the effect of the loss does not fulfill a duty to ameliorate the loss of health by providing health care.

The remedy for the loss sustained is amelioration or restoration. Medical services are probably the only, if not the best, manner in which to provide resto-ration. Because of the nature of the health loss, in many cases discovering the loss may be possible but restoring health impossible. In other cases, some measure of restoration may be possible through medical services. Services begin with discovery through first evaluation and diagnosis and proceed to referral, workup, and care that are medically indicated (Sheridan et al., 2004).

The committee received many written and oral statements, both at its information-gathering meetings and while reviewing the programs of HRSA grantees, that many RECA applicants or patients are uninsured or underinsured for the costs of followup diagnostic or therapeutic interventions. Other populations covered by radiation-compensation programs (REVCA and EEIOCPA) provide medical services; uranium workers covered under RECA are also covered under EEIOCPA for medical expenses.

Hence, for consistency with other similar compensation programs and to overcome any inequities, the committee believes that similar relief is owed to RECA population, and, therefore, **the committee recommends that, if an individual has established eligibility for compensation. ᴿᴱᴄA cover the costs of screening, complications of screening, refeɪ.als (followup), diagnosis (workup), and treatment for the RECA-compensable diseases for which such eligibility has been established.** As stated earlier in this chapter, establishing such eligibility for people exposed to fallout from the US nuclear-weapons program may include obtaining an individually calculated PC/AS for the cancer being proposed for screening.

Screening in an Era of Scarce Resources

Apart from the issues discussed to this point about compensational screening, the cost of expanding RECA is a matter raised by some committee members, by some people who testified at the committee's information-gathering meetings, and by some reviewers of this report. The fundamental question is appropriate allocation of social resources, especially when those funds are limited. Some commentators posed the question in terms of expanding RESEP, in a (present) configuration that focuses so heavily on compensational screening (as well as outreach and education), to a national program. The basic problem is whether resources so directed, if now expanded to cover the country, would be a cost-effective use of HRSA appropriations. More importantly, one must ask whether expanding RECA to additional populations, geographies, or diseases is a potential threat to the "medical commons" of societally shared limited resources (Hiatt, 1975). The committee did not pursue these matters in depth; we view them as beyond both our charge and our time constraints. We note, however, that these matters are subjects that will warrant attention in the larger social debates about the changes in RECA suggested in our report.

CONCLUSION

This chapter has examined the conceptual and practical elements of screening for compensational purposes. We have linked these issues to the committee's adoption of a PC/AS model and we have noted some clinical and epidemiologic issues related to compensable diseases. Some of our recommendations are directed at agents that would need to amend RECA (or RESEP) legislation; others are directed at steps that HRSA can consider without statutory changes. The next chapter addresses on the remaining core elements of the RESEP program: education and outreach.

11

Education and Outreach

This chapter addresses several aspect of the committee's charge from the Health Resources and Services Administration (HRSA) about education, outreach, and screening in the context of the Radiation Exposure Compensation Act (RECA), which is administered by the Department of Justice (DOJ) and HRSA's responsibilities for the Radiation Exposure Screening and Education Program (RESEP). We take into account here the implications of the recommendations set forth in earlier chapters. Because of their inseparable nature in efforts of this sort, we use the terms "education" and "outreach" somewhat interchangeably. We focus on the needs of three broad target audiences: the general public, a variety of specific populations defined by occupational or residential exposure (such as uranium miners and people who may have been exposed to radiation from fallout from the US nuclear-weapons testing), and health care providers.

We briefly review our understanding of current efforts by the six HRSA grantees and the issues and problems with the existing RECA and RESEP programs that participants and HRSA grantees raised at information-gathering meetings in late 2003 and 2004. HRSA originally had six grantees but defunded one after one year and added another, so the number of grantees remains at six.

We examine these issues and their ramifications for HRSA in the context of moving to a "national" RECA program grounded in a probability of causation/ assigned share (PC/AS) approach (see Chapters 5 and 6). We use the term national to refer to expansion of eligibility to persons in all US counties and all US territories whose calculated PC/AS for at least one compensable condition meets or exceeds whatever threshold and credibility limits Congress or other bodies may recommend. As explained earlier, the committee does not intend to imply that all persons, in all counties, are automatically eligible for RECA compensation; rather, the committee's recommendations regarding screening turn in part on proof of administrative eligibility (including calculation of an individual PC/AS for a RECA-compensable disease). Insofar as earlier recommendations have direct ramifications for HRSA and its RESEP program, we make suggestions here for their implementation.

Finally, we outline a planning framework that may help HRSA to strengthen its education and outreach programs. Our recommendations and examples for implementation of proven educational intervention strategies are aimed at overcoming barriers to effective outcomes-based education programs that could be adapted for future RECA and RESEP programs.

CHARACTERISTICS OF THE RADIATION EXPOSURE SCREENING AND EDUCATION PROGRAM

Education is one of nine core activities expected of RESEP grantees. Grantees have told HRSA that they need guidance in public education and outreach mechanisms to those at risk or experiencing symptoms as a result of exposure to radiation (letter from HRSA to committee via Dr. Isaf Al-Nabulsi dated May 6, 2004). In developing its Request for Applications (RFA), HRSA asked potential grantees to develop strategies to expand and enhance public outreach and education in the following six categories:

- The possibility of disease.
- Symptoms.
- The potential need for diagnostic evaluation.
- The availability of screening for disease through RESEP.
- The possibility of compensation through RECA.
- The need for documentation of medical and occupational history if RECA claim is filed.

Although the RFA speaks to symptoms, as explained in Chapter 9 screening is a term that usually applies to asymptomatic individuals. If patients are symptomatic, the perspective is one of diagnosis and treatment, if indicated.

The main elements of RECA were reviewed in the several earlier chapters that dealt with the history of the program, compensation under the current program, and new information and science that will influence future developments. Chapters 9 and 10 examined the screening aspects of the RESEP program. We comment briefly here on the audiences, responsibilities, activities, and funding of the HRSA's RESEP effort.

Audiences

Obvious targets of RECA and RESEP educational and outreach efforts are the public at large, various specific populations defined by occupational or residential exposure (such as uranium miners and downwinders [that is, those who may have been exposed to radiation from fallout from the US nuclear-weapons testing]), and health care providers. The RESEP legislation does not specify populations for education and outreach, but HRSA implicitly adopted this framework, and we apply it here.

Responsibilities

With respect to its scope of services and diseases, RECA and especially the amendments in 2000 that mandated the RESEP program can be read narrowly as focusing on radiogenic diseases (and secondarily on some nonradiogenic pulmonary and renal conditions after occupational uranium exposures). In fact, HRSA has expanded the focus to involve education for persons at risk of or experiencing symptoms of any disease secondary to the radiation or other exposures that made a person eligible for compensation. Thus, HRSA has gone beyond screening per se, although the committee was changed with examining screening. Specifically, HRSA requires its RESEP grantees to convey information about the possibility of disease, symptoms, screening and diagnosis, RECA compensation, and the need to document both medical and exposure (such as occupational) history. The committee agrees that those are, in broad terms, appropriate subjects of education.

Activities

Table 11.1 briefly describes the populations and geographic areas that the current six HRSA grantees cover. It outlines their screening, referral, and other protocols and documents various activities. Table entries were based initially on materials and presentations at a committee information-gathering meeting in Window Rock, Arizona. Grantees later provided further or updated information. Thus, the table has data covering the period spring 2004 to early March 2005.

Grantees carry out a considerable array of outreach and educational efforts. They occur through form letters and various types of mailings, articles in local newspapers, radio and television spots (for example, public service announcements), and education and followup by registered nurses for instruction and health information. The efforts and products vary widely among the grantees, and at least one grantee does little or no broad outreach through broadcast and print media because of costs. We could not determine the extent to which HRSA itself or the grantees standardize messages delivered through different media (print or broadcast) or either centralize or share responsibility for material and message development.

The numbers of people reached through education and outreach efforts (not including medical contacts) apparently vary widely, but stating an overall range is difficult because grantees do not count contacts similarly. One grantee does statewide outreach, so the population contacted (in theory) is in the millions; others count contacts in the hundreds to thousands.

Table 11.1 also notes some clinical outcomes, such as numbers of persons with abnormal tests or possible diagnoses. Those data, too, vary markedly among grantees in both types of variables recorded and numbers (or percentages).

Other than these types of process measures, neither HRSA nor grantees provided evidence of the outcomes of the activities. In particular, we received no information on patient health outcomes as a result of these efforts. In addition, no data on programmatic outcomes, such as better understanding of RESEP or the RECA program or improved access to either program, were provided. Grantees did not appear to be using any outcomes-based educational model.

Finally, the far right column of Table 11.1 presents the many issues that grantees identified in the public meetings or later communications. Some fit well with points that the committee addressed as part of its overall charge. An example is the concern from the Indian Health Service grantee (Navajo Area Health Service) about tests used to document presence of compensable diseases that are not appropriate screening tests, which underscores the distinction made in Chapters 9 and 10 between medical screening and compensational screening. Other issues lay well outside the committee charge per se, but some were heard from more than one grantee. Examples included concerns with attorney fees, lack of feedback to grantees from DOJ on outcomes of compensation claims, and resumption of nuclear weapons testing at the Nevada Test Site. Yet other matters, such as concerns with lack of public or private coverage or other means for paying medical costs of referrals and treatment, were ones that the committee discussed and used in arriving at policy and program recommendations found in other chapters of this report.

Funding

HRSA implements RESEP through a grant mechanism. The RESEP RFA had a fair amount of detail about desired or required activities. Nonetheless, a grant, by definition, permits awardees to fashion programs according to their own preferences and capacities, which are not necessarily related to the populations that they serve or the diseases in question. That the grantees would differ substantially in operations, therefore, is hardly surprising.

Another element of a grant mechanism is a relative lack (at least in comparison with contracts or cooperative agreements) of accountability for activities and allocation of resources except perhaps annually or only upon renewals, which may involve even longer periods of time. HRSA defunded an original grantee after 1 year because of nonperformance. The basic point is that overseeing

TABLE 11.1　HRSA Grantees: Populations and Areas, Screening, Education, and Outreach Activities; Issues

HRSA Grantee, and Population Covered	Screening and Referral Protocols	Individuals Reached
Navajo Service Area Radiation Exposure Screening and Education Program Bruce B. Struminger Navajo Area Health Service Department of Internal Medicine NNMC, Box 160 Shiprock, New Mexico Covers miners, millers, and Nevada Test Site downwinders in New Mexico, Arizona, Colorado, and Utah (essentially Navajo Nation populations) Some post-1971 miners included	Focused on testing for RECA compensation ("disability/ compensable illness") program Send various form letters to primary-care providers (for downwind exposure) and to patients (after chart review to identify RECA diagnoses) Referral and treatment: Indian Health Service (IHS) or outside contracted specialty services Followup: RESEP clinic or patient's IHS primary-care physician	Outreach contacts: ~1,500 Medical encounters: >1400 Educational activities: >100 Eligibility assistance provided: >1200 Of total uranium workers screened (n >1300) since 10/02: 323, positive arterial blood gas 186, positive spirometry; > 181 meet RECA medical qualifications, but not all can prove eligibility because of difficulties documenting work history Have identified > 2000 individuals (Navajo and Hopi) who may meet RECA qualifications as downwinders Have contacted > 100 of these individuals to assist them with identification of the appropriate documentation (from their IHS records) needed for their RECA claims

Issues Identified

Clinical:
- Arterial blood gases and chest x rays are essential for compensation proof, but are not appropriate screening tests
- Latent tuberculosis of concern (reactivation of latent tuberculosis infection related to silicosis)
- Many patients with comorbidities

Other:
- No feedback from Department of Justice on who has been awarded compensation; IHS cannot apply for compensation and patients cannot get transportation covered
- New regulations change the parameters for qualifying FVC and FEV1, using ethnic-specific lower limits of normal rather than previously qualifying FVCs and FEV1s ≤ 80% predicted. This change reduce number of RECA qualifying spirometry examinations
- Necessity of "B" readings—special radiologic interpretations evaluating for evidence of silicosis derived from black lung program—which are increasingly hard to get
- IHS is going digital and no NIOSH guidelines exist for readings of digital images
- Use of unregulated, private consultants or lawyers who overcharge claimants
- Consider amending RECA simply to require proof of exposure without proof of illness or illness severity, both for social justice reasons and given imperfection of current testing regimen, which requires that physicians and nurses compromise Hippocratic Oath to "do no harm" given that potential harm caused by medical testing rarely leads to therapeutic options (no treatment exists for pulmonary fibrosis)
- Consider amending RECA to allow the use of affidavits for proof of presence or residence

continued

TABLE 11.1   Continued

HRSA Grantee, and Population Covered	Screening and Referral Protocols	Individuals Reached
St. Mary's Hospital and Regional Medical Center Teresa Coons, PhD Mercedes Cameron, MD St. Mary's Saccomanno Research Institute 2530 N 8th Street Suite 100 Grand Junction, CO 81506 Covers Colorado and southeastern Utah Navajo areas; Wyoming (uranium miners)	Screening focused on "whole person," not just compensable diseases Extensive medical, occupational, and residential history (but no invasive tests for initial screening unless indicated on work-history or other clinical grounds) Followup within St. Mary's network; Much RN educator instruction and health information and recontact every 6 months	Data as of 11/30/04: 425 individuals screened so far Of those screened: 5% had abnormal oxygen saturation test, but 70% had abnormal arterial blood gasses 69% eligible for Medicare or Medicaid 7% have no insurance Remainder generally have third-party insurance 90% are male ~69% are ≥ 65 years old ~90% are non-Hispanic white ~88% are former uranium industry workers (miners, mill workers, and ore transporters) 327 non-contrast computed tomographic scan of chest completed; 49% abnormal and needing followup 331 patients have been referred for additional diagnostic evaluation or treatment

Issues Identified

Clinical:
- Value of screening these populations just for RECA cancers and other diseases when use of resources might be better justified for providing more complete health care
- More complete examination protocol may lead to identification of radiation exposure-related conditions that are unknown or underestimated; this is important opportunity with RESEP population that should not be missed
- Relationship between renal disease and diabetes for populations that center sees

Other:
- RECA-eligible populations are aging
- Area is very rural, so transportation is issue (although people are accustomed to coming to Grand Junction for other reasons)
- Patient mistrust of "government" studies or services in some cases (Colorado Plateau study or other reasons)
- Expansion of eligibility for RESEP program to post-1971 miners or other groups not included in RECA legislation (but having same exposures) might provide some sense of "justice" to those not included in RECA without amending legislation
- Consider expansion of downwinder category to other geographic regions with documented iodine-131 deposition (NCI study data)—for example, areas in Colorado, New Mexico, and Idaho where deposition levels were similar to those found in current "downwinder" counties

continued

TABLE 11.1   Continued

HRSA Grantee, and Population Covered	Screening and Referral Protocols	Individuals Reached
St. George and Dixie Regional Medical Center	Screening protocols "very similar" to those of other RESEP sites, but screening mostly of downwinders (92%) because of the proximity to geographic area related to Nevada Test Site	Have been seeing patients since March 2004
Rebecca Barlow, RN, BSN, OCN, CPON		Outreach contacts
Carolyn Rasmussen, RN, BS, OCN, CPON		—Newspaper ads/articles—radio spots/interviews: 276
Dixie Regional Medical Center	Provide much cancer screening education, giving written material, to all patients, as well as general healthy life information to those with identified diseases (hypertension, diabetes, and so on)	—Television: 3
544 SO 400 East		—Pamphlets distributed: 2,920
St. George, Utah		—Presentations: 20
		—Fliers distributed: 3,360
		—Interview for documentary on effects of nuclear testing: 1
Covers southwestern Utah and the tristate area of southern Utah, Arizona and Nevada		Number reached by all outreach *(total for all sources):* >29,300,000 people exposed to our outreach attempts through all media
Would like to expand to hold clinics quarterly in Cedar City, Utah, and in Colorado		Total patients scheduled: 716 Total patients screened: 595
		RECA information given to patients and/or general public: about 6,521 Patients sent to RECA specialist for assistance with claims: about 106
		RECA claim form given to potential claimants from office: about 322
		Claimants paid: unable to find out from Department of Justice which claimants have been compensated; will send out 6-month survey to try to capture that information
		Patients served: —Male 46% —Female 54%

Issues Identified

Clinical:

- Screening tests covered by grant monies because Medicare does not reimburse for "screening" examinations
- Uranium miners, millers, ore transporters also get arterial blood gas tests, screening spirometry, two-view chest x ray, and CMP
- Total medical referrals: 1,145

Referrals Made:
— Colonoscopy: 320
— Mammography: 237
— Prostate-specific antigen: 143
— Pelvic exam: 62
— Pap smear: 52
— EGD: 15
— Thyroid ultrasonography: 5
— Testicular ultrasonography: 4
— Breast ultrasonography: 3
— Miscellaneous (for instance, to primary-care physician for non-RECA ailments): 151

Abnormalities Found:
— Rectal mass or positive stool: 49
— Suspicious skin lesion: 47
— Breast Nodules: 24
— Dysphagia: 2
— Prostate nodules: 9
— Thyroid nodule: 5
— Testicular nodule: 5
— Pulmonary nodule: 1
— Prostate cancer: 1

Other:

- Potential of reopening Nevada Test Site (underground "bunker bombs") of huge concern for all residents in area
- Give list of attorneys in area that help people with RECA compensation, but explain to all potential claimants that attorneys are restricted in what they can charge and are not required
- People in tri-state area are used to coming to St. George to receive medical care, do shopping, and so on
- Oncology nurses for 10 years, and have worked with families consumed by cancer and its diagnosis
- Biology of cancer is explained by "two-hit method"—atmospheric exposure was a "hit" exposure to people that does not go away; the younger the patient when exposure occurs, the greater the chance for late effects
- Concerned with interim report about screening not harmful, biggest way to catch cancers early; supported by American Cancer Society
- Feel some monies should be made available to fund followup of abnormal screening findings in high-risk people who have low income and are uninsured

continued

TABLE 11.1   Continued

HRSA Grantee, and Population Covered	Screening and Referral Protocols	Individuals Reached
New Mexico Radiation Exposure Screening and Education Program, University of New Mexico Health Sciences Center Karen Mulloy, DO, MSCH and Elizabeth Kocher Department of Internal Medicine MSC10 5550, 1 University of New Mexico Albuquerque, NM 87131-0001 Covers: all persons in New Mexico except for those members of the Navajo Nation (areas including Grants, Laguna Pueblo, Acoma Pueblo, Gallup, etc.)	Screening protocols are focused on the diseases linked with uranium mining and milling. Complete medical and occupational history taken on all individuals. Chest x-ray with B-reading, spirometry, oximetry, and focused physical exam completed on everyone. BUN, creatinine, UA completed on ore transporter & millers. ABG ordered when medically indicated. CT scan ordered if no PCP and medically indicated Followup with PCP and/or RESEP clinic	Outreach contacts: 398 Medical encounters (screened individuals): 218 Educational activities: 48 Eligibility assistance provided: 115 Medical referrals: 99 Received compensation: 3 Do not hear back on most cases whether accepted or denied
Utah Navajo Health System RESEP Stephanie Singer, MD P.O. Box 130 Montezuma Creek, UT 84534 Covers the Utah Strip of the Navajo Nation through 3 community health centers and collaborates with St. Mary's Northern Navajo Medical Center, and Dixie Regional	Screening protocols are the same as those used by St. Mary's RESEP	~2450 people reached (via presentations, mailings, or direct contact) ~395 people screened, of those screened: 15 positive results 15 people referred Regarding compensation: 3 people received compensation ~16 applications submitted, but no decision yet ~35 applications are in process

Issues Identified

Clinical:
- Other diseases not covered by RECA seen in miners—cases of asbestosis found in miners with no other work history

Other:
- How to reach miners who might have moved away (involves contacting extended families, miner organizations, and use of "word of mouth")

Clinical:
- Little primary or preventive care available internally to these populations

Other:
- No feedback from Department of Justice, so they do not know where applications stand
- Area is very rural and remote, and population does not have telephones or electricity, so followup must be by mail

continued

TABLE 11.1   Continued

HRSA Grantee, and Population Covered	Screening and Referral Protocols	Individuals Reached
The Arizona Radiation Exposure Screening & Education Program (AZRESEP) *"Serving AZ Downwinders and Uranium Mining Industry Workers"* Linda M. Nelson, MPH AZRESEP Director Mountain Park Health Center 635 E. Baseline Rd Phoenix, AZ 85042 Covers uranium miners, millers, ore transporters, downwinders, and onsite participants with focus on the downwinder population	Standardized AZRESEP protocol handbook has been developed. All sites contracted with Mountain Park Health Center are required to follow protocol laid out in handbook. MPHC contracts with three additional main sites (Canyon lands Community Health Care in Page, AZ; North Country Community Health Center in Flagstaff, AZ; and Yavapai County Community Health Services in Prescott, AZ) Basically, protocol is 1. Perform eligibility screening (through questionnaire) 2. Make appointment for patient (unless patient chooses to go to own provider for insurance reasons) 3. Provide patient medical and occupational history, focused physical examination, laboratory tests, and x rays as necessary 4. Provide patient written medical summary with copies of test results in 6 - 8 weeks in mail with recommendations for followup 5. If patient is part of AZRESEP clinical system, followup occurs onsite; if not, patient is referred to his or her own PCP for followup	Outreach contacts (face-to-face and telephone encounters): 2002-2003: 110 2003-2004: 886 2004-2005: 307 *(Quarter 1)* Medical screenings: 2002-2003: 86 2003-2004: 63 2004-2005: 14 *(Quarter 1)* Medical referrals: 2002-2003: 2 2003-2004: 14 2004-2005: 0 *(Quarter 1)* Educational activities (presentations and media): 2002-2003: 48 2003-2004: 107 2004-2005: 14 *(Quarter 1)* RECA eligibility assistance provided: 2002-2003: 78 2003-2004: 298 2004-2005: 28 *(Quarter 1)*

Issues Identified

Arizona has high managed-care penetration, so we encounter problems with insurance plans in this state; most plans require that enrollee be associated with network of providers and have an established primary-care provider; if person with health insurance goes outside his or her network, the services rendered are not reimbursable, and patient is required to pay out of pocket for services

Many Medicare beneficiaries elect to enroll in managed-care plan as secondary coverage for Medicare-covered services. Even though AZRESEP sites use sliding fee schedule for uninsured patients, patients who have health insurance usually do not qualify for it; because AZRESEP sites are required to bill for services, some people have been deterred from coming in for any medical screening that they may have to pay for even if they have health insurance

Other identified issues have been that demand in Maricopa County for medical screening of RECA-eligible people has been lower than expected, and cost of implementing full-fledged marketing campaign in area is beyond realm of funding resources; we initially thought that high number of eligible people would reside in Maricopa County (even though it is not classified as RECA downwinder county) because of out-migration of people from northern Arizona counties, in-migration of people from surrounding states, and sheer number of people who live in Maricopa County (3,396,875) compared with the rest of Arizona (2,232,695); in last past quarter, print advertising has been taking place in Maricopa County with disappointing results; furthermore, prohibitive costs associated with media (especially radio and TV but also print) limit amount of media outreach and education possible with small budget activities and tracking progress are typically not as straightforward for grants as for contracts; to the extent that standardization of messages, materials, and uses of federal funds is desired; we believe a contract mechanism offers more agency control than grant program.

The committee was concerned about standardization and accountability even for the present configuration of RESEP. If RECA moves to a national PC/AS mechanism for determining exposure and potential eligibility, even one involving the preassessment activities proposed earlier, RESEP audiences will expand geographically and change in demographics; similarly, HRSA's responsibilities for RESEP screening, education, and outreach will grow. In our judgment, therefore, HRSA may do well to revamp the RESEP program in several ways. Some recommendations appear below, but we offer the following as a cross-cutting matter: **The committee recommends that HRSA change its RESEP funding mechanism from grants to contracts.**

HRSA might consider using a cooperative agreement mechanism instead of a pure grant program. In our judgment, that would not be a meaningful step, in that cooperative agreements are administered more like grants than contracts. We offer several reasons for our recommendation to move to a contract mechanism.

First, a contract vehicle would enable HRSA to standardize protocols for medical and compensational screening in accordance with the recommendations in Chapters 9 and 10. HRSA's current grantees provide different (or different levels of) services to eligible populations, as documented in Table 11.1, although part of the explanation may be different amounts of grant awards. Nonetheless, to the extent that grantees' goals and approaches differ, they risk inadvertently discriminating against some populations and for others. Such discrimination—insofar as these groups can be considered alike for purposes of RESEP activities, and however unintentional—is not consistent with appropriate use of public funds or with our understanding of the ethical principles that underlie RECA.

The committee would support advances that offered all populations timely access to the services appropriate for realizing RECA and RESEP goals. This is, in part, the implicit motivation for its recommendation about monetary coverage of screening, diagnosis, and treatment services for administratively eligible individuals. The basic activities of the entities that HRSA funds for RESEP need more standardization and supervision to be made consistent. A contract would provide this.

Second, a contract mechanism might help HRSA to expand and centralize attention on the content and literacy level of patient-oriented materials of all types (including non-English versions). Literacy and numeracy are problems even for the current grantees' audiences, and the problems are likely to grow as additional populations not steeped in these issues for the past decade are drawn into the RECA program. We return to literacy and numeracy issues later.

Third, for similar reasons of standardization and efficiency, a contract vehicle is likely to help efforts to inform physicians, other clinicians, and other professional groups about RECA and RESEP (especially in previously uncovered areas of the country). HRSA can work with research or technical assistance firms and with academic institutions and other organizations to develop and test educational messages, in various media, and to create and disseminate appropriate clinical practice guidelines and protocols. HRSA might, for example, partner with the Agency for Healthcare Research and Quality (AHRQ) in commissioning systematic reviews of treatment options for these conditions and in disseminating information from the US Preventive Services Task Force on updated screening protocols for cancer and other relevant disorders.

Fourth, contracts provide a way for HRSA to begin and then maintain ongoing program evaluation of RESEP activities and services beyond what is now possible through annual grant renewals. Program evaluation itself does not appear to be an element of RESEP administration, and this is a matter of some concern. We can envision that Congress and the Department of Health and Human Services might wish to direct more attention to the effectiveness and efficiency of RESEP efforts as RECA itself becomes a more complicated program involving more diverse geographic areas.

In short, we believe that contracts will offer the most efficient and effective mechanism of administration, especially for any national program for which entities already involved in RECA-oriented and RESEP-like activities do not exist. The one possible exception is the Indian Health Service, which in principle could expand its reach to Native American populations outside the Southwest. Efficiency and appropriate oversight are more likely through a contract than a grant mechanism as the RESEP program expands to serve broader RECA purposes.

With respect to funding for HRSA for RESEP, we note that recommendations to move to a PC/AS approach for RECA for the nation as a whole markedly increase the size and scope of RESEP activities beyond the groups in the western and southwestern counties now eligible for RECA compensation. That may have nontrivial implications for RESEP's overall costs. Some observers and policymakers may judge that the administrative costs to expand education and outreach programs to those who are at a very low level of risk (with attendant claims, perhaps in large numbers, from individuals who may not be granted compensation) are not in reasonable proportion to the gain. The committee believed that issues of costs (for RECA or for RESEP) related to its recommendations, although of considerable importance, exceeded its charge for this report, and so we do not explore RESEP programmatic costs further in this chapter.

CORE ISSUES FOR RESEP

At the public meetings in Arizona, New Mexico, and Utah, the committee heard many points that direct attention to issues that HRSA should address. As already introduced, Table 11.1 (rightmost column) briefly note the numerous issues that grantees brought to the committee's attention. Beyond these, many speakers and beneficiaries of RESEP services presented stories and data underscoring what the HRSA grantees said and raising yet other topics.

The major problems arose in four categories: knowledge deficits, inaccessibility, cultural sensitivity, and inequity. Some were discussed in earlier chapters. We refocus on them here in the context of information from the public hearings and other materials that affected groups and individuals forwarded to the committee.

Knowledge Deficits

Knowledge or information deficits in the RECA-RESEP context can lie in several domains: awareness, acceptance, assimilation, and use of materials. The general public and many persons potentially eligible for RECA can, at worst, be said (as of the time of preparation of this report) to be unaware of the RECA and RESEP programs; at best, they probably have not assimilated the details of the programs. Those generalizations are more applicable to a national program than to a program focused nearly exclusively on selected counties in several western and southwestern states. HRSA (with DOJ) will likely need to modify and expand educational efforts to address the awareness domain of knowledge acquisition.

Some members of the general public have difficulty understanding or completing the 25- page RECA application form, especially the necessary residence documentation. Such problems will increase if Congress legislates the committee's recommendation for expanding the program through the PC/AS method and expands coverage of medical services for specified diseases to those approved for eligibility for compensation. As noted in earlier chapters, effective implementation of the PC/AS model may require that formal technical assistance be offered to individuals to use the dose and PC/AS calculators accurately (see Chapters 5 and 6). A single HRSA contractor might provide such assistance nationally either directly to individuals or to patient, volunteer, disease, or other advocacy groups.

HRSA may also need to address the preassessment activities suggested in relation to the committee's first recommendation about moving to a PC/AS approach to compensation. This calls for the National Cancer Institute (NCI) or other appropriate agencies to make two determinations: (1) which diseases are likely to have a high enough probability of satisfying compensation criteria, thereby eliminating scenarios in which the dose to tissues and organs would be universally so low that processing RECA claims would be unwarranted, and (2) which population groups have incurred sufficient risk from fallout-related radiation exposure to warrant consideration for compensation. The committee expects this work to be made widely available to the US public. For that reason, HRSA may need to play an important role in disseminating information for the various audiences that need to be reached and in targeting outreach and education to populations that are especially affected by the determinations (that is, of being at very high risk or at very low risk of exposure).

Inaccessibility

Ease of Understanding

Given the cultural differences, remoteness and broad geographic nature of the regions to be reached, and the limited literacy (reading ability) and numeracy (quantitative literacy) of targeted populations, such traditional outreach methods as pamphlets and public service announcements are of dubious accessibility and thus of questionable value. The impact of the current educational and outreach efforts (such as public service announcements, brochures, and similar media for RESEP and RECA messages) has not been rigorously evaluated. The committee remains uncertain about the net benefits of those efforts in populations now targeted, let alone in much more diverse sets of people who may emerge in a PC/AS-based national program.

Ease of Obtaining Services

Access to health care in the traditional sense (defined a decade ago by the Institute of Medicine as "the timely use of personal health services to achieve the best possible health outcomes" [IOM, 1993]) was often raised as an issue for the current RESEP program grantees and populations. Thus, physical inaccessibility to screening tests, timely referrals, and appropriate health care because of the expanse of the geographic regions covered may continue to pose problems for some populations possibly eligible for RESEP services or RECA compensation. The extent to which the problem may arise if the RECA program is extended through the national application of the PC/AS method is not clear, especially if administrative criteria must first be met. How access to care plays out will depend on the location of health care facilities, which are likely to be far more numerous (and accessible) in many parts of the country than in the present RECA states. It may also hinge on patterns of insurance coverage or eligibility for federal health programs, although the committee has recommended that RECA cover services related to screening (or complications of screening), diagnosis, referral, and treatment. Insofar as some "hot spots" are in relatively rural areas (albeit not as rural as some of the covered areas in the American Southwest), however, geographic inaccessibility may still present an obstacle to high-quality and timely services.

Cultural Sensitivity

In this section and later in this chapter, we use the issue of sensitivity to Native American culture as a prime example because the committee heard much testimony that such sensitivity was not always evident. However, the issue applies to the many different cultures of peoples exposed to radiation by the US nuclear-weapons programs.

Native Americans have a strong sense of being treated as "second-class citizens" at local treatment centers. The committee heard a good deal of testimony about apparent insensitivity to Native American traditions and medical practices. The committee notes that DOJ has apparently improved its practices as regarding documentation of residence, employment, and family relationships to permit more use of traditional (or unconventional) materials appropriate for Native American groups. The committee further addresses the use of affidavits for proof of presence or residence below.

Interactions across DOJ, HRSA, HRSA grantees, and future HRSA contractors with all target populations warrant review of issues of cultural sensitivity (including but not limited to literacy and numeracy questions). Expanding RECA will doubtless increase the need for attention to such concerns.

Inequity

The committee introduced ethical issues in Chapter 8 and wove these considerations into other chapters as well. Generally speaking, inequities are a major concern to the extent that like groups are apparently not treated alike. We note, however, that treating "unlike" groups dissimilarly does not raise an equivalent ethical dilemma.

Many representatives of groups that are not now eligible for RECA compensation spoke strongly about their sense of unfairness about the initial "arbitrary" selection of eligible geographic areas for RECA coverage. The committee understands their perception that a fundamental breach of ethical principles occurred and that it ought to be given great weight in considering their grievances. The basic reason is that they are "like" groups that have not been dealt with similarly. The PC/AS method of extending RECA eligibility nationally is an effort to redress some of these concerns.

DISCUSSION AND RECOMMENDATIONS

In this section, we examine further some of the issues introduced above and offer our recommendations for addressing them. We focus here on advising HRSA about education and outreach. In some cases, however, screening, considered in Chapters 9 and 10, reappears as a matter for further educational efforts and as a part of streamlining HRSA's administration of RESEP. The issues presented here are, in the main, categories used earlier: knowledge deficits, inaccessibility, cultural competence, inequities, and screening. By far the most critical steps will be to overcome knowledge deficits in both patient populations and clinicians; second will be screening questions in the context of education and outreach. **The committee recommends that the Department of Health and Human Services ensure that the content of public and professional educational programs be consistent across all entities that HRSA supports through its RESEP program.**

Knowledge Deficits

Educational Needs Related to Patient Populations

Clinical Issues   Making the RECA and RESEP programs better known to and understood by eligible populations is a major goal. It calls for a larger and better public education effort than we believe has been mounted to date. Although RECA stakeholders appear to be better informed than the general public about radiation issues and present law to remediate harm, one important need is to improve understanding of the nature and meaning of *radiogenic disease* (as dealt with in Chapters 4 and 7). Many conditions may have an etiology related to radiation exposure but also have competing causes that are not related to radiation exposure and may be more likely to be the true cause of a person's illness. Many disorders, including many cancers, have no proven causal relationship with radiation exposure (as discussed in Chapters 4 and 7). In our view, these facts must be laid out before the public at large in language that will be broadly understood, given the literacy competence in this country.

To understand better the RECA eligibility process, potentially eligible persons must have a general understanding of the PC/AS method. This includes an understanding of the "uncertainty" associated with this method (see Chapter 5). Also important will be conveying the reasons that Congress sets criteria of eligibility for compensation at whatever points it eventually adopts. Chapter 5 of this report might be a source for HRSA or others to use in developing better, more standardized documents and materials for educational purposes, although we recognize that it is not written at an average level of literacy for the US population and would need to be simplified for broader use.

Literacy and Numeracy   Low literacy in the United States (in any language, but conventionally considered English) is common. One typical test is the National Adult Literacy Study (NALS), which grades literacy in five levels (1 being lowest and 5 highest). A decade ago, about 90 million adults in the United States scored on level 1 or 2 on the NALS, which means that they have trouble integrating multiple pieces of information from a single document or finding two or more numbers and performing a calculation with them (Berkman et al., 2004; DeWalt et al., 2004). Both skills are critical to comprehending and using a PC/AS method; indeed, many daily activities today require literacy skills above level 1 or 2. Thus, low literacy must be taken into account in a broader RECA program; this requirement poses appreciable challenges for HRSA's RESEP efforts.

Numeracy is a companion construct to literacy, although in the United States it is less well understood and measured. Sometimes deemed *quantitative literacy*, it comprises the knowledge and skills required to apply simple operations of arithmetic, either alone or sequentially, using numbers embedded in printed material. Broadly, numeracy is the aggregate of knowledge, skills, and attitudes needed to perform mathematical calculations and manipulations, grasp measurement units intuitively, estimate known or unknown values, interpret and use mathematical and quantitative information, and think and express oneself effectively in quantitative terms. More prosaically, it is the ability to use mathematics at a level necessary to function effectively in everyday situations of school, work, and society. (For further definitional discussions, see Gal, 1995, and Saskatchewan Education www.sasked.gov.sk.ca/docs/policy/cels/el3.html#e13e10; accessed 12/21/2004).

The general US population is not especially numerate, when judged by questions related to probabilities (how many times will a fair coin flipped 1,000 times come up heads?) and percentages and proportions (converting a 1% chance of winning a lottery to 10 winners in 1,000 people winning that lottery, or converting a chance of winning a raffle from 1 in 1,000 to 0.1%) (Williams et al., 1995; Schwartz et al., 1997; Woloshin et al., 2001; Sheridan and Pignone, 2002). Sheridan and Pignone (2002) reported that 77% of first-year medical students answered three numeracy questions correctly, 18% answered two correctly, and 5% answered one or none correctly. In a study of mostly middle-aged women concerning understanding of screening mammography benefits, Schwartz et al. (1997) reported that 44% answered the coin-flip question incorrectly (generally underestimating the value); moreover, 46% of women were unable to convert 1% to a proportion accurately, and 80% were unable to convert 1 in 1,000 to a percentage.

Such findings call into serious question the ability of RECA populations to grasp the PC/AS concepts easily, let alone to interpret correctly the types of information that come from the dose and PC calculators. This is especially problematic because the population targeted for education in the PC/AS method may, indeed, be locations across the entire nation. In fact, setting the threshold or "cut point" and an acceptable level of uncertainty (for example, credibility intervals) for determining eligibility is a social or political, not a scientific, task. For that reason, affected populations may find themselves motivated to increase their numeracy to have more influence with national representatives on the issue if Congress amends RECA as advocated in this report. Education may be needed to help such claimants to interpret their own (or family members') values correctly. That may in turn call for HRSA to arrange for technical assistance to make it happen.

Benefits of Information Easy-to-understand information about RECA, RESEP, and radiogenic cancers may provide reassurance to some individuals. Specifically, they may be encouraged to learn that even though they had been exposed, not yet having developed a cancer to which the exposure may have contributed many years after exposure may mean that they never will develop it. Issues of age at exposure and the relationship between age and competing causes of cancer are dealt with in Chapter 4. Furthermore, reassurance may flow from educating both the public and health care providers that the relative risk of a radiogenic cancer (and therefore the likelihood that such a cancer will have been the consequence of an exposure to ionizing radiation) declines as time since exposure increases. **The committee recommends that HRSA and all its grantees undertake an appropriately focused educational program explicating the limitations, the benefits, and the risks of medical screening for many RECA diseases.**

In addition, people who know or believe that they were exposed through occupation or fallout from the US nuclear-weapons testing may come to a better appreciation of the disease to which they should be especially attentive. This involves both awareness of symptoms and choices about screening and lifestyle behaviors.

Finally, some people may expend much time, money, and emotion in pursuing campaigns for inclusion of "their" disease among those listed for compensation. If the public improves its understanding of the likelihood that radiation was not likely to have caused those diseases, some frustrating efforts that have little likelihood of payoff might be averted.

Claimant Applications   The RECA-RESEP application process clearly needs to be more user-friendly and easily explained. It is a 25-page application (http://www.usdoj.gov/civil/torts/const/reca/claimform.htm). A question remains as to whether improving and simplifying the application process are responsibilities chiefly of DOJ or HRSA. Nevertheless, if this program is to be implemented effectively, any individuals deemed newly eligible for possible compensation must be made aware of and learn how to use the existing compensation program and its procedures most effectively.

To make the application process more user-friendly, storing all claims and, if possible, clinical information at a central location would be a useful innovation. Those reapplying would then not have to repeat the entire application process. Implicit in this chapter is that HRSA bears substantial responsibility for meeting these informational (educational and outreach) needs, but the committee suspects that many elements of DOJ's administration of RECA could also be streamlined and improved. We do not address them here because DOJ program efforts were not a direct part of the committee's charge.

Multiple Compensation Programs   The multiple compensation programs are confusing to those who may be eligible not only for RECA compensation but also possibly (or alternatively) for other programs. For example, uranium workers are eligible for an additional $50,000 in compensation from the Energy Employees Occupational Illness Compensation Program Act (EEOICPA), but if an onsite participant were to apply to RECA and accept compensation, he or she could not then apply to EEOICPA for even the difference in compensation between the two programs. (See Chapter 2 and Table 2.4 for a more comprehensive review of different compensation programs.)

As part of HRSA's outreach effort, however, and to complement whatever DOJ does in this regard, we believe that HRSA should make more information available to potential claimants on all compensation programs, especially the programs that offer both medical services benefits and monetary payments. HRSA should combine elements of all those programs in one consistent description that helps potential beneficiaries compare programs and understand the full set of benefits to which they might be entitled.

That might be done in several ways. For example, HRSA's RESEP Web site could include more information about and links to relevant sites of other federal departments and agencies, such as the Department of Veterans Affairs and the Department of Labor. HRSA might also attempt to develop a simple triage system for guiding potential applicants to the compensation programs most appropriate for their situations. The committee did not explore these options in detail but did agree that a specific recommendation that HRSA act was justified.

The committee recommends that HRSA provide information to RECA populations about other radiation exposure compensation programs for which they might be eligible. The committee also recommends that an advisory organization should review all federal compensation programs related to radiation exposure to determine similarities and differences and that HRSA periodically convene representatives of all programs to address inconsistencies among programs and determine the effects of developments over time in radiation biology, risk estimates, legislation, and regulations.

Explanations and instructions must be at a level of reading ability appropriate for the target populations. Among the needs are getting and using information about the application process and obtaining help in using PC/AS dose calculators on the Web through the NCI or the National Institute for Occupational Safety and Health (NIOSH, through the Centers for Disease Control and Prevention [CDC]). This effort might be accomplished through some form of technical assistance as mentioned above.

The committee heard repeatedly that simple messages and clear instructions are critical. Complicated or incomplete messages and materials pose difficulties for many of the target populations and may induce frustration, anger, misunderstanding, and even mistrust. A related risk is that people will fail to obtain the benefits to which they are entitled. Thus, taking literacy and numeracy into account is a paramount requirement. Of special concern are materials in languages other than English—most likely Spanish, but also other Native American languages.

The above pertains essentially to print materials: newspaper articles, flyers and brochures, and the like, including Web-based information. HRSA and others should also give similar attention to anything conveyed through the broadcast media.

HRSA's role is to support development of such materials. We believe that the agency should provide guidelines about or assistance in creation and production in accordance with contemporary expectations in the health communications field. Taking the several points above into account, the committee makes several recommendations for HRSA activities. Most important in this context, **the committee recommends that HRSA ensure that all public informational materials are written so that members of target populations can understand their contents.**

All groups from whom the committee heard recommended that HRSA, the agency's grantees or contractors, and other interested parties use all types of media to reach out to potentially affected populations. *Media* in this context appears to mean both traditional print and broadcast channels: newspapers, bulletins, flyers, and similar print vehicles and radio and television media, including both commercial and cable outlets. Mailing materials may be the most direct way to reach members of some populations that lack easy access to broadcast media (at least in their own languages); in some cases, this might need to be accomplished simply through bulk mailing to "resident" or "occupant." We believe, in addition, that populations that have computer access and are computer literate may be well served by expanded use of Web-based outlets. Such Web sites should include all appropriate hyperlinks across sites.

Similarly, engaging all types of community-based organizations is critical. They will differ by locality, but the overall conclusion is unassailable. Among the types of organizations mentioned were local health departments and social-services departments, philanthropic entities (for example, local or regional foundations), and faith-based organizations (for example, church organizations or religious membership groups). Advocacy groups, whether related specifically to the RECA program—such as downwinder societies, mining unions and networks, veterans' networks, and tribal organizations—or to disease organizations, such as local chapters of the American Cancer Society and similar national groups, should be approached. Finally, service organizations, such as groups headed by local commercial leaders, may be useful.partners.

Apart from local efforts or those coordinated through entities other than HRSA, we believe that HRSA can do much more with its Web site. That will be critical if RECA becomes a national program, but even in its present state the agency's Web site is neither informative enough nor user-friendly. This chapter notes many informational messages and materials that HRSA might post on its Web site.

The committee suggests that HRSA expand its RESEP Web pages beyond the minimal elements of information found on the four current pages. Such expansion should include many of the messages and materials issues raised throughout this chapter, taking into account any expansion of RECA with respect to eligible populations and a shift to a PC/AS approach for new or existing localities (including the impact of preassessment findings). The committee also suggests that the Web site be internally hyperlinked more efficiently than it is now. Finally, the committee suggests that HRSA create an expanded set of hyperlinks to other federal agencies (particularly to all the relevant Web sites for DOJ, NCI, CDC, and NIOSH) and national stakeholder organizations in this area.

One element of the committee charge concerned adding new diseases to the current RECA list, and this was explored in Chapter 7. Given currently available epidemiologic, radiobiologic, and dosimetric evidence reviewed there, the committee did not recommend any additional diseases be added to the current list of RECA-compensable conditions. Thus, no issue arises for RESEP as to compensational screening for any new diagnoses. In Chapter 9, however, the committee suggested that HRSA may want to consider screening for depression, in health care settings with appropriate quality-of-care programs and adequate referral or follow through when needed; this is in accord with recommendations of the US Preventive Services Task Force for adult populations. In addition, the committee noted that it considers in utero exposures to be included in determining eligibility. HRSA and DOJ will need to explain and publicize these decisions and changes thoroughly through outreach to organizations, improved Web sites, better print or other materials of appropriate levels of literacy and numeracy, and revised application forms and other protocols.

Yet another challenge arises in relation to the committee's recommendations that RECA cover the costs of screening (and its complications) and appropriate referrals and treatment for RECA-compensable diseases for all individuals who have established eligibility for compensation. If Congress acts on this recommendation, then HRSA almost certainly will need to develop some educational program for both patients and providers to explain how these provisions will work.

Educational Needs Related to Clinicians

Clinical Issues   Facts now emerging about the long-term effects of radiation exposure on mental and emotional health must be emphasized in primary and specialty care fields. Psychologists, psychiatric social workers, and other appropriate community workers also need to be educated on these issues if they are to provide appropriate guidance and care to patients and (potentially) family members.

Another challenge for clinical audiences may be educating them about the applications of PC/AS methods by which their patients may need to determine potential eligibility for compensation. This is especially pertinent to the extent that documenting administrative eligibility for RECA compensation is a prerequisite for requesting compensational screening (as discussed in Chapters 9 and

10). As explained in Chapters 5 and 6, developing the information for a PC/AS calculation for an individual requires the presumption of disease.

We see a high likelihood that patients or their families will turn to physicians for explanations and assistance on PC/AS questions, and we are not convinced that the medical profession, on balance, has a solid understanding of statistical approaches in general or for the RECA program in particular. HRSA may need to take special steps to reach out to the medical profession with technical assistance or educational materials geared to the needs of physicians (or their office staffs) about the meaning and application of PC/AS calculations. The material in Chapter 5 that serves as the PC/AS primer will be an important place for HRSA (or professional societies themselves) to start. As noted earlier, the PC/AS primer information (Chapter 5) is one starting place for HRSA and professional societies. Inasmuch as RECA would not be the first compensation program to use PC/AS, HRSA might investigate effective education efforts on the part of EEOICPA and any other radiation compensation programs, here and abroad, that might be good bases for its own efforts.

Screening Issues   Physicians asked about their own role in RESEP. They asked whether they were expected to use routine screening and checkups with emphasis on radiation-related diseases, to provide examinations with nonroutine tests to determine eligibility for compensation, or both. They also asked about what cancers should be considered for examination.

Clinicians questioned the value of screening populations for RECA cancers when resources might be better used and justified for providing more complete health care. Health care providers—health plans, health care organizations, physicians, nurses, and medical social workers in particular—need reliable, up-to-date information. We were generally impressed with the level of professional knowledge and commitment among the current HRSA grantees on this score, although some questions remain (see below). Any expansion of RECA and RESEP, however, will generate a significant need for provider education on several fronts. The first is simply a better understanding of both programs tailored to the health care interventions and assistance that such professionals may need to give to their patients and families.

As implied in Chapters 9 and 10, screening alone requires that much be done to keep the medical profession up to date, apart from the need for better information about important distinctions between medical and compensational screening. Clinicians should ensure that all persons to be screened are aware of and comprehend the tradeoffs of the benefits of screening and the risks it poses. This is especially important in the context of compensational screening, when few health benefits are likely to accrue (but harms may well) and the likelihood of successful claims is low.

Moreover, as RESEP has been implemented so far, the variation in practices for referral, followup, diagnosis, and treatment is considerable. HRSA may have to spell out more precisely than it has so far what providers outside the southwest grantee areas (or the Indian Health Service) are expected to do. The recommended move to a contract mechanism, in which more procedures can be standardized and centralized, is meant to make it possible to meet such needs more efficiently and effectively. The preassessment recommended by the committee may provide useful guidance as well.

Finally, turnover is one important factor in reaching health professionals who are affiliated with institutions that care for RECA populations. HRSA grantees emphasized difficulties posed by changes in physicians, nurses, or other personnel. Among the intractable problems was providing professional education and training for RECA over and over. We see no obvious way to avoid the problem, but we note it as one that HRSA needs to take into account in future programmatic activities.

Educational Issues and Mental Health

Little or no epidemiologic work appears to have been done on issues related to mental and emotional status and the exposure and continuing concerns of miners and downwinders and their families. Nonetheless, the lack of empirical work cannot disguise the range of psychiatric problems they describe, allude to, or evince. However, the committee could not amass evidence of documented emergence of these conditions as a result of exposure or of sustained symptoms that would provide a direct, causal link between complaints and diagnoses identified today and exposure that occurred decades ago. For that reason, we could assemble no convincing arguments that these conditions should be added to the list of RECA-compensable disorders.

We conclude, however, that prompt provision of accurate information about radiation exposure and its consequences may ameliorate the psychologic consequences (anxiety, depression, post-traumatic stress disorder, chronic environmental stress) of a catastrophic exposure. Although the exposures covered by RECA were not catastrophic accidents, it may now be the case, many decades later, that providing full and truthful information can help to diminish the psychologic burdens that RECA downwinder or other populations may exhibit.

The committee recommends that HRSA undertake an enhanced program of education and communication about the risks posed by radiation exposure for people who may have been exposed to radiation from fallout from US nuclear-weapons testing. If Congress adopts the PC/AS approach to determine eligibility for compensation, education about the nature of the calculations may afford additional opportunities to inform downwinders and others about the relatively small magnitude of the risks. In short, as noted in Chapter 9, the committee emphasizes the need for greater educational efforts to help clinicians, patients, and families recognize mental health problems and obtain appropriate referrals and services for them.

Inaccessibility

The committee heard in several ways that HRSA and others must be especially cognizant of the specific needs or requirements of local populations and the lack of resources in some places. That is true especially for groups that have little access to television (meaning that public service announcements may be unavailable to them). Another barrier is the high geographic dispersion of some of the populations of concern (in rural and frontier areas, including reservations for Native American).

Language is a further complicating issue. Reaching Native American populations now (such as members of the Navajo Nation) may be problematic insofar as English is not their first language. Moreover, expanding RECA nationally raises the probability that messages and materials need to be rendered into Spanish (or perhaps other languages). For complicated subjects—such as RECA, PC/AS concepts and procedures, and services available through RESEP— adequate translations (taking literacy into account) may be problematic.

Cultural Sensitivity

The RECA and RESEP programs should work with local minority populations to understand these types of concerns and to ensure that equitable services are provided in culturally sensitive way. As already noted, the Native American groups in the current RECA areas of the southwest reported many instances in which they felt that the programs were not adequately responsive to their traditions and medical practices. HRSA needs to ensure that clinicians working in the RESEP programs are trained to understand cultural preferences and to develop ways to incorporate into their programs the nontraditional medical practices and life views of the Native Americans and of other groups and cultures (Blackhall et al., 1995; Carrese and Rhodes, 1995; Gostin, 1995). Owing to the complexity of the application process, one step that HRSA might take is to have its contractors engage case managers or triage personnel with special training and capabilities in providing education, information, and services in culturally appropriate ways.

Inequity

The committee heard repeatedly about equity issues. Native Americans stated repeatedly that they received fewer accommodations than the majority population to meet their needs and situations. They and other groups were concerned about the extent of fallout from the nuclear tests under consideration and the arbitrary nature of using geographic boundaries to determine eligibility. If RECA is amended, as recommended earlier, to extend eligibility through a PC/AS mechanism across the nation, a substantial public education effort will be needed in at least the states and counties (localities) that had high levels of fallout. The suggested preassessment effort may help to identify areas warranting early and high-priority attention.

A related public education and information effort should be directed at making the RECA program known to people in these localities. A question remains as to whether this is a responsibility chiefly of DOJ or of HRSA, but one way or another people potentially eligible to apply for compensation—all those outside the southwestern areas now eligible—must be told of the existence of the compensation program. If this is a DOJ duty, such information must include explanation of the existence of the RESEP program. In addition, people must be told how they can find and access needed information, forms, and the like.

To all that must be added explanations and instructions—at a readability level appropriate for the targeted populations—for getting information about the application process. Some potential applicants may need help using the PC/AS dose calculators on the Web through NCI or CDC.

Screening Issues Revisited

Somewhat problematic for the committee was the testimony from physicians regarding screening and their role in it. One issue was confusion about medical vs compensational screening; another was the number and types of tests (for example, arterial blood gases) that were performed on all patients because they were potentially eligible for RECA and whether this constituted good medical practice.

Excessive Testing Concerns

Grantees reported disquiet about some of the diseases being screened for and about the amount and complexity of the diagnostic testing required for potential RECA eligibility. Their concerns are well founded, as laid out in Chapter 9 about the potentially adverse effects of using multiple tests. The committee here reiterates that the several recommendations made in Chapter 9 for addressing these issues are relevant for improving RESEP activities.

Screening for Medical or Compensational Purposes

Clinicians associated with the existing HSRA grants expressed confusion about their screening role. Physicians' traditional view of screening typically does not include examining a patient for compensatory reasons. Chapters 9 and 10 differentiated between medical screening and compensational screening. Compensational screening involves specific tests that HRSA set out for radiogenic disease; these are not traditionally undertaken in routine medical screening but are required to establish eligibility for RECA compensation. The distinction needs to be much clearer to clinicians working in the RECA program. A term other than *screening*, perhaps *compensation examination*, might be developed for examinations that are undertaken strictly to determine RECA eligibility.

Although a potential for harm exists in all screening programs, considering such harm is especially critical for decisions that the public and clinicians make about screening tests for many RECA-compensable diseases. Little or no medical benefit appears to accrue from such screening to offset the risks of harm. In deciding whether to undertake such compensational screening, exposed individuals and their clinicians have to weigh a small medical risk against a perhaps even smaller possibility of receiving compensation under RECA.

In the committee's view, clinicians must inform patients of the possible harms whether they are providing medical screening or screening for compensation. HRSA needs to acknowledge to clinicians that it recognizes that screening tests are not perfect and that not all diagnoses can be medically certain. The committee advocates that, even for compensation examinations, screening activities be undertaken only when the tests and procedures in question are supported by credible scientific evidence and when the expected benefits outweigh the risks. We suggest in Chapter 9, consistent with our concerns in Chapter 8 for the underlying ethical concerns at stake, that shared decision-making models be researched. Here, we advocate that they be understood and implemented by clinicians in helping RECA stakeholder populations to make decisions regarding screening.

EDUCATIONAL PROGRAM PLANNING AND IMPLEMENTATION

So far, this chapter has discussed the issues identified by the general public, grantees, and the committee. We focus here more on the need for a planning framework and an outcomes-based educational model that HRSA could adapt for its future RESEP programs.

Theoretical Background

The discussion and recommendations outlined in this chapter are based on longstanding health-education and social-science concepts and theory (Lewin, 1935). A multidisciplinary body of research and activities labeled *health education* has grown since that time. Numerous groups have applied these concepts to help individuals to improve their own health and to assist health care organizations in raising the health status of their communities.

Health education encompasses a broad range of behavioral and organizational change strategies that are based on research and application by psychologists, sociologists, anthropologists, experts in marketing and communications, clinicians, and health care management professionals. Health education programs typically involve an intense evaluation of a well-defined target population; they examine knowledge, attitudes, goals, perceptions, social status, power structure, and cultural traditions that affect health (Derryberry, 1960). Health educators are concerned not only with individuals and their families but also with the institutions and social conditions that impede or facilitate achievement of optimal health (Griffiths, 1972, p. 13). Health education principles have guided social movements and other broad health-related programs. The work dates back to the pioneering research by Lewin (1935) on group processes in individual change.

Six broad categories of factors—knowledge, personal, interpersonal, institutional, community and public policy—are typically the focus of health educators. The following discussion is derived from this body of research (McLeroy et al., 1988).

Framework

To build a sustainable program, HRSA and its grantees or contractors may usefully consider a planning framework to guide their activities and to evaluate and improve them over time. For example, the planning framework depicted in Table 11.2 uses a seven-step process and is similar to the traditional "diagnosis and treatment" model often used by clinicians and managers.

For each new initiative that a HRSA grantee or contractor undertakes in relation to its RESEP responsibilities, this type of guide specifies clear project objectives, target audiences, and other steps necessary to develop an accurate "diagnosis" of what will need to be done to accomplish specific program objectives. The guide helps users to identify specific knowledge, personal, interpersonal, organizational, community, and public policy barriers that need to be addressed to implement effective outcomes-based, education programs.

Once organizations complete such a plan, their staffs and other users can create grids that focus on the barriers to be addressed and the strategies necessary for effective implementation. Table 11.3 illustrates the approach.

Table 11.4 defines groups of barriers that HRSA and others should evaluate in the RESEP context. They are important to understand because they can interfere with attaining the program's objectives. One general problem with the use of *barrier-driven strategies* is that the full range of barriers is typically not evaluated; a second is that educational deficits are overemphasized. Thus, health educators emphasize that all barrier categories should be examined.

Categorizing barriers provides a method for developing specific corrective interventions that would improve the chances of effective implementation of a program. For example, if knowledge deficit barriers exist among particular physicians, implementing corrective educational strategies for these physicians should be effective. In contrast, if organizational barriers exist, educational interventions and appropriate health care professionals would likely not be helpful or cost-effective, but designing strategies to change the organizational problems would likely be productive.

TABLE 11.2   Framework for Systematic Planning of RESEP Activities

1. Identify specific proposed objectives, suggested changes and measurable outcomes.
2. Identify specific target audiences in terms of RECA-RESEP objectives, including not only those specified by legislation (such as miners, downwinders, and ore transporters) but also clinicians, health-care organizations, and other staff that may need to make changes in methods and operations.
3. Assess target audiences' ability and desire to make changes; "stages of readiness" approach is relevant at this point.
4. Determine potential barriers to accomplishing the outcomes.
5. Recommend specific and customized intervention strategies to remove the barriers and determine the cost effectiveness of doing so.
6. Initiate barrier-specific strategies and remedial interventions.
7. Evaluate results and impact and make changes as needed.   Require assessment plans.

TABLE 11.3   Barriers to Effective Education and Outreach Activities and Implementation of Related Programs

Barrier Classifications	Definitions and Examples of Barriers
Knowledge deficits and skills Sequential steps include awareness, agreement, assimilation, application, and integration	Includes simple "lack of knowledge" or outdated clinical skills. Examples are knowledge deficits about indications or contraindications of medications, about current recommendations or clinical-practice guidelines (such as lack of knowledge of the benefit of treating patients who have silicosis and a positive skin test for tuberculosis with antituberculosis medications such as isoniazid) about use of tests and procedures considered obsolete, and about technical training, skill, or expertise (as in poor surgical or invasive test techniques)
Provider and patient personal barriers	Includes the provider's feelings, beliefs, values and experiences. For example, a patient develops hepatitis while receiving isoniazid therapy, and this affects a physician's decisions regarding use of the therapy for future patients
Provider and patient interpersonal and psychosocial barriers	Includes interpersonal interaction barriers, for example "turf battles" and inability of providers to relate effectively with patients or with other providers
Organizational barriers and lack of organizational support	Includes organizational, structural, and system limitations, including those related to resources and administrative support, for example, lack of standing orders or incomplete standing orders for acute stroke in the emergency room; and process issues with implementing physical-therapy or occupational-therapy orders
Community barriers	Includes existing community resources, public attitudes, and broader general support for a proposed new program
Public policy barriers	Includes existing local, state or federal policies (such as Medicare payment schedules) that may interfere with program implementation

TABLE 11.4 Framework for Addressing Educational and Outreach Barriers

Knowledge Barriers	Potential Effective Intervention Strategies
The general public and persons potentially eligible for RECA and RESEP either lack awareness or have not accepted or assimilated RECA and RESEP information. Therefore, they are not fully using the program.	HRSA grantees are undertaking a variety of knowledge enhancement programs to raise awareness of the RECA program. They include media programs, distribution of flyers and brochures, physician and attorney mailings and a broad array of community wide programs. They also are using not only awareness strategies but also acceptance strategies such as the use of influential local leaders and local community organizations. In many instances group and individual interventions may be necessary. They also are attempting to develop effective programs using influential Native American leaders and organizations. These activities need to be augmented by more customized programs for people with low literacy or who lack access to local media.
The general public has reported difficulty completing the RECA application forms.	DOJ officials need to examine the current application process and work with grantees to address particularly troublesome barriers to efficient claims processing. Personnel are needed to triage potentially eligible claimants to the various programs available to them. Additional personnel with appropriate training in cultural sensitivity are needed to assist claimants. Accommodations are needed for people with low literacy or numeracy, visual or other physical disabilities, or emotional and mental health problems.
Some Native Americans had particular difficulty completing the application form, especially the proof-of-residency documentation. These types of problems will increase if the geographic scope of the program is expanded.	This issue is being addressed by DOJ.
Personal Barriers	**Potential Effective Intervention Strategies**
Some Native Americans prefer nontraditional medical practice and are fearful and suspicious of Western medicine.	Local Native Americans who have influence with these individuals should meet one-on-one to explore these fears, legitimize them, and try to develop a plan that would reduce the fears and allow these individuals to participate in the RECA programs. RECA staff should be involved in this process and try to customize their services to these individuals.

TABLE 11.4 Continued

Interpersonal Barriers	Potential Effective Intervention Strategies
Some Native Americans are suspicious of local health care providers.	RECA and RESEP programs should be designed to recognize the cultural traditions and make accommodations to integrate cultural traditions and local leaders as much as possible in the workup and treatment of Native American and other groups that use non-traditional medical techniques.
The Navajo population does not trust local clinicians and community agencies. They feel that they are treated as second-class citizens in accessibility and equity of local health care services. They also believe that their unique needs and traditions are not addressed or respected.	RESEP organizations should activate influential local Native American and other appropriate community organizations to facilitate an opportunity for groups to identify and express their concerns and to develop specific objectives and plans to improve relationships between Native Americans and local populations.
All three town meetings involved people who felt that they had not been heard and "that's what they wanted."	Similarly, such organizations should activate local organizations and influential leaders to offer people opportunities to identify and express their concerns, have them legitimized, and ensure that an action plan is developed. Simply reiterating concerns can become a perpetual, unproductive process.

Organizational Barriers	Potential Effective Intervention Strategies
Local clinics lack resources and personnel to provide the definitive tests (such as arterial blood gases, chest x rays, and spirometry) apparently required to determine eligibility for compensation.	Federal, state, and local health-care organizations should help to build capacity to provide resources capable of evaluation of covered cancers.

Community Barriers	Potential Effective Intervention Strategies
The targeted populations often live in rural, geographically diverse, and isolated areas. Those factors add complexity to any outreach program, particularly for followup and implementation of clinical and public-health programs.	Contractors cannot rely exclusively on media-based health education. Contact through local community groups and leaders must also be considered. Resources and methods for broader outreach to very small groups need to be found, although the cost effectiveness of such a "small" approach will need to be considered.

Public Policy Barriers	Potential Effective Intervention Strategies
Resources may not be adequate and available to continue and possibly expand the RESEP program.	Involved federal agencies and legislative representatives need to evaluate the RESEP program regularly in light of recommendations in this report.

The committee recommends that HRSA:

a. Use a standardized method to develop outcomes-based goals and objectives for appropriate planning and assessment.

b. Identify and evaluate the cost and effectiveness of steps to remove barriers to program implementation.

c. Train staff to identify specific barriers to implementation and develop strategies to overcome them.

CONCLUSION

The challenges of outreach and education even for the current RESEP program are substantial. If the RECA program is expanded as described in earlier chapters, the target audiences will be larger (for example, clinicians, health care organizations, special exposed populations, and the general public) than expected when the RESEP program was created. The barriers (knowledge deficits, interpersonal, personal factors, organizational, community factors, and public policy) to reaching program objectives are broader than originally understood and will be even more problematic if and when RECA expands nationally but they can be addressed with methods consistent with the health education research literature.

In particular, HRSA and its RESEP grantees or contractors should adopt and implement outcomes-based planning and implementation approaches. The following overarching specific changes are indicated. HRSA must:

1. Provide information about the existence and availability of RECA and RESEP.

2. Explain in clear and simple terms the likelihood of receiving compensation.

3. Put the low risks of radiogenic cancer in context to reassure exposed individuals.

4. Clearly explain the potential risks posed by medical testing and the relative lack of improvement in health outcomes gained by early detection of many RECA-compensable diseases.

5. Explain the proposed PC/AS method both to exposed individuals and to the clinicians who may be guiding their decision-making.

The committee recognizes the expanded nature of such an effort and offers its suggestions and recommendations in the hope that they will be helpful to future RECA and RESEP programs and populations.

References

Adams, R. E., Bromet, E. J., Panina, N., Golovakha, E., Goldgaber, D. Gluzman, S. Stress and Wellbeing in Mothers of Young Children after the Chornobyl Nuclear Power Plant Accident. Psychological Medicine 32(1):143-156. 2002.

ACHRE (Advisory Committee on Human Radiation Experiments). Advisory Committee on Human Radiation Experiments, Final Report. Available at http://www.eh.doe.gov/ohre/roadmap/achre/index.html. Washington, DC: U.S. Government Printing Office. 1995.

ACGIH (American Conference of Govermental Industrial Hygienests). 2004 TLVs and BEIs, ACGIH publication number 0104. Available at http://www.acgih.org/store/. 2004.

Allen vs. United States 588 F. Supp 247 (D. Utah 1984).

Allen, J. E., Henshaw, D. L., Keitch, P. A., Fews, A. P., Eatough, J. P. Fat Cells in Red Bone Marrow of Human Rib: Their Size and Spatial Distribution with Respect to the Radon-Derived Dose to the Haemopoietic Tissue. Int. J. Radiat. Biol. 68(6):669-678. 1995.

American Thoracic Society. Medical Section of the American Lung Association. Single-Breath Carbon Monoxide Diffusing Capacity (Transfer Factor): Recommendations for a Standard Technique-1995 Update. American Journal of Respiratory and Critical Care Medicine 152: 2185-2198. 1995.

Archer, V. E., Wagoner, S. D., Lundin, F. E. Cancer Mortality Among Uranium Mill Workers. J. Occup. Med. 15(1):11-14. 1973.

Aristotle. Nicomachean Ethics, V 1-5, 1129a14-1133b28. Translated by Terence Irwin, Indianapolis: Hackett Publishing Co. 1985.

Ashmore, J. P., Krewski, D., Zielinski, J. M., Jiang, H., Semenciw, R., Band., P. R. First Analysis of Mortality and Occupational Radiation Exposure Based on the National Dose Registry of Canada. American Journal of Epidemiology 148(6): 564-574. 1998.

Astakhova, L. N., Anspaugh, L. R., Beebe, G. W., Bouville, A., Drozdovitch, V. V., Garber, V., Gavrilin, Y. I., Khrouch, V. T., Kuvshinnikov, A. V., Kuzmenkov, Y. N., Minenko, V. P., Moschik, K. V., Nalivko, A. S., Robbins, J., Shemiakina, E. V., Shinkarev, S., Tochitskaya, S. I., Waclawiw, M. A. Chernobyl-Related Thyroid Cancer in Children of Belarus: a Case-Control Study. Radiation Research 150(3):349-356. 1998.

Atkinson, W. D., Law, D. V., Bromley K. J., Inskip, H. M. Mortality of employees of the United Kingdom Atomic Energy Authority, 1946-97. Occup. Environ. Med. 61(7):577-585. 2004.

Austin, D. F., Reynolds, P. J., Snyder, M. A., Biggs, M. W., Stubbs, H. A. Malignant Melanoma Among Employees at the Lawrence Livermore National Laboratory. Lancet 2:712-716. 1981.

Austin, D. F., Reynolds, P. Investigation of an Excess of Melanoma Among Employees of the Lawrence Livermore National Laboratory. American Journal of Epidemiology 145:524-531. 1997.

Banks, D. E., Cheng, Y. H., Weber, S. L., Ma, J. K. Strategies for the Treatment of Penumoconiosis. Occupation Medicine 8:205-232. 1993.

Banks, D. E. Silicosis. In: Text Book of Clinical Occupational and Environmental Medicine. Editors: Rosenstock, L., Cullen, M. R., Brodkin, C. A., Redlich, C. A. Second Edition. Elsevier Saunders. Pp. 380-392. 2005.

Barry, M. J. Health Decision Aids to Facilitate Shared Decision Making in Office Practice. Ann. Int. Med. 136:127-135. 2002.

Bates, D. Y., Gotch, A. R., Brooks, S., Landrigan, P. J., Hankinson, J. L., Merchant, J. A. Prevention of Occupational Lung Disease. Task Force on Research and Education for the Prevention and Control of Respiratory Diseases. Chest 102(3 Suppl):257S-276S. 1992.

Beauchamp, T. L., Childress, J. Principles of Biomedical Ethics. 5th Ed. Oxford University Press (Oxford). 2001.

Beck, H. L. Exposure Rate Conversion Factors for Radionuclides Deposited on the Ground. EML Report-378. 1980.

Beck, H. L., Krey, P. W. Radiation Exposures in Utah from Nevada Nuclear Tests. Science 220(4592):18-24. 1983.

Beck, H. L., Helfer, I. K., Bouville, A., Dreicer, M. Estimates of Fallout Dose in the Continental U.S. from Nevada Weapons Testing Based on Gummed-Film Monitoring Data. Health Physics 59, 565-576. 1990.

Beck, H. L., Bennett, B. J. Historical Overview of Atmospheric Nuclear Weapons Testing and Estimates of Fallout in the Continental United States. Health Physics 82, 591-608. 2002.

Becker, D. V., McConahey, W. M., Dobyns, B. M., Tomkins, E., Sheline, G. E., Workman, J. B. The Results of Radioiodine Treatment of Hyperthyroidism. Preliminary Report of the Thyrotoxicosis Therapy Follow-up Study. Further Advances in Thyroid Research. Vol. 1. Editors: Fellinger, K. and Hofer, R. Vienna. Wien: Verlage der Wiener Medizinischen Akademie. Pp. 603-609. 1971.

Becker, S. M. Psychological Effects of Radiation Accidents. In: Medical Management of Radiation Accidents. Second edition. Editors: Gusev, I. A., Guscova, A. K., Mettler, F. A. Jr. CRC Press, Boca Raton, FL. 2001.

Belmont, J. W., Gibbs, R. A. Genome-Wide Linkage Disequilibrium and Haplotype Maps. Am. J. Pharmacogenomics 4:253-262. 2004.

Bendel, I., Schuttmann, W., Arndt, B. Cataract of the Lens as a Late Effect of Ionizing Radiation in Occupationally Exposed Persons. Late Biological Effects of Ionizing Radiation. Vienna, IAEA. 1978.

Bennett, B. G. Worldwide Dispersion and Deposition of Radionuclides Produced in Atmospheric Tests. Health Physics 82, 644-655. 2002.

Bennington, J. L. Cancer of the Kidney-Etiology, Epidemiology and Pathology. Cancer 32:1017-1029. 1973.

Bentham, J. An Introduction to the Principles of Morals and Legislation. Garden City: Doubleday. Originally published in 1789. 1961.

Beral, V., Inskip, H., Fraser, P., Booth, M., Coleman, D., Rose, G. Mortality of Employees of the United Kingdom Atomic Energy Authority, 1946-1979. Br. Med. J. (Clin. Res. Ed.) 291(6493):440-447. 1985.

Beral, V., Fraser, P., Carpenter, L., Booth, M., Brown, A., Rose, G. Mortality of Employees of the Atomic Weapons Establishment, 1951-82. BMJ 297(6651):757-770. 1988.

Bergonie, J., Tribondeau, L. Interpretation de Quelques Resultants de la Radiotherapie et Essai de Fixation d'une Technique Rationnellle. Compt. Rend. Acad. Sci 143:983. 1906.

Berkman, N. D., DeWalt, D. A., Pignone, M. P., Sheridan, S., Lohr, K., Lux, L., Sutton, S., Swinson, T., Bonito, A. Literacy and Health Outcomes. Evidence Report/Technology Assessment No. 87. AHRQ Publication No. 04-E007-2. Rockville, MD: Agency for Healthcare Research and Quality. 2004.

Berrington, A., Darby, S. C., Weiss, H. A., Doll, R. 100 Years of Observation on British Radiologists: Mortality From Cancer and Other Causes 1897-1997. Br. J. Radiol. 74(882):507-519. 2001.

Blackhall, L. J., Murphy, S. T., Frank, G., Michel, V., Azen, S. Ethnicity and Attitudes Toward Patient Autonomy. Journal of the American Medical Association 274(10):820-825. 1995.

Boice, J. D. Jr. Radiation and Non-Hodgkin's Lymphoma. Cancer Research 52 (19 Suppl):5489s-5491s. 1992.

Boice, J. D. Jr., Miller, R. W. Childhood and Adult Cancer After Intrauterine Exposure To Ionizing Radiation. Teratology 59:227-233. 1999.

Bouffler, S., Lloyd, D. Nuclear Weapons Tests and Human Germline Mutation Rate: A Commentary. NRPB eBulletin, No. 1, National Radiological Protection Board, Chilton (http://www.nrpb.org/publications/bulletin/no1/article5). 2002.

Bouville, A., Simon, S. L., Miller, C. W., Beck, H. L., Anspaugh, L. R., Bennett, B. G. Estimates of Doses from Global Fallout. Health Physics 82(5):690-705. 2002.

Braestrup, C. B. Past and Present Radiation Exposure to Radiologists from the Point of View of Life Expectancy. Am. J. Roentgenol, Rad. Ther. Nucl. Med. 78:988-992. 1957.

Brenner, D. J., Elliston, C. D. Estimated Radiation Risks Potentially Associated With Full-Body CT Screening. Radiology 232(3):735-738. 2004.

Brent, R. L. Utilization of Developmental Basic Science Principles in the Evaluation of Reproductive Risks from Pre- and Postconception Environmental Radiation Exposures. Teratology 59:182-204. 1999.

Bromet, E. J., Parkinson, D. K., Dunn, L. O. Long-Term Medical Health Consequences of The Accident at Three Mile Island. International Journal of Mental Health 19:48-60. 1990.

Bromet, E. J., Gluzman, S., Schwartz, J. E., Goldgaber, D. Somatic Symptoms in Women 11 years after the Chornobyl Accident: Prevalence and Risk Factors. Environmental Health Perspectives 110 Supplement 4:625-629. 2002.

Brugge, D., Goble, R. The History of Uranium Mining and the Navajo People. Am. J. Pub. Health 92(9):1410-1419. 2002.

Brugge, D., Goble, R. The Radiation Exposure Compensation Act: What is Fair? New Solutions 13(4):385-397. 2003.

Caldwell, G. G., Kelley, D. B., Heath, C. W. Jr. Leukemia Among Particpants in Military Maneuvers at a Nuclear Bomb Test: A Preliminary Report. Journal of the American Medical Association 244 :1575-1578. 1980.

Caldwell, G. G., Kelley, D., Zack, M., Falk, H., Heath, C. W. Jr. Mortality and Cancer Frequency Among Military Nuclear Test (Smoky) Participants, 1957 through 1979. Journal of the American Medical Association 250:620-624. 1983.

Cardis, E., Gilbert, E. S., Carpenter, L., Howe, G., Kato, I., Armstrong, B. K., Beral, V., Cowper, G., Douglas, A., Fix, J., Fry, S. A., Kaldor, J., Lavé, C., Salmon, L., Smith, P. G., Voelz, G. L., Wiggs, L. D. Effects of Low Doses and Low Dose Rates of External Ionizing Radiation: Cancer Mortality Among Nuclear Industry Workers in Three Countries. Radiation Research 142(2):117-132. 1995.

Carlson C. S., Eberle, M. A., Kruglyak, L., Nickerson, D. A. Mapping Complex Disease Loci in Whole-Genome Association. Nature 429:446-452. 2004.

Carrese, J. A., Rhodes, L. A. Western Bioethics on the Navajo Reservation: Benefit or Harm. Journal of the American Medical Association 274:826-829. 1995.

CDC-NCI (Centers for Disease Control and Prevention-National Cancer Institute). A Feasibility Study of the Health Consequences to the American Population from Nuclear Weapons Tests Conducted by the United States and Other Nations. Predecisional Draft – For Peer Review and Public Comment. Washington, DC: U.S. Department of Health and Human Services (1997). 2001.

Charpentier, P., Ostfeld, A. M., Hadjimichael, O. C., Hester, R. The Mortality of U.S. Nuclear Submariners. Journal of Ooccupational Medicine 35:501-509. 1993.

Checkoway, H., Pearce, N., Crawford-Brown, D. J., Cragle, D. L. Radiation Doses and Cause-Specific Mortality Among Workers at a Nuclear Materials Fabrication Plant. Am. J. Epidemiol. 127(2):255-266. 1988.

Clarke, R. H., Dunster, J., Nenot, J-C., Smith, H., Voeltz, G. The Environmental Safety and Health Implications of Plutonium. J. Radiol. Prot.16 (2):91-105. 1996.

Clayton, G., Clayton, F. Patty's Industrial Hygiene and Toxicology, 3rd revised edition, New York: Wiley and Sons, Inc. 1981.

Cohen, B. L. Test of The Linear-No Threshold Theory of Radiation Carcinogenesis For Inhaled Radon Decay Products. Health Physics 68(2):157-174. 1995.

Coleman, J. Corrective Justice and Wrongful Gain. Journal of Legal Studies, Vol. 11, No. 2, pp. 421-440. 1982.

Coleman, J. A. Corrective Justice and Property Rights. Social Philosophy and Policy II:2: 124-138. 1994.

Collins, D. L., de Carvalho, A. B. Chronic Stress from the Goiania 137Cs Radiation Accident. Behavior Medicine 18:149-157. 1993.

CIRRPC (Committee on Interagency Radiation Research and Policy Coordination) Science Panel Report No. 6. Use of Probability of Causation by the Veterans Administration in the Adjudication of Claims of Injury Due to Exposure to Ionizing Radiation. Committee on Interagency Radiation Research and Policy Coordination, Office of Science and Technology Policy, Executive Office of the President, Washington, DC. August 1988 (ORAU 88/F-4). 1988.

Craig, R. K. When Daubert Gets Erie. Medical Certainty and Medical Expert Testimony in Federal Court. Univ. of Denver Law Review (77 Denv U L Rev 69). 1999.

Dalager, N. A., Kang, H. K., Mahan, C. M. Cancer Mortality Among the Highest Exposed US Atmospheric Nuclear Test Participants. J. Occup. Environ. Med. 42:798-805. 2000.

Darby, S. C., Kendall, G. M., Fell, T. F., O'Hagan, J. A., Muirhead, C. R., Ennis, J. R., Ball, A. M., Dennis, J. A., Doll, R. A Summary of Mortality and Incidence of Cancer in Men from the United Kingdom Who Participated in the United Kingdom's Atmospheric Nuclear Weapons Tests and Experimental Programmes. British Medical Journal 296:332-338. 1988.

Darby, S. C., Kendall, G. M., Fell, T. P., Doll, R., Goodill, A. A., Conquest, A. J., Jackson, D. A., Haylock, R. G. Further Follow Up of Mortality and Incidence of Cancer in Men from the United Kingdom Who Participated in the United Kingdom's Atmospheric Nuclear Weapon Tests and Experimental Programmes. British Med. J. 307(6918):1530-1535. 1993a.

Darby, S. C., Roman, E. Nuclear Weapons Testing and Childhood Leukaemia. Ann. Med. (5):429-430. 1993b.

Darby, S. C.,Whitely, E., Howe, G. R., Hutchings, S. J., Kusiak, R. A., Lubin, J. H., Morrison, H. I., Tirmarche, M., Tomasek, L., Radford, E. P., Roscoe, R. J., Samet, J. M., Shu, X. Y. Radon and Cancers Other Than Lung Cancer in Underground Miners: A Collaborative Analysis of 11 Studies. J. Nat. Cancer Inst. 87:378-384. 1995.

Davis, S., Stepanenko, V., Rivkind, N., Kopecky, K. J., Voilleque, P., Shakhtarin, V., Parshkov, E., Kulikov, S., Lushnikov, E., Abrosimov, A., Troshin, V., Romanova, G., Doroschenko, V., Proshin, A., Tsyb, A. Risk of Thyroid Cancer in the Bryansk Oblast of the Russian Federation after the Chernobyl Power Station Accident. Radiation Research 162(3):241-248. 2004a.

Davis, S., Kopecky, K. J., Hamilton, T. E., Onstad, L. Thyroid Neoplasia, Autoimmune Thyroiditis, and Hypothyroidism in Persons Exposed to Iodine 131 from the Hanford Nuclear Site. Journal of the American Medical Association 292(21):2600-2613. 2004b.

Decisioneering 2001. Crystal Ball® 2000.2, distributed by Decisioneering, Inc., Denver, Colorado. 2001.

Delongchamp, R. R., Mabuchi, K., Yoshimoto, Y., Preston, D. L. Cancer Mortality Among Atomic Bomb Survivors Exposed in Utero or as Young Children, October 1959-May 1992. Radiation Research 147:385-395. 1997.

Denman, A. R., Eatough, J. P., Gillmore, G., Phillips, P. S. Assessment of Health Risks to Skin and Lung of Elevated Radon Levels in Abandoned Mines. Health Physics 85(6):733-739. 2003.

DOE (Department of Energy). United States Nuclear Tests July 1945 through September 1992. US DOE publication DOE/NV—209-Rev 15, December 2000.

DOJ (Department of Justice). Final Report of the Radiation Exposure Compensation Act Committee, submitted to the Human Radiation Interagency Working Group, July 1996.

DOJ (Department of Justice). 28 CFR Part 79. Claims under the Radiation Exposure Compensation Amendments of 2000; Final Rule and Proposed Rule. Federal Register Vol. 67, No. 152, pp. 51422-51439. Washington, DC: U.S. Government Printing Office. August 7, 2002.

DOJ (Department of Justice). 28 Part 79. Civil Division: Claims Under the Radiation Exposure Compensation Act Amendments of 2000; Amendments Contained in the 21st Century Department of Justice Appropriations Authorization Act of 2002; Final Rule. Federal Register Vol. 69, No. 56, pp. 13628-13676. Washington DC: U.S. Government Printing Office. March 23, 2004.

Derryberry, M. Health Education—Its Objectives and Methods. Health Education Monographs 8:5-11. 1960.

DeWalt, D. A., Berkman, N. D., Sheridan, S., Lohr, K. N., Pignone, M. P. Literacy and Health Outcomes: A Systematic Review of the Literature. J. Gen. Intern. Med. 19:1228-1239. 2004.

Doody, M. M., Mandel, J. S., Lubin, J. H., Boice, J. D. Jr. Mortality Among United States Radiologic Technologists, 1926-90. Cancer Causes Control 9(1):67-75. 1998.

Dubrova, Y. E., Bersinbaev, R. I., Djansugurova, L., Tankimanova, M. K., Mamyrbaeva, Z. Z., Mustonen, R., Lindholm, C., Hulten, M., Salomaa, S. Nuclear Weapons Tests and Human Germline Mutation Rate. Science 295:1037. 2002a.

Dubrova, Y. E., Grant, G., Chumak, A. A., Stehka, V. A., Karakasian, A. N. Elevated Minisatellite Mutation Rate in the Post-Chernobyl Families from Ukraine. Am. J. Hum. Genet. 71:801-809. 2002b.

Dubrova, Y. E., Nesterov, V. N., Krouchinsky, N. G., Ostapenko, V. A., Neumann, R., Neil, D. L., Jeffreys, A. J. Human Minisatellite Mutation Rate after the Chernobyl Accident. Nature 380:683-686. 1996.

Dubrova, Y. E., Nesterov, V. N., Krouchinsky, N. G., Ostapenko, V. A., Vergnaud, G., Giraudeau, F., Buard, J., Jeffreys, A. J. Further Evidence for Elevated Human Minisatellite Mutation Rate in Belarus Eight Years after the Chernobyl Accident. Mutat. Res. 381:267-278. 1997.

Dworkin, G. Autonomy and Behavior Control. Hastings Center Report, Fall 76; 6:23-28. 1976.

Eatough, J. P. Alpha-Particle Dosimetry for the Basal Layer of the Skin and the Radon Progeny 218-Po and 214-Po. Phys. Med. Biol. 42(10):1899-1911. 1997.

Eatough, J. P., Henshaw, D. L. Radon and Monocytic Leukaemia in England. J. Epidemiol. Community Health 47(6):506-507. 1993.

Eckerman, K. F., Ryman, J. C. External Exposure to Radionuclides in Air, Water, and Soil. Washington, DC: U.S. Environmental Protection Agency, Federal Guidance Report No. 12, EPA-402-R-93-081. 1993.

Eckerman, K. F., Wolbarst, A. B., Richardson, A. C. B. Limiting Values of Radionuclide Intake and Air Concentration and Dose Conversion Factors for Inhalation, Submersion, and Ingestion. Washington, DC: U.S. Environmental Protection Agency, Federal Guidance Report No. 11, EPA 520/1-88-020. 1988.

Eddy, D. How to Think about Screening. In Screening for Diseases. Snow V, ed. American College of Physicians. Philadelphia, ix-xxv. 2004.

Epstein, R. A. A Theory of Strict Liability, 2 J. Legal Stud. 151. 1973.

Federal Registry Vol. 67, No. 83, pages 21256-21258. Department of Health and Human Services. Health Resources and Services Administration. Fiscal Year 2002 Competitive Application Cycle for the Radiation Exposure Screening and Education Program 93.257. Washington, DC: U.S. Government Printing Office. April 30, 2002.

Finch, S. C. Myelodysplasia and Radiation. Radiation Research 161:603-606. 2004.

Fletcher, G. Fairness and Utility in Tort Theory, 85 Harvard Law Review (Harv. L. Rev.) 537. 1972.

Frame, P. S., Carlson, S. J. A Critical Review of Periodic Health Screening Using Specific Screening Criteria. J. Fam. Practice 2:29-36. 1975.

Franklyn, J.A., Maisonneuve, P., Sheppard, M., Betteridge, J., Boyle, P. Cancer Incidence and Mortality after Radioiodine Treatment for Hyperthyroidism: A Population-Based Cohort Study. Lancet 353:2111-2115. 1999.

Freedman, D. M., Sigurdson, A., Rao, R. S., Hauptmann, M., Alexander, B., Mohan, A., Doody, M. M., Linet. M. S. Risk of Melanoma Among Radiologic Technologists in the United States. Int. J. Cancer 103(4):556-562. 2003.

Fricchone, G. Generalized Anxiety Disorder. New England Journal of Medicine 351:675-682. 2004.

Gal, I. Big Picture: What Does "Numeracy" Mean? GED Items 12(4/5). 1995. Available at: http://mathforum.org/teachers/adult.ed/articles/gal.html.

Galen, R. S., Gambino, R. S. Beyond Normality: The Predictive Value and Efficiency of Medical Diagnosis. John Wiley & Sons (New York), p. 237. 1975.

Gavrilin, Y., Khrouch, V., Shinkarev, S., Drozdovitch, V., Minenko, V., Shemiakina, E., Ulanovsky, A., Bouville, A., Anspaugh, L., Voilleque, P., Luckyanov, N. Individual Thyroid Dose Estimation for a Case-Control Study of Chernobyl-Related Thyroid Cancer Among Children of Belarus-Part I: 131I, Short-Lived Radioiodines (132I, 133I, 135I), and Short-Lived Radiotelluriums (131MTe and 132Te). Health Physics 86(6):565-585. 2004.

Gembicki, M., Stozharov, A. N., Arinchin, A. N., Moschik, K. V., Petrenko, S., Khmara, I. M., Baverstock, K. F. Iodine Deficiency in Belarusian Children as a Possible Factor Stimulating the Irradiation of the Thyroid Gland during the Chernobyl Catastrophe. Environ. Health Perspect. 105 Suppl 6:1487-1490. 1997.

GAO (General Accounting Office). Radiation Exposure Compensation: Analysis of Justice's Program Administration, GAO-01-1043. Washington, DC: U.S. General Accounting Office. 2001.

GAO (General Accounting Office). Funding to Pay Claims May be Inadequate to Meet Projected Needs, GAO-03-481. Washington, DC: U.S. General Accounting Office. 2003.

Gilbert, E. S. Invited Commentary: Studies of Workers Exposed to Low Doses of Radiation. Am. J. Epidemiol. 153:319-322. 2001.

Gilbert, E. S., Marks, S. An Analysis of the Mortality of Workers in a Nuclear Facility. Radiation Research 79:122-148. 1979.

Gilbert, E. S., Omohundro, E., Buchanan, J. A., Holter, N. A. Mortality of Workers at the Hanford Site: 1945-1986. Health Physics 64(6):577-590. 1993.

Gilbert, E. S., Land, C. E., Simon, S. S. Health Effects from Fallout. Health Physics 82:726-735. 2002.

Gill, K. A. The Moral Functions of an Apology. In Roberts, R. C. (ed) Injustice and Rectification. New York: Peter Lang. 2002.

Ginzberg, H. M. The Psychological Consequences of the Chernobyl Accident—Findings from the International Atomic Energy Agency. Public Health Reports108:184-192. 1993.

Goans, R. E., Holloway, E. C., Berger, M. E., Ricks, R. C. Early Dose Assessment in Criticality Accidents. Health Physics 81:446-449. 2001.

Gold, M. R., Siegel, J. E., Russell, L. B., Weinstein, M. C. Cost Effectiveness in Health and Medicine. New York: Oxford University Press; p. 28. 1996.

Gorry, G. A., Pauker, S. G., Schwartz, W. B. The Diagnostic Importance of the Normal Finding. New England Journal of Medicine 298:486-489. 1978.

Gostin, L. O. Informed Consent, Cultural Sensitivity, and Respect for Persons. Journal of the American Medical Association 274(10):844-845. 1995.

Griffiths, W. Health Education Definitions: Problems and Philosophies. Health Education Monographs 31:12-14. 1972.

Gutheil, T. G. The Psychiatrist as Expert Witness. Washington, DC: American Psychiatric Press, p. 6. 1998.

Hall, E.J. Radiation Carcinogenesis. In: Radiobiology for the Radiologist. Fourth Edition. J.B. Lippincott Company, Philadelphia. Pp. 323-350. 1994.

Hall, E. J. Radiation Carcinogenesis. In: Radiobiology for the Radiologist. Fifth Edition. J. B. Lippincott Company, Philadelphia. 2000.

Hall, P., Mattsson, A., Boice, J. D. Jr. Thyroid Cancer after Diagnostic Administration of Iodine-131. Radiation Research 145(1):86-92. 1996a.

Hall, P., Furst, C. J., Mattson, A., Holm, L. E., Boice, J. D., Inskip, P. D. Thyroid Nodularity after Diagnostic Administration of Iodine-131. Radiation Research 146:673-682. 1996b.

Hamilton, T. E., van Belle, G., LoGerfo, J. P. Thyroid Neoplasia in Marshall Islanders Exposed to Nuclear Fallout. Journal of the American Medical Association 258(5):629-635. 1987.

Harris, R., Donahue, K., Rathore, S. S., Frame, P., Woolf, S. H., Lohr, K. N. Screening Adults For Type 2 Diabetes: A review of the evidence for the U.S. Preventive Services Task Force. Annals of Internal Medicine 138(3):215-229. 2003.

Harris, R. P., Helfand, M., Woolf, S. H., Lohr, K. N., Mulrow, C. D., Teutsch, S. M., Atkins, D. For the Methods Work Group, Third US Preventive Services Task Force. Current Methods of the US Preventive Services Task Force: A Review of the Process. American Journal of Preventive Medicine 20 (3S):21-35. 2001.

Harris, R., Lohr K. N. Screening for Prostate Cancer: An Update of the Evidence for the U.S. Preventive Services Task Force. Annals of Internal Medicine 137:917-929. 2002.

Henderson, R. W., Smale, R. F. External Exposure Estimates for Individuals Near the Nevada Test Sites. Health Physics 59:715-721. 1990.

Hiatt, H. H. Protecting the Medical Commons: Who is Responsible? New England Journal of Medicine 293:235-241. 1975.

Holm, L. E., Wiklund, K. E., Lundell, G. E., Bergman, N. A., Bjelkengren, G., Cederquist, E. S., Ericsson, U. B., Larsson, L. G., Lindberg, M. E., Lindberg, R. S., et al. Thyroid Cancer after Diagnostic Doses of Iodine-131: A Retrospective Cohort Study. J. Natl. Cancer Inst. 80(14):1132-1138. 1988.

Holm, L. E., Hall, P., Wiklund, K., Lundell, G., Berg, G., Bjelkengren, G., Cederquist, E., Ericsson, U. B., Hallquist, A., Larsson, L. G. Cancer Risk after Iodine-131 Therapy for Hyperthyroidism. J. Natl. Cancer Inst. 83(15):1072-1077. 1991.

Hornung, R. W., Deddens, J. A., Roscoe, R. J. Modifiers of Lung Cancer Risk in Uranium Miners from the Colorado Plateau. Health Physics 74(1):12-21. 1998.

Houlston, R. S., Peto, J. The Search for Low-Penetrance Cancer Susceptibility Alleles. Oncogene 23:6471-6476. 2004.

Howard, J. E., Heotis, P. M., Scott, W. A., Adams, W. H. Medical Status of Marshallese Accidentally Exposed to 1954 BRAVO Fallout Radiation; January 1988 Through December 1991. Upton, NY: Brookhaven National Laboratory; Report DOE/EH-0493 PB95. 1995.

Howard, J. E., Vaswani, A., Heotis, P. Thyroid Disease Among the Rongelap and Utirik Population— An Update. Health Physics 73(1):190-198. 1997.

Howe, G. R., Nair, R. C., Newcombe, H. B., Miller, A. B., Burch, J. D., Abbatt, J. D. Lung Cancer Mortality (1950-80) in Relation to Radon Daughter Exposure in a Cohort of Workers at the Eldorado Port Radium Uranium Mine: Possible Modification of Risk by Exposure Rate. J. Natl. Cancer Inst. 79(6):1255-1260. 1987.

Howe, G. R. Studies of Nuclear Workers: the Pros and Cons. Lancet 364:1199-1200. 2004.

Howe, G. R., Zablotska, L. B., Fix, J. J., Egel, J., Buchanan, J. Analysis of the Mortality Experience Amongst U.S. Nuclear Power Industry Workers After Chronic Low-Dose Exposure to Ionizing Radiation. Radiation Research 162(5):517-526. 2004.

Hughes, J. S., O'Riordan, M. C. Radiation Exposure of the UK Population: 1993 review. Chilton: National Radiological Protection Board. 1993.

Hursh, J. B., Spoor, N. L. Data on Man. Chapter 4 In: Uranium, Plutonium and Transplutonic Elements: Handbook of Experimental Pharmacology, New Series, Vol 34. Editors: Hodge, H. C., Stannard, J. N., Hursh, J. B. New York: Spinger-Verlag. 1973.

IOM (Institute of Medicine). Clinical Practice Guidelines: Directions for a New Program. Editors: Field, M. J., Lohr, K. N. Washington, DC: National Academy Press. 1990a.

IOM (Institute of Medicine). A Strategy for Quality Assurance. Volumes I and II. Editor: Lohr, K. N. Washington, DC: National Academy Press. 1990b.

IOM (Institute of Medicine). Guidelines for Clinical Practice: From Development to Use. Editors: Field, M. J., Lohr, K. N. Washington, DC: National Academy Press. 1992.

IOM (Institute of Medicine). Access to Health Care in America. M. Millman, ed. Committee on Monitoring Access to Personal Health Care Services. Washington DC: National Academy Press. 1993.

IOM (Institute of Medicine). Potential Radiation Exposure in Military Operations: Protecting the Soldier Before, During and After. Committee of Battlefield Criteria. Washington, DC: National Academy Press. 1999.

IOM (Institute of Medicine). The Five Series Study: Mortality of Military Participants in U.S. Nuclear Weapons Tests. Medical Follow-up Agency. Washington, DC: National Academy Press. 2000.

IOM (Institute of Medicine). Crossing the Quality Chasm: A New Health System for the 21st Century. Washington, DC: National Academy Press. 2001.

IOM (Institute of Medicine). Preparing for the Psychological Consequences of Terrorism. A Public Health Strategy. Washington, DC: The National Academies Press. 2003a.

IOM (Institute of Medicine). Fulfilling the Potential of Cancer Prevention and Early Detection. Washington, DC: The National Academies Press. 2003b.

IOM-NRC (Institute of Medicine-National Research Council). Exposure of the American People to Iodine-131 from Nevada Nuclear-Bomb Tests. Review and Assessment of the Public Health Implications of the National Cancer Institute's "Estimated Exposures and Thyroid Doses Received by the American People from Iodine-131 in the Fallout Following Nevada Atmospheric Nuclear Bomb Tests." Washington, DC: National Academy Press. 1999.

IAC (International Advisory Committee). The International Chernobyl Project: Assessment of Radiological Consequences and Evaluation of Protective Measures. Technical Report. Vienna, Austria: International Atomic Energy Agency. 1991.

IARC (International Agency for Research on Cancer). Working Group on the Evaluation of Carcinogenic Risks to Humans 1999. Monographs on the Evaluation of Carcinogenic Risks to Humans. Vol 75 Ionizing Radiation, Part 1: X- and Gamma (G - Symbol) - Radiation and Neutrons. World Health Organization, IARC Press, Lyon, France. 2000.

IARC (International Agency for Research on Cancer). Working Group on the Evaluation of Carcinogenic Risks to Humans 2001. Monographs on the Evaluation of Carcinogenic Risks to Humans. Vol 78 Ionizing Radiation, Part 2: Some Internally Deposited Radionuclides. World Health Organization, IARC Press, Lyon, France. 2001.

ICRP (International Commission on Radiological Protection). Limits for Intakes of Radionuclides by Workers. ICRP Publication 30, Part 1. Annals of the ICRP 2(3/4). Oxford, U.K.: Pergamon Press. 1979.

ICRP (International Commission on Radiological Protection). Nonstochastic Effects of Ionizing Radiation. ICRP Publication 41. Annals of the ICRP 14(3). Oxford, U.K.: Pergamon Press. 1984.

ICRP (International Commission on Radiological Protection). 1990 Recommendations of the International Commission on Radiological Protection. ICRP Publication 60. Annals of the ICRP 21(1-3). Elmsford, NY: Pergamon Press. 1991.

ICRP (International Commission on Radiological Protection). Human Respiratory Tract Model for Radiological Protection. ICRP Publication 66. Annals of the ICRP 24(1-3). Elmsford, NY: Pergamon Press. 1994.

ICRP (International Commission on Radiological Protection). Basic Anatomical and Physiological Data for Use in Radiological Protection: the Skeleton. ICRP Publication 70. Annals of the ICRP 26(2). Elmsford, NY: Pergamon Press. 1995.

ICRP (International Commission on Radiological Protection). Conversion Coefficients for Use in Radiological Protection Against External Radiation. ICRP Publication 74. Annals of the ICRP 26/(3). Elmsford, NY: Pergamon Press. 1996.

ICRP (International Commission on Radiological Protection). Genetic Susceptibility to Cancer (Ed. J Valentin). Annals of the ICRP, Publication 79. Oxford, U.K.: Pergamon Press. 1998.

ICRP (International Commission on Radiological Protection). Pregnancy and Medical Radiation. Volume 31/1. Annals of the ICRP, Publication 84. Pergamon Press. 2001a.

ICRP (International Commission on Radiological Protection). Doses to the Embryo and Fetus from Intakes of Radionuclides by the Mother. Annals of the ICRP, Publication 88. Pergamon Press. 2001b.

ICRU (International Commission on Radiation Units and Measurements). Retrospective Assessment of Exposure to Ionising Radiation. ICRU Report 68, Volume 2, No 2. Bethesda, MD: International Commission on Radiation Units and Measurements. 2002.

Ivanov, V. K., Chekin, S. Y., Parshin, V. S., Vlasov, O. K., Maksioutov, M. A., Tsyb, A. F., Andreev, V. A., Hoshi, M., Yamashita, S., Shibata, Y. Non-Cancer Thyroid Diseases Among Children in the Kaluga and Bryansk Regions of the Russian Federation Exposed to Radiation following the Chernobyl Accident. Health Physics 88(1):16-22. 2005.

Iwasaki, T., Murata, M., Ohshima, S., Miyake, T., Kodo, S., Inoue, Y., Narita, M., Yoshimura, T., Akiba, S., Tango, T., Yoshimoto, Y., Shimizu, Y., Sobue, T., Kusumi, S., Yamagishi. C., Matsudaira, H. Second Analysis of Mortality of Nuclear Industry Workers in Japan, 1986-1997. Radiation Research 159: 228-238. 2003.

Izumi, S., Koyama, K., Soda, M., Suyama, A. Cancer Incidence in Children and Young Adults did not Increase Relative to Parental Exposure to Atomic Bombs. British Journal of Cancer 89:1709-1713. 2003.

Jablon, S., Miller, R. W. Army Technologists: 29-year Follow up for Cause of Death. Radiology 126:677-679. 1978.

Jablon, S., Boice, J. D., Jr., Hrubec, Z. Cancer in Populations Living Near Nuclear Facilities: A Survey of Mortality Nationwide and Incidence in Two States. Journal of the American Medical Association 265:1403-1408. 1991.

Jeffreys, A. J., Dubrova, Y. E. Monitoring Spontaneous and Induced Human Mutations by RAPD-PCR: a Response to Weinberg et al. (2001). Proc. R. Soc. Lond. B. Biol. Sci. 268:2493-2494. 2001.

Jemal, A., Tiwari, R. G., Murray, T., Ghafoor, A., Samuels, A., Ward, E., Feuer, E. J., Thun, M. J. Cancer Statistics, 2004. Cancer Journal for Clinicians 54:8-29. 2004.

Johnson, J. C., Thaul, S., Page, W. F., Crawford, H. Mortality of Veteran Participants in the CROSS-ROADS. Washington, DC: National Academy Press. 1996.

Jones, C. R. Radiation Protection Challenges Facing the Federal Agencies. Health Physics 87(3): 273-281. 2004.

Jonsen, A. R., Siegler, M., Winslade, A. J. Clinical Ethics. 5th Ed. McGraw Hill, New York. 2002.

Junk, A. K., Kundiev, Y., Vitte, P., Wogal, B. V. Ocular Radiation Risk Assessment in Populations Exposed to Environmentatal Radiation Contamination. Proceedings of the Advanced Research Workshop Kiev, Ukraine, 19 July-1 August, 1997. NATO Science Partnership Sub-Series: Kluwer Academic Publishers 2, October 1998.

Kemeny, J. G. The Need for Change: The Legacy of TMI, Report of the President's Commission on The Accident at Three Mile Island, Report of the Emergency Preparedness and Response Task Force. Government Printing Office, Washington, DC. 1979.

Kerber, R.A., Till, J.E., Simon, S.L., Lyon, J.L., Thomas, D.C., Preston-martin, S., Rallison, M.L., Lloyd, R.F., Stevens, W. A Cohort Study of Thyroid Disease in Relation to Fallout from Nuclear Weapons Testing. Journal of the American Medical Association 270:2076-2082. 1993.

Kiuru, A., Auvinen, A., Luokkamaki, M., Veidebaum, T., Tekkel, M., Rabu, M., Hakulinen, Y., Servomaa, K., Rytomaa, T., Mustonen, R. Hereditary Minisatellite Mutation Frequencies Among Pre- and Post-Chernobyl Children of Estonian Cleanup Workers. International Symposium on Radiation and Homeostasis, July 13-16, 2001, Kyoto, Japan. 2001.

Kodaira, M., Satoh, C., Hiyama, K., Toyama, K. Lack of Effects of Atomic Bomb Radiation on Genetic Instability of Tandem-Repetitive Elements in Human Germ Cells. Am. J. Hum. Genet. 57:1275-1283. 1995.

Kuletz, V. H. The Tainted Desert. Routledge, New York. pp. 336. 1998.

Lafata, J. E., Simpkins, J., Lamerato, L., Poisson, L., Divine, G., Johnson, C. C. The Economic Impact of False-Positive Cancer Screens. Cancer Epidemiol. Biomarkers Prev. 13(12): 2126-2132. 2004.

Lagakos, S. W., Mosteller, F. Assigned Share in Compensation for Radiation-Related Cancers. Risk Analysis 6(3):345-357. 1986.

Land, C. E., Tokunaga, M., Koyama, K., Soda, M., Preston, D. L., Nishimori, I., Tokuoka, S. Incidence of Female Breast Cancer Among Atomic Bomb Survivors, Hiroshima and Nagasaki, 1950-1990. Radiation Research 160(6):707-717. 2003.

Land, C., Zhumadilov, Z., Gusev, B., Hartshorne, M., Crooks, L., Wiest, P., Carr, Z., Luckyanov, N., Fillmore, C-M., Weinstock, B., Bouville, A., Simon, S. Thyroid Nodule Prevalence and Radiation Dose Among Childhood Residents of Villages Near the Semipalatinsk Nuclear Test Site in Northeastern Kazakhstan. Poster Presentation. International Congress of the Radiation Research Society, Brisbane, Australia. 2004.

Laurier, D., Valenty, M., Tirmarche, M. Radon Exposure and the Risk of Leukemia: A Review of Epidemiological Studies. Health Physics 81(3):272-88. 2001.

Lee, T. R. Environmental Stress Reactions Following the Chernobyl Accident. In: One Decade After the Chernobyl Accident. International Atomic Energy Agency, Vienna. Pages 283-310. 1996.

Lessard, E. T., Miltenberger, R. P., Cohn, S. H., Musolino, S. V., Conard, R. A. Protracted Exposure to Fallout—The Rongelap and Utirik Experience. Health Physics 46(3):511-527. 1984.

Lessard, E. T., Brill, A. B., Adams, W. H. Thyroid Cancer in the Marshallese: Relative Risk of Short Lived Internal Emitters and External Radiation Exposure. Proceedings of Fourth International Radiopharmaceutical Symposium. Oak Ridge, TN., BNL-37335; CONF-851113-3. 1985.

Levey, A. S., Coresh, J., Balk, E., Kausz, A. T., Levin, A., Steffes, M. W., Hogg, R. J., Perrione, R. D., Lau, J., Eknoyan, G. National Kidney Foundation Practice Guidelines for Chronic Kidney Disease: Evaluation, Classification, and Stratification. Ann. Intern Med. 139:137-147. 2003.

Lewin, J. L. The Genesis and Evolution of Legal Uncertainty About Reasonable Medical Certainty, 57 Maryland L. Rev. 380, 498. 1998.

Lewin, K. A Dynamic Theory of Personality. New York: McGraw-Hill. 1935.

Likhtarov, I., Kovgan, L., Vavilov, S., Chepurny, M., Bouville, A., Luckyanov, N., Jacob, P., Voillequé, P., Voigt, G. Post-Chornobyl Thyroid Cancers in Ukraine. Report 1: Estimation of Thyroid Doses. Radiation Research 163(2):125-136. 2005.

Lindberg, S., Karlsson, P., Arvidsson, B., Holmberg, E., Lunberg, L. M., Wallgren, A. Cancer Incidence after Radiotherapy for Skin Hemangiomas During Infancy. Acta Oncol. 34: 735-740. 1995.

Little, J. B. Radiation-Induced Genomic Instability. Int. J. Radiat. Biol. 74:663-671. 1998.

Little, M. P. The Proportion of Thyroid Cancers in the Japanese Atomic Bomb Survivors Associated With Natural Background Radiation. J. Radiol. Prot. 22(3):279-291. 2002.

Little, M. P., Muirhead, C. R. Curvature in the Cancer Mortality Dose Response in Japanese Atomic Bomb Survivors: Absence of Evidence of Threshold. Int. J. Radiation Biology 74: 471-480. 1998.

Livshits, L. A., Malyarchuk, S. G., Kravchenko, S. A., Matsuka, G. H., Lukyanova, E. M., Antikin, Y. G., Arabskaya, L. P., Petit, E., Giraudeau, F., Gourmelon, P., Vergnaud, G., Le Guen, B. Children of Chernobyl Cleanup Workers do not Show Elevated Rates of Mutations in Minisatellite Alleles. Radiation Research 155:74-80. 2001.

Lloyd, R. D., Gren, D. C., Simon, S. L., Wrenn, M. E., Hawthorne, H. A., Lotz, T. M. Stevens,W., Till, J. E. Individual External Exposure from Nevada Test Site Fallout for the Utah Leukemia Cases and Controls. Health Physics 59:723-737. 1990.

Lohr, K.N. Quality of Health Care. In Encyclopedia of Bioethics (2nd ed.). Editor: Reich, W.T. New York: Macmillan Publishing Company. 1995.

Lohr, K.N. Clinical Practice Guidelines: Historical Perspectives. In: Guidelines for Clinical Practice. Editors: Mattison, N., Tilson, H. The International Challenge. Proceedings of the Marlow Workshop. London: Pharmaceutical Partners for Better Healthcare. 1998.

Lohr, K.N. The Strengths and Limitations of Evidence-based Medicine and CPGs. In: Scripting a Future for Clinical Practice Guidelines. Editor: Usher, S. Proceedings from a Multidisciplinary Summit. Montreal, QC. Health Policy Forum. 1999.

Lohr, K.N., K. Eleazer, K., Mauskopf, J. Review Article. Health Policy Issues and Applications for Evidence-based Medicine and Clinical Practice Guidelines. Health Policy Vol. 46, No. 1, pp. 1-19. 1998.

Loomis, D. P., Wolf, S. H. Mortality of Workers at a Nuclear Materials Production Plant at Oak Ridge, Tennessee, 1947-1990. American Journal of Industrial Medicine 1:121. 1996.

Lundell, M., Hakulinen, T., Holm, L. E. Thyroid Cancer after Radiotherapy for Skin Hemangioma in Infancy. Radiation Research 140(3):334-333 1994.

Lyon, J. L., Klauber, M. R., Gardner, J. W., Udall, K. S. Childhood Leukemias Associated with Fallout from Nuclear Testing. New England Journal of Medicine 300(8):397-402. 1979.

Machado, S. G. Cancer Mortality and Radioactive Fallout in Southwestern Utah. American Journal of Epidemiology 125(1):44-61. 1987.

MacMahon, B., Hutchison, G. B. Prenatal X-Ray and Childhood Cancer: A Review. Acta. Unio. Int. Contra. Cancrum. 20:1172-4. 1964.

Marcinowski, F., Lucas, R. M., Yeager, W. M. National and Regional Distributions of Airborne Radon Concentrations in U.S. Homes. Health Physics 66:699-706. 1994.

Martland, H. S. Some Unrecognized Dangers in The Use and Handling of Radioactive Substances. Journal of the American Medical Association 85:1769-1776. 1929.

Matanoski, G. M., Seltser, R., Sartwell, P. E., Holland, E. L., Elliott, E. A. The Current Mortality Rates of Radiologists and Other Physician Specialists: Deaths from All Causes and from Cancer. American Journal of Epidemiology 101:188-198. 1975a.

Matanoski, G. M., Seltser, R., Sartwell, P. E., Holland, E. L., Elliott, E. A. The Current Mortality Rates of Radiologists and Other Physician Specialists: Specific Causes of Death. American Journal of Epidemiology 101:199-210. 1975b.

Matanoski, G. M., Sartwell, P., Elliott, E., Toniscia, J., Sternberg, A. Cancer Risks in Radiologists and Radiation Workers. In Radiation Carcinogenesis: Epidemiology and Biological Significance. Edited by Boice, J. D. Jr., Fraumeni, J. F. Jr. Raven Press. New York. 1984.

Matanoski, G. M. Health Effects of Low-Level Radiation in Shipyard Workers. Final Report. Johns Hopkins University, Baltimore, MD. DOE Report DE-AC02-79 EV 100095. 1991.

May, C. A., Tamaki, K., Neumann, R., Wilson, G., Zagars, G., Pollack, A., Dubrova, Y. E., Jeffreys, A. J., Meistrich, M. L. Minisatellite Mutation Frequency in Human Sperm Following Radiotherapy. Mutat. Res. 453:67-75. 2000.

McGale, P., Darby, S. C. Low Doses of Ionizing Radiation and Circulatory Diseases: A Systematic Review of the Published Epidemiological Evidence. Radiation Research 163: 247-257. 2005.

McLeroy, K. R., Bibeau, D., Steckler, A., and Glanz, K. An Ecological Perspective on Health Promotion Programs. Health Education Quarterly 15:351-377. 1988.

McGeoghegan, D., Binks, K. The Mortality and Cancer Morbidity Experience of Workers at The Capenhurst Uranium Enrichment Facility 1946-1995. J. Radiat. Prot. 20, 381-401. 2000a.

McGeoghegan, D., Binks, K. The Mortality and Cancer Morbidity Experience of Workers at The Springfields Uranium Production Facility 1946-1995. J. Radiat. Prot. 20, 111-137. 2000b.

McMichael, A. J. Standardized Mortality Ratios and the "Healthy Worker Effect": Scratching beneath the surface. Journal of Occupational Medicine 18 (3):168-187. 1975.

Meltzer, P. S., Kallioniemi, A., Trent, J. M. Chromosome Alterations in Human Solid Tumors. In: The Genetic Basis of Human Cancer (Eds. Vogelstein, B., and Kinzler, K. W.) McGraw-Hill: New York. 2002.

Mettler, F. A. Jr., Upton, A. C. Medical Effects of Ionizing Radiation. W.B. Saunders, Philadelphia. 1995.

Mill, J. S. Utilitarianism, edited with an introduction by Roger Crisp. New York: Oxford University Press. Originally published in 1861. 1998.

Miller, R. W., Jablon, S. A Search for Late Radiation Effects Among Men Who Served as X-Ray Technologists in the U.S. Army During World War II. Radiology 96:269-274. 1970.

Mohan, A. K., Hauptmann, M., Freedman, D. M., Ron, E., Matanoski, G. M., Lubin, J. H., Alexander, B. H., Boice, J. D. Jr., Doody, M. M., Linet, M. S. Cancer and Other Causes of Mortality Among Radiologic Technologists in the United States. Int. J. Cancer 103(2):259-267. 2003.

Mole, R. H. Childhood Cancer after Prenatal Exposure to Diagnostic X-Ray Examinations in Britain. Br. J. Cancer 62:152-168. 1990.

Morgan, W. F. Non-Targeted and Delayed Effects of Exposure to Ionizing Radiation-Induced Genomic Instability and Bystander Effects in vivo, Clastogenic Factors and Transgenerational Effects. Radiation Research 159: 581-596. 2003.

Mortensen, J. D., Woolner, L. B., Bennet, W. Gross and Microscopic Findings in Clinically Normal Thyroid Glands. J. Clin Ed. and Metab. 15:1270-1280. 1955.

Mothersill, C., Seymour, C. B. Review: Radiation-Induced Bystander Effects: Past History and Future Perspectives. Radiation Research 155:759-767. 2001.

Muller, H. J. Measurement of Gene Mutation Rate in Drosophila, Its High Variability, and Its Dependence Upon Temperature. Genetics 13:279-357. 1928.

Mulrow, C. D., Williams, J. W., Trivedi, M., Chiquette, E., Aguilar, C., Cornell, J. E. Treatment of Depression: Newer Pharmacotherapies. Evidence Report/Technology Assessment No. 7. Rockville, MD: Agency for Healthcare Research and Quality. AHRQ Publication No. 00-E003. 1999.

Muirhead, C. R., Goodhill, A. A., Haylock, R. G., Vokes, J., Little, M. P., Jackson, D. A., O'Hagan, J. A., Thomas, J. M, Kendall, G. M., Silk. T. J., Bingham, D., Berridge, G. L. Occupational Radiation Exposure and Mortality: Second Analysis of the National Registry for Radiation Workers. Journal of Radiation Protection 19(1):3-26. 1999.

Nagataki, S., Shibata, Y., Inoue, S., Yokoyama, N., Izumi, M., Shimaoka, K. Thyroid Diseases Among Atomic Bomb Survivors in Nagasaki. Journal of the American Medical Association 272(5):364-370. 1994.

Najarian, T., Colton, T. Mortality from Leukaemia and Cancer in Shipyard Nuclear Workers. Lancet 1:1018-1020. 1978.

NCI (National Cancer Institute). Estimated Exposures and Thyroid Doses Received by the American People from Iodine-131 in Fallout Following Nevada Atmospheric Nuclear Bomb Tests. NIH Publication No. 97-4264. Washington, DC: U.S. Department of Health and Human Services. 1997.

NCI-CDC (National Cancer Institute-Centers for Disease Control and Prevention). Land, C., Gilbert, E., Smith, J. M. Report of the NCI-CDC Working Group to Revise the 1985 NIH Radioepidemiological Tables. NIH Publication No. 03-5387. National Institutes of Health, Bethesda, MD. 2003.

NCRP (National Council on Radiation Protection and Measurements). Induction of Thyroid Cancer by Ionizing Radiation. NCRP Report No. 80. NCRP: Bethesda, MD. 1985.

NCRP (National Council on Radiation Protection and Measurements). Exposure of the Population in the United States and Canada from Natural Background Radiation. NCRP Report No. 94. NCRP: Bethesda, MD. 1987.

NCRP (National Council on Radiation Protection and Measurements). Misadministration of Radioactive Material in Medicine. NCRP Commentary No. 7. NCRP: Bethesda, MD. 1991.

NCRP (National Council on Radiation Protection and Measurements). The Probability that a Particular Malignancy may have been Caused by a Specific Irradiation. NCRP: Bethesda, MD. 1992.

NCRP (National Council on Radiation Protection and Measurements). A Guide for Uncertainty Analysis in Dose and Risk Assessments Related to Environmental Contamination. NCRP Commentary No. 14. NCRP: Bethesda, MD. 1996.

NCRP (National Council on Radiation Protection and Measurements). Uncertainties in Fatal Cancer Risk Estimates Used in Radiation Protection. NCRP Report No. 126. NCRP: Bethesda, MD. 1997.

NCRP (National Council on Radiation Protection and Measurements). Biological Effects and Exposure Limits for "Hot Particles". NCRP Report No 30, NCRP: Bethesda, MD. 1999.

NCRP (National Council on Radiation Protection and Measurements). Evaluation of the Linear-Non-Threshold Model for Ionizing Radiation. NCRP Report No 136, NCRP: Bethesda, MD. 2001.

NIH (National Institutes of Health). Rall, J. F., Beebe, G. W., Hoel, D. G., Jablon, S., Land, C. E., Nygaard, O. F., Upton, A. C., Yalow, R. S., Zeve, V. H. Report of the National Institutes of Health Ad Hoc Working Group to Develop Radioepidemiology Tables. National Institutes of Health, Bethesda, MD, NIH Publication No. 85-2748 (U.S. Government Printing Office, Washington). 1985.

National Kidney Foundation. K/DOQI Clinical Practice Guidelines for Chronic Kidney Disease: Evaluation, Classification and Stratification. Am. J. Kidney Dis. 39:S1-S266 (Suppl 1). 2002.

NRC (National Research Council). Committee on the Biological Effects of Ionizing Radiation. The Effects on Populations of Exposure to Low Levels of Ionizing Radiation: BEIR III. Washington, DC: National Academy Press. 1980.

NRC (National Research Council). Oversight Committee on Radioepidemiological Tables. Assigned Share for Radiation as a Cause of Cancer—Review of Radioepidemiological Tables Assigning Probabilities of Causation (Final Report). Washington, DC: National Academy Press. 1984.

NRC (National Research Council). An Assessment of the New Dosimetry for A-bomb Survivors, Panel on Reassessment on A-bomb Dosimetry. Washington, DC: National Academy Press. 1987.

NRC (National Research Council). Committee on Health Effects of Exposure to Radon. Health Effects of Exposure to Radon and Other Internally-Deposited Alpha Emitters: BEIR IV. Washington, DC: National Academy Press. 1988.

NRC (National Research Council). Committee on the Biological Effects of Ionizing Radiation. The Effects on Populations of Exposure to Low Levels of Ionizing Radiation: BEIR V. Washington, DC: National Academy Press. 1990.

NRC (National Research Council). Report of the Committee on Health Effects of Exposure to Low Levels of Ionizing Radiations (BEIR VII) Phase I. Health Effect of Exposure to Low Levels of Ionizing Radiation. Time for Reassessment? Washington, DC: National Academy Press. 1998.

NRC (National Research Council). Health Effects of Exposure to Radon: BEIR VI. Washington, DC: National Academy Press. 1999.

NRC (National Research Council). Committee on an Assessment of Centers for Disease Control and Prevention Radiation Studies from DOE Contractor Sites: Subcommittee to Review the Radioepidemiology Tables. A Review of the Draft Report of the NCI-CDC Working Group to Revise the "1985 Radioepidemiological Tables." Washington, DC: National Academy Press. 2000.

NRC (National Research Council). Status of the Dosimetry for the Radiation Effects Research Foundation (DS86). Washington, DC: National Academy Press. 2001.

NRC (National Research Council). Assessment of Scientific Information for the Radiation Exposure Screening and Education Program: Interim Report. Washington, DC: The National Academies Press. 2003a.

NRC (National Research Council). Review of the Dose Reconstruction Program of the Defense Threat Reduction Agency. Washington, DC: The National Academies Press. 2003b.

NRC (National Research Council). Exposure of the American Population to Radioactive Fallout from Nuclear Weapons Tests. A Review of the CDC-NCI Draft Report on a Feasibility Study. Washington, DC: The National Academies Press. 2003c.

Nozick, R. Anarchy, State and Utopia New York: Basic Books. 1974.

NEA/OECD (Nuclear Energy Agency/Organisation for Economic Co-operation and Development). Chernobyl: Assessment of Radiological and Health Impacts. 2002 Update of Chernobyl: Ten Years On. Nuclear Energy Agency/Organisation for Economic Co-operation and Development. Paris, France. 2002.

Olsen, D. M., Kane, R. L., Proctor, P. H. A Controlled Trial of Multiphasic Screening. New England Journal of Medicine 294:925-930. 1976.

Pauker, S. G., Kassirer, J. P. The Threshold Approach to Clinical Decision Making. New England Journal of Medicine 302(20):1109-1117. 1980.

Pearce, N. Mortality and Cancer Incidence in New Zealand Participants in United Kingdom Nuclear Weapons Tests in The Pacific. Supplementary Report. Wellington, New Zealand. Wellington School of Medicine. 1996.

Pearce, N., Winkelmann, R., Kennedy, J., Lewis, S., Purdie, G., Slater, T., Prior, I., Fraser, J. Further Follow-Up of New Zealand Participants in United Kingdom Atmospheric Nuclear Weapons Tests in The Pacific. Cancer Causes and Control 8:139-145. 1997.

Pierce, D. A., Shimizu, Y., Preston, D. L., Vaeth, M., Mabuchi, K. Studies of the Mortality of Atomic Bomb Survivors. Report 12, Part 1. Cancer 1950-1980. Radiation Research 146:1-27. 1996.

Pierce, D. A., Preston, D. L. Radiation-Related Cancer Risks at Low Doses Among Atomic Bomb Survivors. Radiation Research 154(2):178-186. 2000.

Pignone, M. P., Gaynes, B. N., Rushton, J. L., Burchell, C. M., Orleans, C. T., Mulrow, C. D., Lohr, K. N. Screening for Depression in Adults: A Summary of the Evidence for the U.S. Preventive Services Task Force. Annals of Internal Medicine 136:765-776. 2002.

Pinkerton, L. E., Bloom, T. F., Hein, M. J., Ward, E. M. Mortality Among a Cohort of Uranium Mill Workers: An Update. Occup. Environ. Med. 61:57-64. 2004.

Polednak, A. P., Frome, E. L.   Mortality Among Men Employed Between 1943 and 1947 at a Uranium-Processing Plant. J. Occup. Med. 23(3):169-178. 1981.

Posner, R. The Concept of Corrective Justice, 10 J. Legal Studies 187. 1981.

Preston, D. L., Kusumi, S., Tomonaga, M., Izumi, S., Ron, E., Kuramoto, A., Kamada, N., Dohy, H., Matsuo, T., Matsuo, T.   Cancer Incidence in Atomic Bomb Survivors. Part III. Leukemia, Lymphoma and Multiple Myeloma, 1950-1987. Radiation Research 137 (2 Suppl):S68- S97. 1994.

Preston, D. L., Ron, E., Yonehara, S., Kobuke, T., Fujii, H., Kishikawa, M., Tokunaga, M., Tokuoka, S., Mabuchi, K.   Tumors of the Nervous System and Pituitary Gland Associated with Atomic Bomb Radiation Exposure. J. Nat. Cancer Inst. 94:1555-1563. 2002.

Preston, D. L., Shimizu, Y., Pierce, D. A., Suyama, A., Mabuchi, K.   Studies of Mortality of Atomic Bomb Survivors. Report 13: Solid Cancer and Non Cancer Disease Mortality: 1950-1997. RERF Report No. 24-02. Radiation Research 160:381-407. 2003.

Preston, D. L., Pierce, D. A., Shimizu, Y., Cullings, H. M., Fujita, S., Funamoto, S., Kodama, K.   Effect of Recent Atomic Bomb Survivor Dosimetry Changes on Cancer Mortality Risk Estimates. Radiation Research 162:377-389. 2004.

Prise, K. M., Folkard, M., Michael, B. D.   Experimental Studies of Bystander Responses: Challenging Fundamental Mechanisms. Abstract for 4th International Conference on Health Effects of Low-level Radiation, Keble College, Oxford, UK. 2002.

Prosser, W. L.   Handbook on the Law of Torts 4th Edition, St. Paul: West Publications. 1971.

Rallison, M. L., Dobyns, B. M., Keating, F. R., Rall, J. E., Tyler, F. H.   Thyroid Disease in Children. A Survey of Subjects Potentially Exposed to Fallout Radiation. Am. J. Med. 56(4):457-463. 1974.

Rallison, M. L., Lotz, T. M., Bishop, M., Divine, W., Haywood, K., Lyon, J.L., Stevens, W.   Cohort Study of Thyroid Disease Near the Nevada Test Site: A Preliminary Report. Health Physics 59(5):739-746. 1990.

Rawls, J.   A Theory of Justice. Cambridge, Mass., Belknap Press of Harvard University Press. 1971.

Read, C. H., Tansey, M. J., Menda, Y.   A 36 Year Retrospective Analysis of the Efficacy and Safety of Radioactive Iodine in Treating Young Graves' Patients. J. Clin. Endocrinol. Metab. 89(9):4229-4233. 2004.

Reynolds, P., Austin, D. F.   Cancer Incidence Among Employees of the Lawrence Livermore National Laboratory, 1069-1980. Western Journal of Medicine 142:214-218. 1985.

Ricks, R. C., Berger, M. E., O'Hara, F. M. Jr.   The Medical Basis for Radiation-Accident Preparedness III: The Psychological Consequences. Elsevier Science Publishing Co. Inc. New York, NY. 1991.

Rinsky, R. A. Zumwalde, R. D., Waxweiler, R. J., Murray, W. E. Jr, Bierbaum, P. J., Landrigan, P. J., Terpilak, M., Cox, C.   Cancer Mortality at a Naval Nuclear Shipyard. Lancet 1:231-235. 1981.

Rinsky, R. A., Melius, J. M., Hornung, R. W., Zumwalde, R. D., Waxweiler, R. J., Landrigan, P. J., Bierbaum, P. J., Murray, W. E. Jr.   Case-Control Study of Lung Cancer in Civilian Employees at the Portsmouth Naval Shipyard, Kittery, Maine. Am. J. Epidemiol. 127(1):55-64. 1988.

Ritz, B., Morgenstern, H., Froines, J., Young, B. B.   Effects of Exposure to External Ionizing Radiation on Cancer Mortality in Nuclear Workers Monitored for Radiation at Rocketdyne/ Atomics International. American Journal of Industrial Medicine 35:21-31. 1999a.

Ritz, B., Morgenstern, H., Moncau, J.   Age at Exposure Modifies the Effects of Low-Level Ionizing Radiation on Cancer Mortality in an Occupational Cohort. Epidemiology 10(2):135-140. 1999b.

Roberts, R. C.   Injustice and Rectification. New York: Peter Lang. 2002.

Robins, J., Greenland, S.   Estimatibility and Estimation of Expected Years of Life Lost Due to a Hazardous Exposure. Stat. Med. 10:79-93. 1991.

Robins, J. M. Should Compensation Schemes Be Based on the Probability of Causation or Expected Years of Life Lost? Journal of Law and Policy 12(2):537-548. 2004.

Robbins, J. Thyroid Cancer: A lethal endocrine Neoplasm. NIH Conference. Ann. Int. Med. 115:133-147. 1991.

Robbins, J., Adams, W. H. Radiation effects in the Marshall Islands. Radiation and the Thyroid. S. Nagataki. Amsterdam, Excerpta Medica. Pp. 11-24. 1989.

Roedler, H. D. Assessment of Foetal Activity Concentration and Fetal Dose for Selected Radionuclides Based on Human and Animal Data. In: Age-related Factors in Radionuclide Metabolism and Dosimetry. Editors: Gerber, G. B., Metivier, H., Smith, H. Martinus Nijhoff, Dordrecht, pp. 327-337. 1987.

Ron, E., Modan, B., Preston, D., Alfandary, E., Stovall, M., Boice, J. D. Jr. Thyroid Neoplasia Following Low-Dose Radiation in Childhood. Radiation Research 120(3):516-531. 1989.

Ron, E., Preston, D. L., Mabuchi, K., Thompson, D. E., Soda, M. Cancer Incidence in Atomic Bomb Survivors. Part IV: Comparison of Cancer Incidence and Mortality Radiation Research 137(2Suppl):S98-S112. 1994.

Ron, E., Lubin, J. H., Mabuchi, K., Modan, B., Pottern, L. M., Schneider, A. B., Tucker, M. A., Boice, J. D. Jr. Thyroid Cancer after Exposure to External Radiation: A Pooled Analysis of Seven Cohort Studies. Radiation Research 141:259-277. 1995.

Ron, E., Doody, M. M., Becker, D. V., Brill, A. B., Curtis, R. E., Goldman, M. B., Harris, B. S. 3rd, Hoffman, D. A., McConahey, W. M., Maxon, H. R., Preston-Martin, S., Warshauer, M. E., Wong, F. L., Boice, J. D. Jr. Cancer Mortality Following Treatment for Adult Hyperthyroidism. Cooperative Thyrotoxicosis Therapy Follow-up Study Group. Journal of the American Medical Association 280(4):347-355. 1998a.

Ron, E., Preston, D. L., Kishikawa, M., Kobuke, T., Iseki, M., Tokuoka, S, Tokunaga, M., Mabuchi, K. Skin Tumors Risk Among Atomic Bomb Survivors in Japan. Cancer Causes and Control 9:393-401. 1998b.

Rooney, C., Beral, V., Maconochie, N., Fraser, P., Davies, G. Case-Control Study of Prostatic Cancer in Employees of the United Kingdom Atomic Energy Authority. British Medical Journal 307:1391-1397. 1993.

Roscoe, R. J. An Update of Mortality from all Causes Among White Uranium Miners from the Colorado Plateau Study Group. Am. J. Ind. Med. 31(2):211-222. 1997.

Ross, R. K., Schottenfeld, D. Prostate Cancer. In: Cancer Epidemiology and Prevention. Second edition. Editors: Schottenfeld, D., Fraumenis, J. F. Jr. Oxford University Press, New York, NY. 1996.

Roth, B. J., Nichols, C. R., Einhorn, L. H. Neoplasms of the Testis. In: Cancer Medicine. Eds Holland, J. F., Frei, E., Bast, R. C. Jr., Kufe, D. W., Morton, D. L., Weichselbaum, R. Third Edition, Vol 2. Lea and Febiger, Philadelphia. pp. 1593-1619. 1992.

Rothman, K. J. Clustering of Disease. American Journal of Public Health 77(1):13-15. 1987.

Rudner, R. The Scientist Qua Scientist Makes Value Judgments. Philosophy of Science, XX. 1953.

Russell, W. L., Russell, L. B., Kelly, E. M. Dependence of Mutation Rate on Radiation Intensity. In: Immediate and Low Level Effects of Ionizing Radiations. Editor: Buzzatti-Traverso Taylor and Francis, Ltd., London. 1960.

Santora, M., Cary, B. Depressed? New York City Screens for People at Risk. New York Times, April 13. Pp. A1, A16. 2005.

Saskatchewan Education. Understanding the Common Essential Learnings. Regina, SK:Curriculum and Instruction Branch, Saskatchewan Education. Available at: http://www.sasked.gov.sk.ca/docs/policy/cels/el3.html#e13e10; accessed 12/21/2004.

Schlenger, W. E., Jernigan, N. E. Mental Health Issues in Disasters and Terrorist Attacks. Ethn. Dis. 13 (Supplement 3):89-93. 2003.

Schneider, A. B., Ron, E., Lubin, J., Stovall, M., Gierlowski, T. C. Dose-Response Relationships for Radiation-Induced Thyroid Cancer and Thyroid Nodules: Evidence for the Prolonged Effects of Radiation on the Thyroid. J. Clin. Endocrinol. Metab. 77(2):362-369. 1993.

Schottenfeld, D., Winawer, S. Cancers of the Large Intestine. In: Cancer Epidemiology and Prevention. Second edition. Editors: Schottenfeld, D. and Fraumeni. JF., Jr. Oxford University Press, New York, NY. 1996.

Schubauer-Berigan, M. K., Wenzl, T. B. Leukemia Mortality Among Radiation-Exposed Workers. Occupational Medicine 16: 271-287. 2001.

Schull, W. J., Otake, M., Neel, J. V. Genetic Effects of the Atomic Bombs: A Reappraisal. Science 213:1220-1227. 1981.

Schwartz, L. M., Woloshin, S., Black, W. C., Welch, H. G. The Role of Numeracy in Understanding the Benefit of Screening Mammography. Ann. Intern. Med. 127(11):966-72. 1997.

Seltser, R., Sartwell, P. E. The Influence of Occupational Exposure to Radiation on the Mortality of American Radiologists and Other Medical Specialists. Am. J. Epidemiol. 81:2-22. 1965.

Senate Debate, 106th Congress, 2nd Session 146 Cong. Rec. S 6036, Vol. 146, No. 84. Wednesday, June 28. 2000.

Sevcova, M., Svec. J., Thomas, J. Alpha Irradiation of the Skin and the Possibility of Late-Effects. Health Physics 35:803-806. 1978.

Sharma, S. K., Pande, J. N., Verma, K. Effect of Prednisolone Treatment in Chronic Silicosis. Am. Rev. Respir. Dis. 143(4 Pt 1):814-821. 1991.

Shearer, S. D., Jr., Sill, C. W. Evaluation of Atmospheric Radon in the Vicinity of Uranium Mill Tailings. Health Physics 17:77-88. 1969.

Sheridan, S. L., Pignone, M. Numeracy and the Medical Student's Ability to Interpret Data. Effective Clinical Practice 5(1):35-40. 2002.

Sheridan, S. L., Harris, R. P., Woolf, S. H. Shared Decision Making about Screening and Chemoprevention. A Suggested Approach from the U.S. Preventive Services Task Force. Am. J. Prev. Med. 26(1):56-66. 2004.

Shimizu, Y., Kato, H., Schull, W. J., Hoel, D. G. Studies of the Mortality of A-Bomb Survivors. 9. Mortality 1950-1985; Part 3 Noncancer Mortality Based on Revised Doses (DS 86). Radiation Research 130:249-266. 1992.

Shintani, T., Hayakawa, N., Hoshi, M., Sumida, M., Kurisu, K., Oki, S., Kodama, Y., Kajikawa, H., Inai, K., Kamada, N. High Incidence of Meningioma Among Hiroshima Atomic Bomb Survivors. J. Radiat. Res. (Tokyo) 40:49-57. 1999.

Shore, R. Overview of Radiation-Induced Skin Cancer in Humans. International Journal of Radiation Biology 57:809-827. 1990.

Shore, R. E., Hildreth, N., Dvoretsky, P., Andresen, E., Moseson, M., Pasternack, B. Thyroid Cancer Among Persons Given X-Ray Treatment in Infancy for an Enlarged Thymus Gland. Am. J. Epidemiol. 137(10):1068-1080. 1993.

Sigurdson, A. J., Doody, M. M., Rao, R. S., Freedman, D. M., Alexander, B. H., Hauptmann, M., Mohan, A. K., Yoshinaga, S., Hill, D. A., Tarone, R., Mabuchi, K., Ron, E., Linet, M. S. Cancer Incidence in the US Radiologic Technologists Health Study, 1983-1998. Cancer 97(12):3080-3089. 2003.

Silver, S. R., Daniels, R. D., Taulbee, T. D., Zaebst, D. D., Kinnes, G. M., Couch, J. R., Kubale, T. L., Yiin, J. H., Schubauer-Berigan, M. K., Chen, P. H. Differences in Mortality by Radiation Monitoring Status in an Expanded Cohort of Portsmouth Naval Shipyard Workers. J. Occup. Environ. Med. 46(7):677-690. 2004.

Simon, M. A. Causation, Liability and Toxic Risk Exposure. Journal of Applied Philosophy, Vol. 9, No. 1, pp. 35-44. 1992.

Simon, S. L., Till, J. E., Lloyd, R. D., Kerber, R. L., Thomas, D. C., Preston-Martin, S., Lyon, J. L., Stevens, W. The Utah Leukemia Case-Control Study: Dosimetry Methodology and Results. Health Physics 68(4):460-471. 1995.

Simon, S. L., Bouville, A. Radiation Doses to Local Populations Near Nuclear Weapons Test Sites Worldwide. Health Physics 82(5):706-725. 2002.

Smith, P. G., Doll, R. Mortality from Cancer and All Causes Among British Radiologists. British Journal of Radiology 54:187-194. 1981.

Smith, P. G., Douglas, A. J. Mortality of Workers at the Sellafield Plant of British Nuclear Fuels. British Medical Journal 293:845-854. 1986.

Stannard, J. N. Radioactivity and Health: A History. Pacific Northwest Laboratory. (DOE/RL/ 01830-T-59 UC-408). 1988.

Stebbings, J. H., Lucas, H. F., Stehney, A. F. Mortality from Cancers of Major Sites in Female Radium Dial Workers. Am. J. Ind. Med. 5(6):435-459. 1984.

Stepanenko, V. F., Voilleque, P. G., Gavrilin, Y. I., Khrouch, V. T., Shinkarev, S. M., Orlov, M. Y., Kondrashov, A. E., Petin, D. V., Iaskova, E. K., Tsyb, A. F. Estimating Individual Thyroid Doses for a Case-Control Study of Childhood Thyroid Cancer in Bryansk Oblast, Russia. Radiat. Prot. Dosimetry 108(2):143-160. 2004.

Stern, F. B., Waxweiler, R. A., Beaumont, J. J., Lee, S. T., Rinsky, R. A., Zumwalde, R. D., Halperin, W. E., Bierbaum, P. J., Landrigan, P. J., Murray, W. E. Jr. A Case-Control Study of Leukemia at a Naval Nuclear Shipyard. Am. J. Epidemiol. 123(6):980-992. 1986.

Stevens, W., Thomas, D. C., Lyon, J. L., Till, J. E., Kerber, R. A., Simon, S. L., Lloyd, R. D., Elghany, N. A., Preston-Martin, S. Leukemia in Utah and Radioactive Fallout from the Nevada Test Site: A Case Control Study. Journal of the American Medical Association 265:585-591. 1990.

Takahashi, T., Trott, K. R., Fujimori, K., Nakashima, N., Ohtomo, H., Shoemaker, M. J., Simon, S. L. Thyroid Disease in the Marshall Islands: Findings from 10 years of Study. Tohoku University Press, Sendai, Japan. 2001.

Tan, G. H., Gharib, H. Thyroid Incidentalomas: Management Approaches to Nonpalpable Nodules Discovered Incidentally on Thyroid Imaging. Ann. Intern. Med. 126(3):226-231. 1997.

Telle-Lamberton, M., Bergot, D., Gagneau, M., Samspm, E., Giraud, J. M., Neron, M. O., Hubert, P. Cancer Mortality Among French Atomic Energy Commission Workers. American Journal of Industrial Medicine 45:34-44. 2004.

The Atomic Energy Act of 1946 (42 U.S.C. 2012(d)-(e), 2013(d), 2051(d).

Thomas, G., Karaoglu, A., Williams, E. D. Radiation and Thyroid Cancer. World Scientific Publishing Co. 1999.

Thompson, D. E., Mabuchi, K., Ron, E., Soda, M., Tokunaga, S., Ochikubo, S., Sugimoto, S., Ikeda, T., Terasaki, M., Preston, D. L. Cancer Incidence in Atomic Bomb Survivors. Part II: Solid Tumors, 1958-1987. Radiation Research 137(Suppl.) S17-S67. 1994.

Thompson, I. M., Pauler, D. K., Goodman, P. J., Tangen, C. M., Lucia, M. S., Parnes, H. L., Minasian, L. M., Ford, L. G., Lippman, S. M., Crawford, E. D., Crowley, J. J., Coltman, C. A. Jr. Prevalence of Prostate Cancer Among Men With A Prostate-Specific Antigen Level. N. Engl. J. Med. 350(22):2239-2246. 2004.

Thun, M. J., Baker, D. B., Steenland, K., Smith, A. B., Halperin, M., Berl, T. Renal Toxicity in Uranium Mill Workers. Scandinavian Journal of Environmental Health 11:81-90. 1985.

Till, J. E., Simon, S. L., Kerber, R., Lloyd, R. D., Stevens, W., Thomas, D. C., Lyon, J. L., Preston-Martin, S. The Utah Thyroid Cohort Study: Analysis of the Dosimetry Results. Health Physics 68:472-483. 1995.

Tomasek, L., Darby, S. C., Swerdlow, A. J., Placek, V., Kunz, E. Radon Exposure and Cancers Other Than Lung Cancer Among Uranium Miners in West Bohemia. Lancet 341:919-923. 1993.

Tronko, M. D., Howe, G. R., Bogdanova, I., Bouville, A. C., Epstein, O. V., Brill, A. B., Likhtarev, I., Fink, D. J., Markov, V. V., Grenebaum, E., Olijnyk, V. A., Masnyk, I. J., Shpak, V. M., McConnell, R. J., Tereshchenko, V. P., Robbins, J., Zvinchuk, O. V., Zablotska, L. B., Hatch, M., Luckyanov, N. K., Ron, E., Thomas, T. L., Voillequé, P. G., Beebe, G. W. A Cohort Study of Thyroid Cancer and Other Thyroid Disease After the Chernobyl Accident: Thyroid Cancer in Ukraine Detected During First Screening. In Press.

Tucker, M. A., Jones, P. H., Boice, J. D. Jr., Robison, L. L., Stone, B. J., Stovall, M., Jenkin, R. D., Lubin, J. H., Baum E. S., Siegel, S. E. Therapeutic Radiation at a Young Age is Linked to Secondary Thyroid Cancer. The Late Effects Study Group. Cancer Research 51(11):2885-2888. 1991.

Tversky, A., Kahneman, D. Judgment under Uncertainty: Heuristics and Biases. Science 185:1124-1131. 1974.

UNSCEAR (United Nations Scientific Committee on the Effects of Atomic Radiation). Ionizing Radiation: Sources and Biological Effects. Report E, 82, IX, 8. New York: United Nations. 1982.

UNSCEAR (United Nations Scientific Committee on the Effects of Atomic Radiation). Sources, Effects and Risks of Ionizing Radiation, No. E.94.IX.2, United Nations, New York. 1993.

UNSCEAR (United Nations Scientific Committee on the Effects of Atomic Radiation). Sources and Effects of Ionizing Radiation, Report to the General Assembly, with Scientific Annexes. New York: United Nations. 1994.

UNSCEAR (United Nations Scientific Committee on the Effects of Atomic Radiation). Exposures to the Public from Man-Made Sources of Radiation. Annex C In: Sources and Effects of Ionizing Radiation. Volume I: Sources. New York: United Nations. 2000.

UNSCEAR (United Nations Scientific Committee on the Effects of Atomic Radiation). Hereditary Effects of Radiation, Report to the General Assembly, with Scientific Annex. New York: United Nations. 2001.

United States Congress. House. Committee on the Judiciary. Discretionary Function Exemption of the Federal Tort Claims Act and the Radiation Exposure Compensation Act. Hearing, November 1, 8, 1989. 101st Congress, 1st Session. CIS Legislative Histories, 90 CIS H 52129.

United States Congress. Senate. Committee on Labor and Human Resources. Radiation Exposure Compensation Act of 1979. Hearing, June 10, 1980. 96th Congress, 2nd Session. CIS Legislative Histories, 81 CIS S 5414.

USEPA (United States Environmental Protection Agency). Limiting Values of Radionuclide Intake and Air Concentration and Dose Conversion Factors for Inhalation, Submersion, and Ingestion. Federal Guidance Report No. 11. EPA 520/1-88-020. 1988.

USEPA (United States Environmental Protection Agency). External Exposure to Radionuclides in Air, Water, and Soil. EPA Federal Guidance Report No. 12. EPA 402-R-93-81. 1993.

USEPA (United States Environmental Protection Agency). Cancer Risk Coefficients for Environmental Exposure to Radionuclides. Federal Guidance Report No. 13 supplemental CD, 2000. EPA 402-R-99-001. 1999.

Upton, A. C., Wilson, R. Compensating Government Workers Exposed to Radiation. Harvard Center for Risk Analysis. Risk in Perspective Volume 8 (7). 2000.

Ursano, R. J., McCaughey, B. G., Fullerton, C. S. The Structure of Human Chaos. In: Individual and Community Responses to Trauma and Disaster. Editors: Ursano, R. J., McCaughey, B. G., Fullerton, C. S. Cambridge University Press, Cambridge, MA. Pages 403-410. 1994.

van Roosmalen, M. S., Stalmeier, P. F., Verhoef, L. C., Hoekstra-Weebers, J. E., Oosterwijk, J. C., Hoogerbrugge, N., Moog, U., van Daal, W. A. Randomized Trial of A Shared Decision-Making Intervention Consisting of Trade-Offs and Individualized Treatment Information For BRCA1/2 Mutation Carriers. J. Clin. Oncol. 22(16):3293-3301. 2004.

Voelz, G. L., Lawrence, J. N. P., Johnson, E. R. Fifty Years of Plutonium Exposure to the Manhattan Project Plutonium Workers: An update. Health Physics 73:611-619. 1997.

Wakeford, R., Antell, B. A., Leigh, W. J. A Review of Probability of Causation and its Use in a Compensation Scheme for Nuclear Industry Workers in the United Kingdom. Health Physics 74(1):1-9. 1998.

Wakeford, R. The Cancer Epidemiology of Radiation. Oncogene 23:6404-6428. 2004.

Walchuk, M. Radiation Compensation Programs. Health Physics News, Vol. XXX Number 11. 2002.

Watanabe, K. K., Kang, H. K. Military Service in Vietnam and The Risk of Death from Trauma and Selected Cancers. Ann. Epidemiol. (5):407-412. 1995a.

Watanabe, K. K., Kang, H. K., Dalager, N.A. Cancer Mortality Risk Among Military Participants of a 1958 Atmospheric Nuclear Weapons Test. Am. J. Public Health 85(4):523-527. 1995b.

Waxweiler, R. J., Archer, V. E., Roscoe, R. J., Watanabe, A., Thun, M. J. Mortality Patterns Among a Retrospective Cohort of Uranium Mill Workers. In Proceedings of the 16th Midyear Topical Meeting of the Health Physics Society (Albuquerque, NM), Health Physics Society, pp. 428-435. 1983.

Weaver, V., Lelievre, S., Lakins, J., Chrenek, M., Jones, J., Giancotti, F., Werb, Z., Bissell, M. Beta4 Integrin-Dependent Formation of Polarized Three-Dimensional Architecture Confers Resistance to Apoptosis in Normal and Malignant Mammary Epithelium. Cancer Cell 2:205-216. 2002.

Weinberg, H. S., Korol, A. B., Kirzhner, V. M., Avivi, A., Fahima, T., Nevo, E., Shapiro, S., Rennert, G., Piatak, O., Stepanova, E. I., Skvarskaja, E. Very High Mutation Rate in Offspring of Chernobyl Accident Liquidators. Proc. R. Soc. Lond. B. Biol. Sci. 268:1001-1005. 2001.

Weinrib, E. J. Causation and Wrongdoing, 63 Chi.-Kent L. Rev. 407; at p. 409. 1987.

Whitney, S. N., McGuire, A. L., McCullough, L. B. A Topology of Shared Decision Making, Informed Consent, and Simple Consent. Ann. Int. Med. 140:54-59. 2004.

Wilde, G., Sjostrand, J. A Clinical Study of Radiation Cataract Formation Assessed in Adult Life Following Gamma Irradiation of the Lens in Early Childhood. Brit. J. Ophth. 81:261-266. 1997.

Williams, E. D. Biological Effects of Radiation on the Thyroid. In Braverman, L. E., and Utiger, R. D. (eds). Werner and Ingbar's, The Thyroid. Philadelphia, Lippincott. Pp. 421-436. 1991.

Williams, M. V., Parker, R. M., Baker, D. W., Parikh, N. S., Pitkin, K., Coates, W. C., Nurss, J. R. Inadequate Functional Health Literacy Among Patients at two Public Hospitals. Journal of the American Medical Association 6;274(21):1677-1682. 1995.

Wing, S., Richardson, D., Wolf, S., Mihlan, G. Plutonium-Related Work and Cause-Specific Mortality at the United States Department of Energy Hanford Site. Am. J. Ind. Med. (2):153-164. 2004.

Woloshin, S., Schwartz, L. M., Moncur, M., Gabriel, S., and Anna N. A. Tosteson, A. N. A. Assessing Values for Health: Numeracy Matters. Med. Decis. Making 21:380-388. 2001.

Yamada, M., Izumi, S. Psychiatric Sequellae in Atomic Bomb Survivors in Hiroshima and Nagasaki Two Decades After the Explosions. Society of Psychiatry and Psychiatric Epidemiology 37:409-415. 2002.

Yamada, M., Wong, F. L., Fujiwara, S., Akahoshi, M., Suzuki, G. Noncancer Disease Incidence in Atomic Bomb Survivors, 1958-1998. Radiation Research 161(6):622-632. 2004.

Yonehara, S., Brenner, A. V., Kishikawa, M., Inskip, P. D., Preston, D.,L., Ron, E., Mabuchi, K., Tokuoka, S. Clinical and Epidemiologic Characteristics of First Primary Tumors of the Central Nervous System and Related Organs Among Atomic Bomb Survivors in Hiroshima and Nagasaki, 1958-1995. Cancer 101(7):1644-1654. 2004.

Yoshinaga, S., Mabuchi, K., Sigurdson, A. J., Doody, M. M., Ron, E. Cancer Risks Among Radiologists and Radiologic Technologists: Review of Epidemiologic Studies. Radiology 233: 313-321. 2004.

Yoshinaga, S., Hauptmann, M., Sigurdson, A. J., Doody, M. M., Freedman, D. M., Alexander, B. H., Linet, M. S., Ron, E., Mabuchi, K. Nonmelanoma Skin Cancer in Relation to Ionizing Radiation Exposure Among U.S. Radiologic Technologists. Int. J. Cancer. February 9, 2005.

Young, R. W. Acute Radiation Syndrome. Military Radiobiology. Editors: Conklin, J. J., Walker, R. I. Academic Press, Inc., Orlando, FL. 1987.

Zablotska, L. B., Ashmore, J. P., Howe, G. R. Analysis of Mortality Among Canadian Nuclear Power Industry Workers after Chronic Low-Dose Exposure to Ionizing Radiation. Radiation Research 161(6):633-641. 2004.

Appendixes

Appendix A

Invited Speakers and Public Comment

INVITED SPEAKERS

We would like to cordially thank all the invited speakers listed here for their presentations that provided background material related to our study. Their contributions to our study are greatly appreciated.

All the information gathered at the open meetings is part of the National Research Council's public-access file and is available on request to anyone interested.

Dr. Rebecca Barlow, Dixie Regional Medical Center
Louise Benson, Chairwoman, Hualapai Indian Tribe
Dr. Alfred Berg, University of Washington
Commissioner Bruce Blackham of Sanpete County, Utah
Jeffrey Bradshaw
Ed Brickey, Uranium Industry Workers Representative
Dr. Evelyn J. Bromet, SUNY Stony Brook
Dr. Douglas M. Brugge, Tufts University School of Medicine
John Cathey
Dr. Teresa Coons, Saccomanno Research Center at St. Mary's Hospital
Tom Coughlin, Bureau of Primary Health Care
Dr. Regan Crump, Bureau of Primary Health Care
Brad Ence
Eleanore Fanire, Mohave County Downwinders
Gerard Fischer, U.S. Department of Justice

Gloria O. Gustin
Commissioner Ira Hatch, of Emery County, Utah
Senator Orrin G. Hatch, United States Senator, Utah
Dr. Richard Kerber, University of Utah
Kelly King
Verr D. Leavitt
Dr. Lynn Lyon, University of Utah Health Sciences Center
Dr. Kiyo Mabuchi, National Cancer Institute
Dr. Parthiv Mahadevia, MEDTAP International
Congressman Jim Matheson, Congressman, Utah 2nd District
Councilman Jerry McNeely of Grand County, Utah
Harvey W. Merrell
Hazel Merrit, Utah Navajo Downwinders
Dr. Karen Mulloy, New Mexico Radiation Exposure Screening and Education
    Program
Linda Nelson, Mountain Park Health Center
Dr. Lynne Pinkerton, National Institute for Occupational Safety and Health
Dr. Neil R. Powe, Johns Hopkins University
Dr. Stephen Prescott, Executive Director, Huntsman Cancer Institute
Carolyn Royce
William E. Seegmiller
Dr. Steve Simon, National Cancer Institute
Dr. Stephanie Singer, Utah Navajo Health System, Inc.
Dianne Spellberg, US Department of Justice
Commissioner Robert Steele of Juab County, Utah
Shauna Stout
Dr. Bruce Struminger, Navajo Area Indian Health Service
Kathleen Taimi, U.S. Department of Energy
Laura J. Taylor, Mohave County Downwinders advocate
Iris P. Tolley
Peter M. Turcic, U.S. Department of Labor
Dr. Robert J. Ursano, Uniformed Services University
Carol Williams
Dr. Steven H. Woolf, Virginia Commonwealth University

STATEMENTS FROM MEMBERS OF THE PUBLIC

We also would like to thank the following representatives and individuals for their oral and written statements related to our study. We greatly appreciate the time and effort given to our study through their testimony and comments. All statements are part of the National Research Council's public-access file and are available on request to anyone interested.

Joyce Marie Jensen Abernathy
JoEtta Abo
Lynette M. Adams
Michael S. Adams
Becky Adamson
Sherry Adderly
Alice Addington
Pearl Ahnstedt
Linda S. Aiken
Diane Albinola
Mona Alldredge
Marva G. Allen
Ms. Allen
Xan G. Allen
Tiffany Allgood, Coeur d'Alene
    Tribe
Joyce E. Allie
Loralee Alt
Carol Alworth
Kathy Amidon
Lonny R. Amen
Clifford C. Amundsen
Ross C. Anderson
Lynn R. Anspaugh
Diane Applegate
Don and Corrine Applegate
Opel Aprell
C. Tom Arkoosh
Cathy Armacost
Janet M. Arnel
Corinna Arnell
Charlotte L. Arnold
Janie Arnzen
Dorothy M. Ash
James E. Asher
Margaret J. Asher
Sara Ashley
Ray Aspiri
Lou Atoigue
John H. Austin Jr.
Chris Axford
Craig Axford
Dianna Babbitt

Todd Babbitt
Patricia Baggerly
Margaret Linville Bailey
Cynda Bain
Netti Bain
Guy Taft Baker
Joyce Vieselmeyer Baker
Kathleen Baker
Robert L. Baker
Deloris Loveland Balch
Jack B. Baldwin
Muriel J. Baldwin
Wendy Ball
Crystal Babbitt Banning
Juanita Barkdull
Cheryl F. Barker
Bonnie Barkley
Melinda K. (Thirlwell) Barnes
Theodore and Neoma Barr
Beverly Barrett
Leslie Batt-Corbet
Senator Max Baucus (Montana)
Barbara C. Walker Baune
Mrs. Charlie Ross Beale
Kathy Bean
Jennifer Beasley
Beverly Becenti-Pigman
David W. Becker, Jr.
Douglass A. Becker
L. W. Becker
Alvin Beers
Cynda S. Beeson
Karen Beeson
Robert Begay
Sammie Begaye
Cheryl Behr
Sue Beitia
Don Benally
Evelyn L. Bennett
Randy Benson
Shirley Benson
Shirley Colburn Benson
Jay Bentley

Patricia Y. Beonde
Teresa M. Bergin
Christine Bergin-Werner
Margaret Bergin-Satterlee
Marnie Bernard
Shirley A. Berry
Wallace E. Betzold
Lori Bevan-Gardiner
GeorgeAnn Bevill
Larry D. Bidiman
Kerry A. Bidiman
Tami Billman
Senator Jeff Bingaman
Michelle Bird
Don Black
Kitty Blake
Jill L Blanton
Judy L. Blanton
Suzie Blatz
Virginia Blessinger
Ginger Blevins
Cassandra J. Bloedel
Theodore F. Blume
Board of Washington County Commissioners Diana Thomas, Roy Mink, and Rick Michael
Frances Eileen Day and Ruby Bockman
Jeanne Boehm
Neil Bolich
Kathleen C. Bongiovi
Crystal Book
Mirinda J. Booker
Janet S. Boor
Charlene Booth
Congresswoman Madeleine Z. Bordallo
Karen Bork
Darline Bowers
Leonard Bowers
Marjorie Bowman
Laralee Steinfeldt Boyd
Christina (Lancaster) Brackney

Jessie Mae Bradford
Connie Hopkins Brandau
Travis Brandon
Doris Gissel Brannan
Chrystine Eiguren Brassey
Bennie Bray
Connie L. Breener
Patricia A. Bricker
Eugene Bridges
Lee V. Brinkerhoff
Bob Brister
Beulah Brooks
Chuck Broscious
    Environmental Defense Institute
Violet M. P. ners
Adare brown
Carmen W. Brown
Eldene Brown
Gerald Brown
John Brown
Nancy Bailey Smith Brown
Norma Francis Brown
Norman Patrick Brown
Susan Rae Brown
Valerie Brown
Beverly Browne
Ruth Browning
Connie L. Bruner
Lisa Bruner
Pansy Bryson
Stuart Bryson
Colby Buchanan
Tracy Buchanan
Lynn Buck
Sandy Perkins Buffington
Sandra Bunch
Hazel Bungor
Mary E. Burket
Jean Burkhart
Senator Conrad Burns (Montana)
Connie Burrell
Kathleen Burrup
Susan Burstedt

Gae Burton
V. Faye Burton
Sharon Bussert
Velma Butler
Sandra E. Cady
Kimberley Calder
Becky Caldwell
Christopher Caldwell
Linda Calhoun
Mary Jane Calhoun
Luis Caloca
Bill A. Campbell
Norma Frances Snodgrass Campbell
Patsy L. Campbell
Sharla Campbell
Deon L. Canaday
Marolyn Stevens Canaday
Michael R. Canaday
Neal Canaday
Victor Leon Cannon
Paulette Caprai-Findlay
Karen L. Carley
Elmer Carlock
Gale Carlock
Leonard Carlock
William R. Carlock Sr.
Joye C. Carlson
Bert E. and Carol P. Carpenter
C C Carpenter
James Carpenter
Robert and Eileen Carrier
Judy Caruso
Lorraine Carvener
Lynette Case
Glenalee Casper
Nancy Caster
Elizabeth Bruhn Catalan
Lorraine Cavener
Bonnie Cawthra
Robert Celestial
Bob and Georgia Lee Chapin
Greg Chapin
Steve Chapman

Anita J. Chase
Debra (Dubrava) Chase
Darlene Cheney
Zola June Cheney
Debbie A. Childers
Dr. Herbert E. Childress
William R. Chivers
Alice I. Choat
Carolyn Christensen
Diane R. Christensen
Dorene Wise Christensen
Kris Christensen
Anne Christie
Ethel Cignar
Alyce Clark
Arlene Clark
Dayna Clark
Jill Claxon
Diane Clayton
Jeanne Clifford
Patricia Cluff
Stephen R. Cobbley
Loretta Cochran
Kathryn Cock
Jack Colburn
Janet Colburn
Juanita Colburn
Dave Cole
Ada Ruth Colwell
John Commander
Karen Compton
Vicki de la Concepcion
Cathy Conley
Pam Conley
Nancy Reynolds Connell
Jeff Conner
Katherine Contway
Kerry Cooke
Wannetta Cooke
Dr. Teresa Coons
Myrna Copley
Dianne Christensen Corbin
Donald R. Corley

Melissa Corp
Gretchen Cotrell
George W. Cox
Dr. Timothy R. Cox
Clara Cox-Wood
Maxine V. Crandall
Vicki L. Crank
Dave Crawforth
Robert W. Cressey
Addie M. Cristy
Nancy Crookston
Barbara Cross
Myra Dean Crouch
Bonnie Currin
Sue Curtis
Ruth Leora Cutbirth
Sharon Ann Morse Daly
Paul Van Dam
Eletha S. Daniels
Renae Dastrup
Linda Davidson
Carol Davis
James H. Davis
Jody Davis
Gary D. Davison
Mardelle E. Davison
Bernard Dawson
Dan Dawson
Dr. Susan Dawson
Leslie L. Dean
Luann Dean
Dora DeBoard
Ardis Deen
Freda Deen
Della Demerjian
Louis Denetsosie
Shirley Patrick Derbins
Barbi Welch DeSalvo
Tami Dickerson
Mary Dickson
Ed Dilkes
Dietra E. Dixon
Dorothy Dodson

Nikki Doll
Linda Lee Dombeck
Peyton Donatelli
John L. Donovan
April J. Dorey
Bonnie M. Dorris
Ann Dorsey
Anne Dougan
Kee Dougi
Boyd E. Draper
Kristin Scott Draper
Gayla Richardson Dreith
Kay Drerup
Joyce Drexler
Cecil A. Driscoll
Alan Droegemeier
Doris R. Drooger
Sue Dubrava
Betty Duffy
Judy Dykman
Edie Easterbrook
Michael P. Easterbrook
Bonny Easton
Zelma Eberlein
John C. Eckroat
Carla Ecland
Dean Edgar
Eddie A. Edwards
Linda L. Edwards
Marlene Edwards
Nancy C. Edwards
Rhonda Edwards
Lynn Howard Ehrle
June Eiguren
Camellia Elder
Ora Ellis
Rena Harrison Ellis
Jon L. Elsberry
Debra Erickson
Steve Erickson
Florence Ellen Ethington
Clois Evans
Karen Evans

Tim Evans
Stanley Everett
Dennis and Rennee Faatz
Betty and Bill Fackler
LaFaye Skaggs Fahl
Mary Fain
Eleanor Fanire
Shari Farrington
Carmen Fernandez, Senator of
    Guam
Debra Fewkes
Pamela L. Field
Marypat Fields
Wes Fields
Emma Hafen Fife
Nancy Fifer
Royce N. Fifer
Deborah L. Durham Fisher
Katie Fite
Sharon Fitzgerald
Edward Fleenor
Lola Flemmer
Violet M. Flowers
Samuel A. Fluetsch
Mark Forbes (Office of the Speaker,
    Guam)
Pearl L. Ford
Jennifer Foss
Kristi Blaylock Foster
Mabel Foster
William E. Foster
Wynoma R. Foster
Barooze Terrance Fouldapour
Tony Fouldapour
Judy A. Follbaum
Rose M. Fowler
Wendy Fox
Charla Francis
Gary Franklin
Vickie Franklin
Congressman Trent Franks
Andrew R. Fraser
Richard Frasier

Ann Marie Frazier
Pamela L. Pinn Frield
Betty S. Frisbee
Billy M. Frisbee
Judy Frisbee
Michael C. Fritz
Janet G. Fry
Janet (Waltman) Fry
Gerald L. Fullmer
Teresa Fullmer
Albert G. Funk
Patricia Ineas Gabica
LoRee Christensen Gang
Jessica Gardetto
Barbara Gardiner
Geri Susan Gardner
Larry Gardner
Max and Leola Gardner
Mildred E. Gardner
Sherry Gardner
Janice Garlock
Apryl J. Garmon
Donald L. Garmon
Sharon Garmon
Layne Garner
James M. Garrett
Cynthia Gearhard
Dr. Martin Gelman
Gem County Commissioners (Ed
    Mansfield, Sharon Pratt, and
    Michele Sherrer)
Patricia George
Doris Getty
Olva Getty
Walter Getty
Dana Geuther
Glennea Gibson
Margaret Giddeon
Madeline Gilbert
Vicki Gilbert
Ellen Glaccum
Larry Glandon
Thelma (Hutchinson) Glasgow

Charolette Gleave
Josephine Gline
Louise Godby
Aubrey Godwin
Vern M. Goehring
David L. Goff
Cindy Goffinet
Jo Gooch
Donn C. Goode
Lori Goodman
Robert Goodwin
Cynthia L. Gorino
George A. Gorino
Betty Griffith
Vivian Graig
Arenett Grant
Cynthia Grarhard
Robert D. Gray
Nancy Greaves
Jeff Green
Thomas Olen Green
Cathy Gregory
Lyle Greig
Betty Griffith
Mark Gritton
Elna H. Grover
Betty L. Gull
Fred Gunter
Sue A. Brown Gunter
Laraine K. Gurney
R. Lynn Gurney
Esther Gutierrez
Sierra Guy
Pablo S. Guzman
Roger Hadley Jr.
Sonja K. Hadley
Karen E. Hall
Lanita Hall
RaeAnn Hall
Kelly Cannon Hamilton
Carol D. Hampton
Janet K. Hankins
Elaine Hanna

Barbara S. Hansen
David Hansen
Eve Hansen
Jean Hansen
Ralph Hansen
Tamera Hansen
Lois J. Hanson
Marilyn J. Hanson
Michael L. and Lory Richardson
    Hanson
Sherrie J. Harberton
Donald E. Hardin
Janet Hardin
Mary Hardisty
Mary Hardy
John Harms
Sharon Stevens Dilkes Harmsen
Ila Harnar
Morgan C. Haroldsen
Elizabeth L. Harris
Gerald C. Harris
Jerry Harris
Donald D. Harrison
Phil Harrison
Leroy Harry
Kim Richard Hartnett
Terry Hartnett
Ginger Harvey
Katie Hass
Charlotte Hatch
Janet Hatcher
Verily Hatfield
Nellie L. Hawill
Richard D. Hazen
Melinda Heath
Deborah Stevens Heazle
Margaret Jean Heazle
Donna E. Hebert
Julie Richardson Heckart
Ann Heilman
Kim Heinke
Lester Helms
Barbara Henderson

Eltona L. Henderson
Robert Henderson
Dianne Henley
Ivan Hepworth
Rae Ann Hepworth
Betty O. Hess
Rose Heuett
Scott Higbee
Camellia C. Higgins
Gayle C. Hightower
Paula Washburn Hill
Sandra K. Hill
Susan Hill
William F. Hill
Patricia Hines
Lin F. Hintze
Diana Lee Hirschi
Gwen Hirschi
Stan Hitesman
Calvin F. Hoagland
Robert Hobart
Jean S. Hobbs
Helen Hobdey
Dean Hodges
Richard D. Hodgins
Anita Hoebelheinrich
Kristi Hoebelheinrich
Arthur M. Hoffman
Jeanette H. Hoffman
Ilene Hoisington
Marvin and Evelyn Holbrook
Patti Holbrook
Ardelle Holderness
Greg Hollingshead
Johnny Hollingshead
Stanley Hollingshead Jr.
Pam Holloway
Lorissa Wilfong Holt
Louis Hooban
Rand Hooban
Lewis Hoopan
Jerry E. Hoopes
Delbert R. Hopkins

Donna B. Hopkins
Jay Hormel
Lucinda Hormel
Lowell and Gloria Horne
Nikki Horner
Anita Louise Horton
Kenneth Horton
Maria T. Horton
Wendy Hosman
Judy Jones Houde
Helen Houghton
Eileen R. Hoverson
Jim Hovren
Blaine N. Howard
Dan and Dorothy Howard
Darell D. Howard
Steff Howard
Vicki Howard
Estelle Howe
Charles L. Howell
Clara F. Howell
Dave Howell
David K. Howell
Faye Howell
Phyllis C. Howell
Elaine Howen
Pamela Howerton
Martin F. Huebner
Viola M. Huber
Cari R. Hug
Norma Huggins
Carolyn Hull
Della Hull
Neile Hulsey
Brad Hungerford
Ray Hungerford
H. C. Hunt
Darlene B. Hunter
Dorothy L. Hunter
Glenda Hurd
Ephia Hussey
Karl Graff Hutchings
Lester Hutchinson

Vera Hutchinson
Idaho Congressional Delegation
    (Senator Larry E. Graig, Sena-
    tor Mike Crapo, Congressman
    Mike Simpson, and Congress-
    man C.L. "Butch" Otter)
Idaho State Minority Leaders (Sena-
    tor Clint Stennett and Represen-
    tative Wendy Jaquet)
Gladys E. Ingram
Jani C. Ingram
Sondra Ingram
Vicki Inselman
Mark Jackson
Elinor S. Jacobson
Jane Jacobson
Tammy Jacoby
Cindy A. James
Wendy Jaquet
Bellie Jean
Lois Jeffs (Weimer)
Grace C. Jenkins
Vivian L. Jenkins
Donald L. Jennings
Don L. Jensen
Eric Jensen
Homer D. Jensen
Joseph Winston Jensen
Josyph W. Jensen
Gary Jenson
Leslie Gene Jeppson
Jerry John
Ella Mae Johns
Billy Johnson
Darlene Johnson
Gay Johnson
Louella (Lou) Johnson
Neal L. Johnson
Norma Johnson
Rebecca S. Johnson
Ruth Johnson
Shauri Johnson
Troessean N. Johnson

Carrie Jones
Edward Jones
John D. Jones
Kathy Jones
Lori Jones
Opal E. Jones
Phyllis Jones
Taunya Jones
Patricia G. Jordan
Thomas R. Jordan
Pamela Jorstad
Keith N. Joseph
Marigold Joseph
Sheila Joseph
Connie Joslin
Grace Kaiser
Marcia Kaiser
Thomas Kalac
Lydia Kamara
Cindy Kawano
Harold J. Keaton
LouEtta Keef
Janice Kegel
Pamela J. Kelley
Teresa Kelly
Annabelle Kelson-Wood
Dirk Kempthorne, Idaho Governor
James A. Kennedy
John Keogh
Sheila Kerbs
LaRae Kerr
Marea Kettler
Esther A. Kevan
Marci A. Keyes
Anne Kiefer
Alice Kincaid
Brenda G. King
Pam King
Robert Vern King
Carol A. Kinzer
Bonnie Kirk
Valli Kirk
Stanley S. Kirkpatrick

Dave Kissner
Katie Klein
Patricia M. Kletke
Karen B. Knudsen
Peggy Joyce Jensen Kole
Marjorie G. Konrad
Mae Kopp
Linda J. Korell
Loreen Korell
Royce L. Korell
Janeen Jensen Krambule
Verna L. Krizenesky
Robert E. Kroush
Hazel Krull
Nancy Kunes
Bret Kuntz
Rex N. Labrie
Susan J. Lacy
John Reynolds LaFordge
Charles P. Lamm
Douglas P. Lamm
Georgia M. Lamm
Marie Lancaster
Catherine Lane
Lillie Lane
Caren Langford
David Langhorst
Hazel Lanham
Blanche Lanier
Cindy Sue Lankford
Janette Travis Laroless
Geraldine Larribeau
Dona Martin Lathbury
Bernie Lawerence
Don R. Lawrence
Richard D. Laws
Elise Lazar
Allen LeDelle
Lisa Ledwidge
Albert Lee
Dorothy E. Lee
Elaine F. Lee
Jim Lee

Pamela Lee
Roy Lee
Faye Legg
AJ Lehman
Edward L. Leissring
Joan C. Lemon
Dale E. Lemmons
Phyllis Newman Lenker
Gwendolyn Tolmie Leth
George M. Van Leuven
Gary Leuzinger
Judi Leuzinger
James I. Levers
Wilma Levtzow
Johnnye Lewis
Lea J. Lewis
Suzanne Lewis
Vickie M. Lewis
Keith Lincoln
Andrea Lindberg
Jill M. Link
Thomas R. Linville
Holly Lloyd
Lilly Logue
Eva Lord
Shirley Loreen
Robin Lorentzen
Marlene Lovejoy
Thomas Gerald Loveland
Kathleen R. Lovell
Cynthia A. Lovendahl
Rhonda Low
Barbara Blake Lubeke
Dayle C. Lundy
Clarisse Lunt
Juli Lynch
Teri M. Lynch
Lennis Mabee
Raymond and Christy Mabee
Ella and Thomas Mackley
Laurie MacMillan
Nina Madry
Carol Madsen

Dr. Gary Madsen
Susan Magnan
Kristeen L. Main
Arjun Makhijani
Dennis L. Mallory
James R. Mallory
Jim Mallory
Margie Mallory
Marjean Malstrom
Evelyn M. Maltbie
Nancy Marker
Bessie Jolene Marshall
Don and Peggy Marshall
Vickie Jo Martell
Glen Martin
Karen Turner Martin
Suesan Martin
Carol Martinez
Elton Martinez
Israel Martinez
Melton Martinez
Michael A. Martinez
Theresa Martinez
James R. Martz
Helen Masten
Jacque Matson
Judy Matthews
VaNeta Mattulat
Jeremy Maxand
Joyce May
Linda May
John B. Maycock
Shirley Mayfield
Mary I. Mays
Barbara L. McBride
Don E. McBride
Lana Rae McBride
Senator John McCain
Loree McCain
Joan McCall
Shirley McCartney
Carmen McClish
Donna Mcclure

Ben McCoy
David B. McCoy
Laurie McCrea
David McDonald
Dylan McDonald
Kayla D. McDonald
Winfred McDonald
Bess McDoniel
Mike McDonough
Linda McErlean
Cherie McFarland
Beatrice McGavin
Zell A. McGee
Becky McGowan
Linda McGraw
Beverly N. McGuire
Lynnette McHenry
John McKinney
Miriam Faulks McKinney
Bonnie McMullen
Janet L. McNamer
Brent L. McNealy
Frank L. McNealy
Von Zae McNelis
Debi McReynolds-Farm
Luke McThias (Yazzie Jr.)
Debra Sue Pinn Mecham
Karen Black Medlin
Iris Memmott
Joe Merrick
Verla Merrick
Kathleen Meyer
Nancy L. Middlebrook
Brain Miles
H. Richard Miles
Dr. Paul V. Miles
Patsy Ann Sorenson Miles
Brenda Miller
Charlotte Miller
Clay Miller (Mrs.)
Darlene Miller
Delores A. Miller
Gary L. Miller

Jenny Miller
Jerry Miller
Lorraine Miller
Penelope E. Miller
Richard Miller
Shari N. (Dilkes) Miller
Terry R. Miller
Connie Millman
Mohave County Downwinders
    (Miller, Wright, Wise, Akins,
    Langdon, King, Conneally,
    Bishop, Davis, Keller and
    Haywood, Voss, Bee, Marter,
    Blum, Maberry, Inman, etc)
Jeanna Mitchell
Lynnette R. Mitchell
Samuel W. Mitchell
Sherry L. Moffitt
Sherie Mohr
Jim Monroe
James Montgomery
Dr. Les Montgomery
Phyllis Rainey Montgomery
Sally Montgomery
Monticello Community Foundation
    and the City of Monticello
Group from Monticello
Jerald Moon
David S. Moore
William and Patty Moore
Leo Morales
Ed Mordhorst
Ruben Morfin
June S. Morgan
Floyd R. Morley
Ilene Morley
Julia J. Morley
Jim Morphey
Judith Morris
Kara L. Morris-Gimlin
Mrs. William D. Morris
Rob Morrison
Glenn Morrow

Lorene Morrow
Georgia F. Mosbrucker
Katie Moser
Mrs. Roy Motichka
Traci Mousetis
Paul D. Moyer
Billie Munger
Wesley J. Murphy
Greg Murray
Cindy L. Myers
Elaine (Cloughton) Myers
Irene Mylar
Ron Nachtigall
Dale Nash
Gerald H. Nasker
Ronald E. Nasker
Ihla Nation
Jessie Nau
Sharon Paul Nault
Bonnie L. Nebeker
Carmelita Nedrow
June Neibaur
Shelly Neibaur
Arlene Nelson
Dennis Nelson
Leila Nelson
Leslie Nelson
Linda Nelson
Marian Nelson
Marty Pickett Nelson
Philip Nelson
Gwen Nettleton
Virginia Newsom
Beverly A. Nichol
Carmen Nicholls
Audrey Nichols
Barbara Nichols
Monna Nichols
Carrie L. Nichols
Dianne Nichols
Harry E. Nichols, Sr.
James E. Nikirk
Michael J. Nikirk

Randall Nilson
Sharon K. Nilson
Maxine Johnson Ninas
Shirley Noland
Phyllis M. Norwood
Judith Norwood-Parsons
Rose Marie Nutsch
Linda Nygaard
Rebecca C. Obletz
Ardyth O'Connor
Cynthia Ofthedesert
Judith L. Ogawa
Walter S. Okamoto
Terri O'Keeffe
Shirley Olberding
Gedge Oliver
Billie Olson
Curtis Olson
Patricia Crank Olson
Jocelyn Orr
Martin Orr
Phyllis Osborn
Arlene Rekow Osborne
Kathryn Packard
Daniel Packham
Sandra Pace
Kris Packer
Gilbert Padoni
Rena M. Perry Van Paepedgem
Koy Page
Maiera Page
Patricia A. Page
Patricia J. Page
Jane M. Paller
Connie Parker
Tara Parker
Richard Parks
Lola Hereth Parsons
Karen Parton
Randall Patton
Gary B. Peak
Kathleen G. Perkins
Melvin R. Potter

Louise Patterson
Teresa Lynn Patterson
Deanna Patton
Gloria Payne
John A. Paynter
Carol Peacock
LaRae Rawson Peacock
Richard C. Peacock
Clayne Pearson
Joanne Whitworth Pederson
Noela Pegram
Kathleen G. Perkins
Nadean Perkins
Shirley I. Perry
Terri Person
Dale H. Petersen
Claudia Peterson
Donna and Larry Peterson
Gail Peterson
Kalvin E. Peterson
Kenneth M. Peterson
Mendel L. Peterson
Roy C. Peterson
Peter and Drinda Petroni
Sylvia E. Pfeiffer
Barbara E. Phillips
Cora Maxx Phillips
Darlene Phillips
Gene Phillips
Warren Phillips
Barbara Philson
Tony Pickering
Katherine Pierce
Linda J. Pinkert
Manuel Pinto
Barbara L. Pipkin
Fritz Pipkin
Evelyn Pitman
S. Loreen Poff
John C. Pointer
Cindy Pollard
Bethene Pook
Julia Pook

Debbie Poole
Grover Edward Poppleton
Mike Poppleton
Virginia Posey
Marilyn Post
Mary Potter
Dixie S. Poulson
Cindy Povlsen
Doris Power
John Powers
Teena Poxleitner
Henrietta Prater
Diane Worley Pride
Trisha Pritikin
Traci Proctor
Carolyn A. Proshek
Azell and Dwayn Pruett
Paula Pruitte-Paul
Gwendolyn J. Pughes
Bethene Pulliam
Hazel Purdue
Mary Purket
Eric Puschmann
Carol M. Qualman
Ardyce L. Quenzer
Sandra L. Rabehl
Michael G. Rachiele
Jerry Radandt Jr.
Susan Ramos
Larry T. Ranstrom
Jeanette Rauch
Maria Rebollozo
Alicia Records
Max D. Rector
Garth Reece
Tommy Reed
Patricia Ann Reed-Wood
Lorna B. Reeder
Linda R. Reich
Norman and Shirley Reich
Mel Reimers
Darrell R. Reinig
Colleen Rekow

Richard C. Renstrom
Dwaine J. Reusser
Debbie K. Reyes
Betty Reynolds
Lynn Rhodes
John M. Rhoton
Carmen A. Riccardi
Sandra F. Rice
Cheri L. Kimmel Richardson
Connie Sue Richardson-Estes
Danny Lee Richardson
Jerry Richardson Jr.
Jerry Ray Richardson
Tony Richardson
Trina Richardson
Marilyn Richer
Rachel Richert
Gloria Richhart
Dr. Peter Rickards
Michael Ricken
Phyllis Rickert
Robert M. and Lupe V. Ricks
Richard L. Ridgway
Eleanor Riggin
Paul Riggins
Judy (Owen) Riley
Theodore R. Riley
LaVonne Rinaldo
Mary E. Riker-Rivers
Robert and Rene Robertson
Lorraine Limb Robinson
Becky Rockwell
Charles A. Roesberry
Carol Rogers
Glenda L. Rogers
Leonard Rogers
Karla Rogerson
Larry F. Rollins
Diane Ronayne
Marci L. Rood
Bonnie L. Rose
Debbie Rotchford
Sandra P. Rountree

Venice Kate Allred Routsong
Ronald and Joyce Rovetto
Kodie A. Roy
Clay Leland Rudd
Geraldine Rudd
Owen Darrell Rudd
Judy Rule
Phyllis H. Russell
Susan E. Ryder
Ken Rynearson
Richard Rynearson
Maria Salazar
Christine Morris Salstrom
Earl Saltes Jr.
Huldah J. Sanders
Margie Sanders
Ilene Sandifer
Shirley Savaria
Lynn Sawyer
Terri Schaffner
Miriam Schenk
Bonnie Scheuffele
Beverly Schevling
Valerie Schlader
Saundra Schmidt
Shawn Garcia Schow
A. Charles Schroeder
John H. Schrom
Judy Schultz
Mr. Schultz
Don Schweitzer
Richard Leon Scoffield
Clay and Margie Scofield
Helen M. Scott
Janet Eve Scott
Joyce D. Sedlacek
Holly Seefried
Ellen Seibolt
Dan Sellers
Gaye Selover
Emily Severance
Iris H. Seyfried
Kathryn Shadduck

Robert W. Shaffer
Ruth Johannesen Shane
JoAnn Shaul
Doris Shaw
Heidi J. Perkins Sheets
Valerie Shell
Mike Sheppard
Cornelia Sheridan
Denna Sherwood
Janet Shira
Joe Shirley Jr., President of Navajo
    Nation
Evelyn Shoemaker
Lenore Mitchell Short
Tamara S. Showers
Chris Shuey
Cheryl K. Shurtleff
Carolee Shurts
Belen Vallejos Sickles
Mickey Lee Sickles
Elsworth Lee Sievers
Johnson Silvers
Jami S. Simmons
Cheryl Simpson
Elva Simpson
Philip Earl Simpson
Charles Slode
Nancy L. Skidnore
Colleen E. Skinner
Judy Skinner
Tawna Skinner
Kathy Skippen, Idaho
    Representative
Drusilla Small
Laina Smalley
Margaret Smelcer
Alice Smith
B. Gary Smith
Carol Rae Smith
Eldon E. Smith
Hyrum Smith
James Smith
James A. and Eunice Eckroat Smith

Jimmie Smith
John Smith
Kathleen Smith
Kathy Smith
Lisa Louise Smith-Salelana
Mary L. Smith
Ralph Eugene Smith
Ruth E. Smith
Sharon G. Smith
Tena M. Smither
Sue Benjamin Snell
Marjorie E. Sobers
Hazel Solomon
Ramona Sommer
Barton T. Sonner
Dolores (Dee) Southard
Karen Spaulding
Walter B. Spear
Boyd E. Spence
Larinda Spencer
Shirl W. Spencer
Judy Spottswood
Helen Spradling
Brent Sprague
Sandra B. Sprague
Evelyn Spreier
Barbara Squires
Vaughn J. Squires
Cheryl McNarie Stackhouse
Michael Stambulis
Ruby Standerfer
Diana Stanford
Debra J. Stansell
Nancy H. Stathis
Jean Steele
Noel C. Steele
Robert L. Steele
Charles W. Steiner
Florence W. Steiner
Mildred Steiner
Willene Griffin Steitz
Jode P. Stella
Mary M. Stemen

Senator Gary Stennett
Ernest L. Stensgar, Coeur d' Alene
    Tribal Chairman
Danielle Stephens
Larry and Sue Stephenson
Shirley J. Sterling
Boyce G. Stevens
Kimberly Stevens (Babbitt)
Jane Stevens-Thompson
Dr. Stewart
Clella Stiles
Shari Stiles
Barbara E. Stipp
Carlen Stitt
Judith Stockham
Jerry Stocks
Mardie Stone
Patricia Tracy Stowell
Zella Strickland
Dixie Stroud
Gayle Sarriugrate Stroud
Mary Jane Stroup
Linda Jo Struthers
Melissa Sullivan
Lynne B. Summers
Rita M. Sussex
Gloria J. Sutton
Lou Ann Sutton
Nancy Swainston
Eldon and Paula Swapp
Lea Sweat
Betty Sweet
William E. Sweet
Linda R. Swickard
Kathern Syrett
Irvin V. Syrie
Patricia Ann Harral Tabet
Richard H. Taddiken
Lilas L. Talley
Cindy R. Tappe
Minnie Tappe
Vincent E. Tappe
Eleanor Taylor

Dr. LaVern Taylor
M. Bonnie Taylor
Shane Taylor
Bev Teichert
Marlee Terry
Coleen Tertelgte
Frances Tester
Glennis Tester
James M. Thielges
Jeane Thomas
Judy Thomas
Keith Thomas
M. K. Thomas
Mike Thomas
Lori K. Thomason
Kay Erickson-Thompson
Keith R. Thompson
Linda Thompson
Monte O. Thompson
Neal and Faye Thompson
Regie Thompson
Sylvia A. Thompson
Virginia K. Thompson
Kathy Thomsen
Alexander Thorne
Dale E. Thornton
Darryl Thornton
Katherine Thornton
Penny Thorpe
Don V. Tibbs
Dave Timothy
Donna Tinker-Mitchell
Lori O'Keefe Titus
Lucy M. Todecheenie
Karen Draper Tollefson
Kathleen (Welch) Tolmie
Marie C. Tomchee
Janet Tomita
Rebecca Tommack
Kristi Johnson Tompkins
Amber Toole
JoAnne Torrez
Sharon Traughber

Beth A. Travis
Tawna L. Travis
Tanette Travis-Lawless
Joyce L. Treat
Fred Trenkle
Kathleen E. Trever—Department of
    Environmental Quality
Larry B. Trimble
Carolyn Troutner
Patricia M. Fuller Truitt
Preston J. Truman
Renee Truscott
Andrew Tso
Kathleen Tsosie
Arita Griffin-Turner
Susan M. Tyrer
Ivan Ulmer
Loma E. Ulmer
Deana L. Ulrich
Orville L. Updike
Daniel M. Urruttia
Dave Vahlberg
Ruth Vahlberg
Cecilia Vallejos
Dr. Arthur A. Vandenbark
Eve Mary Verde
Katherine Heimbuck Viken
Heather Villanueva
Sharon Villanueva
Tony Villanueva
Candace L. Villarreal (Mooney)
Kathleen Vopat
Al Waconda
Clint Wadsworth
Kriss Wagner
Nola Walden
Jonathan Waldman
Nancy Waldner
Don E. Walker
George E. Walker
LaMar R. Wall
James Wallace
Dennis R. Waller

Larry R. Waller
Sandra E. Walsh
Tracy Walton, President of Gem
    County Farm Bureau
Renee M. Wankier
RaVae Warmenhoven
Arlea W. Washburn
Lark L. Washburn
Stephen Washburn
Betty J. Watson
Bonnie Watson
Norman and Betty Watson
Jayne Watterlin
Steven Weaver
Verna L. Webb
Darrell L. Webster
Billy DeRay Welch
Teresa May Welch
Christine Welch-Galvan
Ralph Wells
Nicole Wend
Connie L. Wensman
James Wensveen
Steve Westfall
Carole Wheeler
Colleen M. Wheeler
Deborah Whipple
William Whitaker
Dennis R. White
Frank K. White
Janet R. White
Judy May White
Winnogne White
Dick Whiting
Clair M. Whitlock
Donna Whitmire
Jack R. Whitney
Karrel P. Whitsell
Joan Wilber
Wilma Wilhite
Corrine Wilkening
Alice S. Will
Carol Williams

Dwight Williams
Logan and Linda Williams
Shauna Williams
Terry Tempest Williams
Ilene Williamson
Susan Fleming Williamson
Donna Willmorth
Christopher Wilma
George A. Wines
Norta Winkler
Charlene Wiscombe
Perry N. Wise
Joy Withers
James T. Wolcott
Sarah L. Wolfe
Mildred Wolfkiel
Ann and John Wood
Frank G. Wood
John W. Wood
Joyce M. Wood
Lorna Wood
Waunita Wood
Carolyn R. Woodall
Cherry Woodbury
Teresa Bussert-Woods
Louise Worley
Darla J. (Shaeffer) Wright
Howard E. Wright
Laurel Wright
Mary J. Wurst-Church
Chris Wylie
Cheryl L. Yancey
Milton Yazzie
Geraldine E. Yeakel
Mathew Yocum
Norma Osborn Yoneda
Andrew Young
Donna M. Young
Glade Young
Martha Young
Wesley Youngberg
Diane Youngstrom
Antoinette Young-Talker

Timothy Yuen
Maria Zapata
Teena Zemke
cottontail0604@aol.com
cybercalm@yahoo.com
danwood1@yahoo.com
Hollilm@aol.com

mohavedownwinders@yahoo.com
JimATP@aol.com
pjwill@netutah.com
rainey_in_id@yahoo.com
tawnyb@earthlink.net
Bob R.

Appendix B

A Comparison of the Risk of Skin Cancer with the Risk of Lung Cancer from Exposure to Radon Decay Products in Underground Mines

It has been suggested that ambient conditions in underground mines might cause cancers at sites other than the lung (Tomasek et al., 1993; Denman et al., 2003), specifically, that radon in the ambient air might be responsible for an excess risk of skin cancer and leukemia. The committee has reviewed the relevant literature and has compared the risk of skin cancer with the risk of ling cancer.

The nobel gas radon-222 (^{222}Rn) is produced naturally in rocks and underground formations. Because of its mobility, it can escape from the geologic matrix and accumulate in the atmosphere. It can reach high concentrations particularly in caves and underground mines. Radon is radioactive; when it decays, it initiates a series of radioactive transitions. The short-lived radioactive descendents of ^{222}Rn—polonium-218 (^{218}Po), lead-214 (^{214}Pb), bismuth-214 (^{214}Bi), and polonium-214 (^{214}Po)—have historically been called radon daughters, radon progeny, or radon decay products. Their kinetic behavior is complicated. They are not inert and do adhere to other objects when contact is made. Many of the daughters "plate out" on surfaces, where they become immobile and decay in situ. Others attach to aerosols and remain suspended in the atmosphere until they decay.

Risks of lung cancer are related to inhalation of radon daughters, not radon itself, because almost all radon inhaled is immediately exhaled, but inhaled radon daughters mostly become resident in the airways of the lung. The radioactive half-lives of the daughters are shorter than the respiratory clearance time, so they decay in the lung. The alpha particles released by ^{218}Po and ^{214}Po have sufficient penetration to reach and deliver high doses to sensitive cells in the bronchial epithelium.

The concentration of radon daughters suspended in air is measured in working levels (WL) or equilibrium equivalent concentration (EEC). Human exposure to radon daughters depends on the concentration at a specific location and the time that the person spends at that location; it is measured in working level months (WLMs). The risk increases as the exposure in WLM increases. Numerical estimates of risk have been reviewed extensively by the committee on health risks of exposure to radon (BEIR VI) (NRC, 1999) and International Commission on Radiation Protection (ICRP, 1981; 1991; 1993).

The risk of skin cancer in radon-rich atmospheres is not associated with inhalation or other intake of radioactivity. It is related to the plate out of radon daughters on the skin. The dose to sensitive cells in the skin depends on the concentration of radioactivity on the surface skin and the ability of alpha particles from ^{218}Po and ^{214}Po to penetrate to the location of the sensitive cells. Previous studies of that process have used a nominal value of 70 mm as the depth of the sensitive basal cells at the base of the epidermis. Recent measurements have indicated large variation in the depth of the basal cells, and this could lead to higher doses than previously expected (Eatough, 1997). We have assessed the pathway to determine whether the projected risk of skin cancer is comparable with the risk of lung cancer posed by the same exposure in WLM.

The dose delivered by an alpha particle as it passes through tissue depends on the stopping power, dE/dx. Values of that quantity have been complied by the National Institute of Standards and Technology and can be downloaded directly from its Web site (http://physics.nist.gov/PhysRefData/contents-radi.html). Those data can be used to compute the dose as a function of depth for radioactivity uniformly deposited on the surface. Figure B.1 shows the results of the computation for ^{218}Po (6 MeV alpha particle) and ^{214}Po (7.7 MeV alpha particle). The data clearly illustrate that the alpha particle from ^{214}Po will not reach basal cells at a depth of more than 50 μm.

The epidermis is not a flat organ located uniformly below the skin. There are undulations that are responsible for large variations in depth. Eatough has published distributions of epidermal thickness in various locations on the body (Eatough, 1997). We have simulated those data as a lognormal distribution (median = 50, σ_g = 1.5) to represent the epidermal thickness on the exposed face of a person in Figure B.2.

The mean dose to basal cells per disintegration on the basis of the data in Figures B.1 and B.2 is shown in Table B.1.

The next step is to determine the amount of radioactivity that will plate out on the skin for a given concentration of radon daughters suspended in the air. Denman et al. (2003) and Sevkova et al. (1978) have reported measurements of plate out on exposed skin. We have adopted a value of 2 (Bq/m^2)/(Bq/m^3) which is consistent with the reported results. Combining those results and assuming a steady-state condition with equal concentrations of ^{218}Po and ^{214}Po on the skin, we obtain a result of 200 mSv/WLM for the effective dose to the skin.

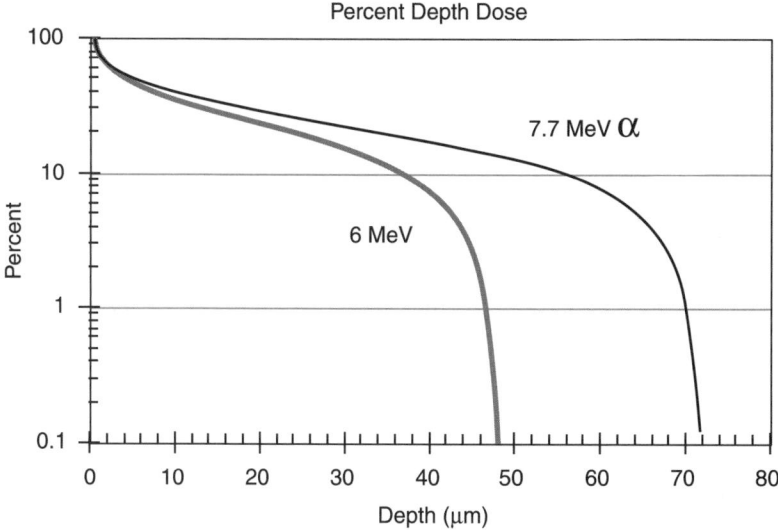

FIGURE B.1   Percent depth dose distribution for 6 MeV and 7.7 MeV alpha particles
that emitted isotropically emission from a source that is uniformly distributed on surface.

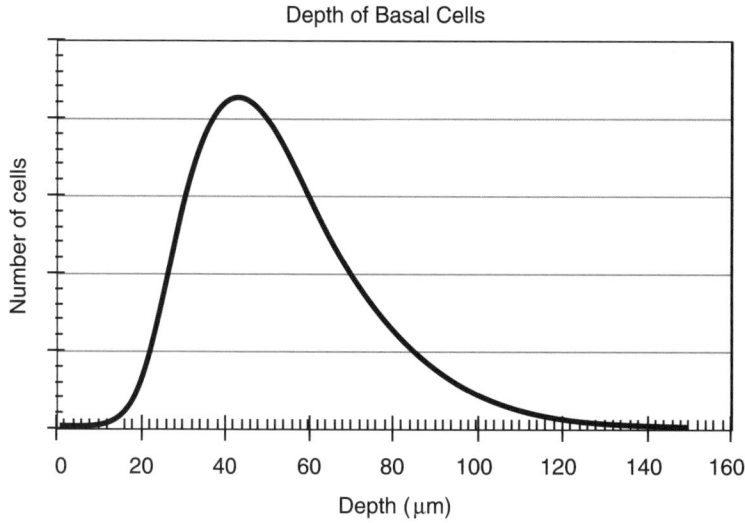

FIGURE B.2   The distribution of basal cells as a function of depth below surface of
skin. Diagram has been normalized such that the area under the curve is equal to 1.0.

TABLE B.1   Mean Dose to Basal Cells per Disintegration

Isotope	Gy·cm^2/dis	μSv·cm^2/dis
^{218}Po	1×10^{-8}	0.2
^{214}Po	4×10^{-8}	0.8

Mining activity is physically rigorous and some perspiration is expected while underground. That adds a layer of water on the skin and increases the thickness of material between the radioactivity and the basal cells.

We assumed that prolonged activity in the mine results in a water loss of 5 liters per day through the skin. The water will be released uniformly over a surface area for a standard adult of 1.7 m^2. It will evaporate from exposed surfaces at about 10% per minute. The result is a steady-state layer of water that is about 20 mm thick. In addition to perspiration, fumes and dust will accumulate on the surface of the skin, and we have assumed that this will provide a thickness of material equivalent to another 20 mm of water.

We have computed the reduction of dose from alpha particles as a function of the thickness of water-equivalent material on the surface of the skin, assuming that the activity is on the top surface of the water. The attenuation factor is shown in Figure B.3.

It can be seen that 40 μm of material will reduce the dose to the basal cells by a factor of about 100. Thus, the equivalent dose is reduced to 2 mSv/WLM.

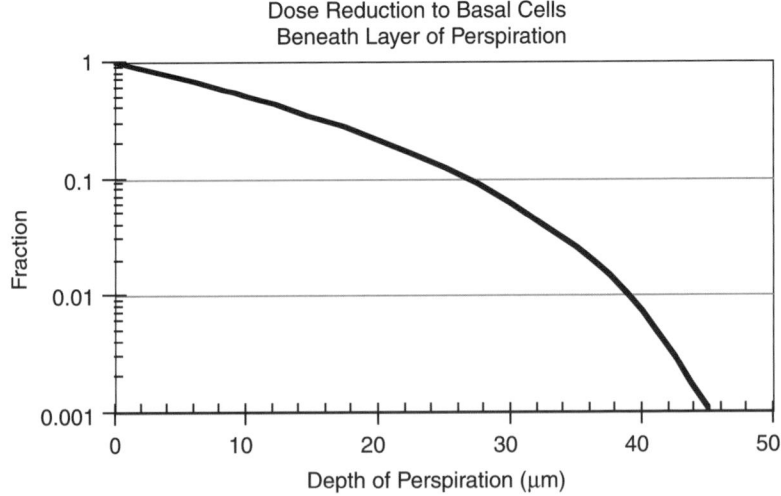

FIGURE B.3   Reduction of dose to basal cells from water-equivalent material on skin. Activity is assumed to be on surface of water above skin.

Uranium miners wear protective clothing while underground, so only their arms, hands, heads, and necks are directly exposed to the atmosphere. That represents less than 25% of the surface area of the skin. Using that value, the effective dose to the exposed portion of a person's body 0.5 mSv/WLM.

The lifetime risk of fatal skin cancer after whole-body exposure to the skin is 2×10^{-4} /Sv (ICRP, 1991 p 132). The risk from radon decay products deposited on the skin would be about 1×10^{-7}/WLM, that is, $(5 \times 10^{-4}$ Sv/WLM) or $(2 \times 10^{-4}$ /Sv). The lifetime risk for lung cancer from inhalation of radon daughters is $\sim 2 \times 10^{-4}$/WLM (ICRP, 1991 p 139). From this dosimetric analysis, the risk of lung cancer is greater than 2000 times the risk of skin cancer from the same exposure to radon daughters.

We have repeated the calculation assuming that the radon decay products are uniformly distributed throughout the layer of perspiration on the skin. The average attenuation factor for this case increased from 0.01 to 0.32. The effective dose to the skin for the exposed portion of the body is then 16 mSv/WLM. The associated risk of lung cancer in this case is more than 60 times the risk of skin cancer.

Epidemiologic studies have evaluated the incidence of skin cancer in underground miners (Darby et al., 1995) and have concluded that high concentrations of radon are not associated with excess mortality from cancers other than lung cancer. A cohort of underground miners in the Czech Republic did show some increased incidence of skin cancer (Sevcova et al., 1978). It was not related to accumulated radon exposure, but it did seem to be associated with length of exposure (i.e., 10 years or longer). The pathologic evaluation diagnosed the lesions as basal cell carcinomas. In all cases, the tumors were excised surgically, and there were no signs of recurrence.

On the basis of dosimetry and epidemiology, the risk of lung cancer is significantly higher than the risk of fatal skin cancer in persons working in an environment where radon decay products are suspended in air.

REFERENCES

Darby, S. C., Whitley, E., Howe, G. R., Hutchings, S. J., Kusiak, R. A., Lubin, J. H., Morrison, H. I., Tirmarche, M., Tomasek, L., Radford, E. P., , R. J., Samet, J. M., Shu, X. Y. Radon and Cancers other than Lung Cancer in Underground Miners: A Collaborative Analysis of 11 Studies. J. Natl. Cancer Inst. 87(5):378-384. 1995.

Denman, A. R., Eatough, J. P., Gillmore, G., Phillips, P. S. Assessment of Health Risks to Skin and Lung of Elevated Radon Levels in Abandoned Mines. Health Physics 85(6):733-739. 2003.

Eatough, J. P. Alpha-Particle Dosimetry for the Basal Layer of the Skin and the Radon Progeny 218-Po and 214-Po. Phys. Med. Biol. 42(10):1899-1911. 1997.

ICRP (International Commission on Radiological Protection). Limits for Inhalation of Radon Daughters by Work. ICRP Publication 32, Annals of the ICRP 6(1). Elmsford, NY: Pergamon Press. 1981.

ICRP (International Commission on Radiological Protection). 1990 Recommendations of the International Commission on Radiological Protection. ICRP Publication 60, Annals of the ICRP 21(1-3). Elmsford, NY: Pergamon Press. 1991.

ICRP (International Commission on Radiological Protection). Protection Against Radon-222 at Home and at Work. ICRP Publication 65, Annals of the ICRP 23(2). Elmsford, NY: Pergamon Press. 1993.

NRC (National Research Council). Health Effects of Exposure to Radon: BEIR VI. Washington, DC: National Academy Press. 1999.

Sevcova, M., Sevc, J., Thomas, J. Alpha Irradiation of the Skin and the Possibility of Late Effects. Health Physics 35(6):803-806. 1978.

Tomasek, L., Darby, S. C., Swerdlow, A. J., Placek, V., Kunz, E. Radon Exposure and Cancers Other than Lung Cancer among Uranium Miners in West Bohemia. Lancet 341(8850): 919-923. 1993.

Appendix C

Radioactivity in Guam After Nuclear-Weapons Testing in the Pacific

RADIOACTIVE FALLOUT IN GUAM

Guam is about 1,200 miles west of the Marshall Islands at about 10° north of the equator. The trade winds are predominantly from east to west at that latitude. Atmospheric testing began in the Marshall Islands with Operation Crossroads in July 1946. Operation Ivy began in October 1952. On October 31, 1952, the first thermonuclear device, with the code name Mike, was detonated. It had a total yield of 10.4 Mt of which about half was fission energy.

A radiation-fallout monitoring program for Operation Ivy was established in 1952 and coordinated by the New York Operations Office of the Atomic Energy Commission. It consisted of a worldwide network of 111 monitoring stations with at least one in every continent, but it was concentrated in the Northern Hemisphere. One of those monitoring stations was located in Guam. In addition to stationary ground-based monitoring, aerial surveys using low flying aircraft were conducted from several Air Force bases.

Instrumentation during aerial surveys included a gamma-ray detector and a recording unit. It was calibrated at the Nevada Proving Ground to measure exposure rate at 1 meter from the ground. It was capable of measuring dose rates in air from external gamma radiation in the range of 0.0001-10 mGy/h.

An aerial survey over Guam began on November 1, 1952. The resulting data are shown in Figure C.1 (Eisenbud, 1953).

Integrating the dose rate over time produced a total effective dose to persons on the ground of about 0.6 mSv (w_r and w_T = 1.0) from external gamma rays as a result of fallout from Mike. To put that into perspective, we have compared the

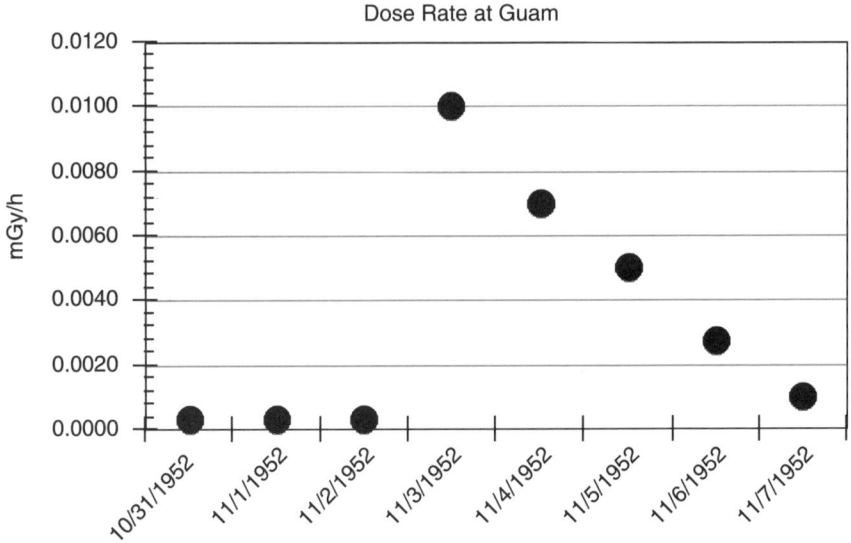

FIGURE C.1   Data from aerial surveys in Guam before and after detonation of nuclear test Mike in Marshall Islands during Operation Ivy.

result with the annual effective dose received from natural background radiation today, as shown in Figure C.2.

As seen in Figure C.2, the external dose received by residents of Guam from the test was about 20% of the annual effective dose received from natural background radiation in the continental United States and about 50% of the annual effective dose received from current values of natural background in Guam.

To gain an appreciation of the fallout received from other tests, we used data from the ground-based monitoring stations. At each station, 24 hour samples of airborne dust were collected on 30×30-cm sheets of adhesive (gummed film). All samples were mailed to the Health and Safety Laboratory in New York City for analysis, where they were ashed and counted for gross beta activity. The total activity on the sample at the time of collection was determined; a power function decay with a coefficient of 1.2 was assumed.

Operation Castle began in 1954. There were 16 tests in 1954, 17 tests in 1956, and 33 tests in 1958. No tests were conducted in 1955 and 1957. Results from the gummed-film data collected at Guam are shown in Figure C.3. The ordinate is the sum of monthly data reported as the deposition of strontium-90 (^{90}Sr) on the surface of the gummed film (Harley et al., 1960).

As mentioned above, there were more than 100 gummed-film stations around the globe. Monthly data from each station have been compiled for the 5-year period of atmospheric testing from 1954 to 1958. The 5 year accumulation of

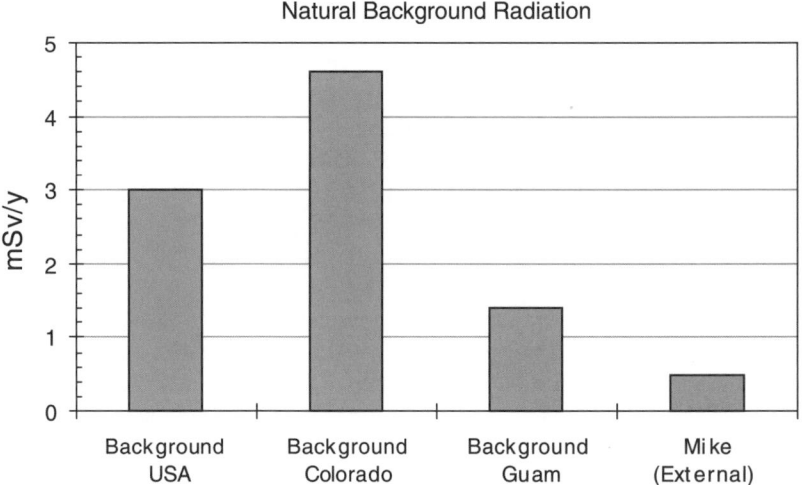

FIGURE C.2   Annual effective dose rates from natural background radiation in US mainland and Guam and external effective dose in Guam from nuclear test Mike in Marshall Islands during Operation Ivy.

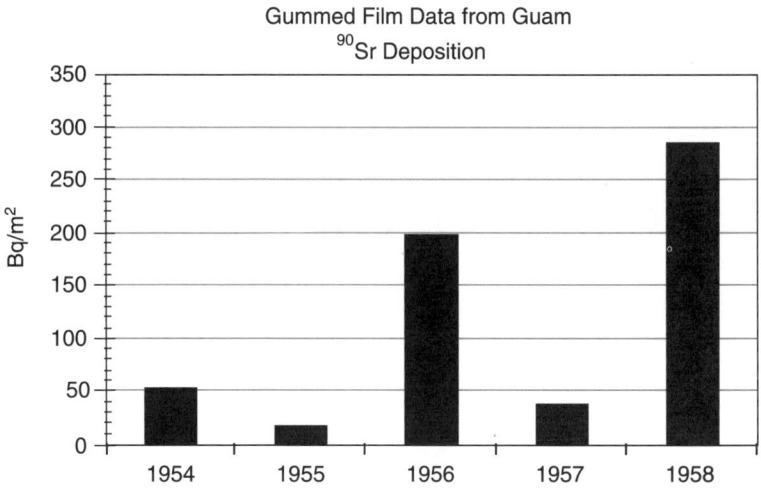

FIGURE C.3   Annual deposition of ^{90}Sr based on data collected at gummed-film station in Guam.

[90]Sr during that period for several locations, including Guam, is displayed in Figure C.4.

Figure C.4 shows that Guam did receive radioactive debris from fallout during the nuclear-weapons testing in the Pacific Ocean. The vertical error bars are used in an attempt to show the uncertainty in the gummed-film measurements. They represent the 95% confidence limits based on a lognormal distribution with a geometric standard deviation of 1.5. Uncertainty in the gummed-film method for measuring fallout was probably much greater. The analysis demonstrates that fallout in Guam during that period was similar to that in other parts of the US and its territories.

An extensive radiologic monitoring program was conducted in Micronesia, including Guam. A report was published in 1975 (Nelson, 1975). In general, the data did not indicate that the concentrations of fission-product radioactivity in samples of soil or biota in Guam were greater than the concentration of naturally occurring radionuclides.

The pathway that is responsible for the largest collective doses from radioactive fallout is the intake of iodine-131 (^{131}I) through consumption of fresh milk. The risk to persons in Guam during this period will depend on iodine deposition

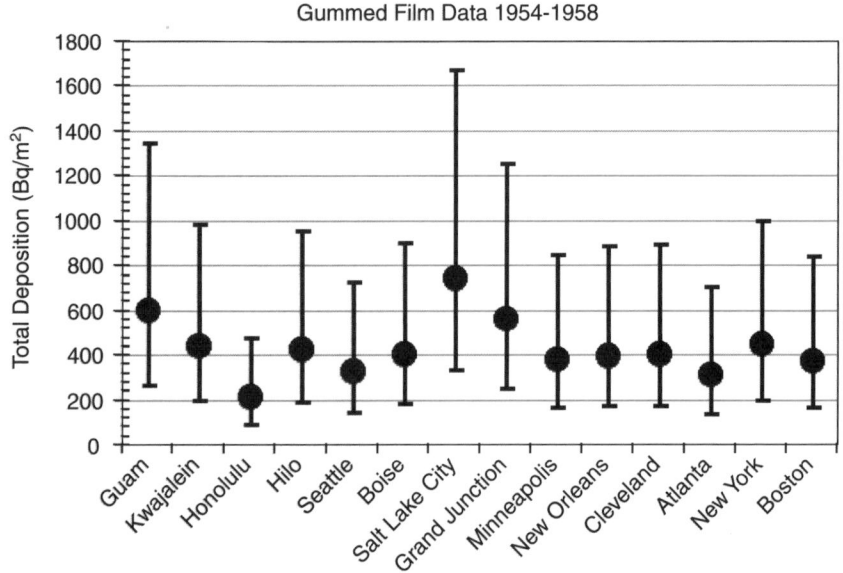

FIGURE C.4   Deposition of ^{90}Sr from 1954 to 1958 based on data collected at gummed-film stations in Pacific and locations in continental United States. Locations are ordered by longitude to provide rough estimate of distance from Guam.

and consumption of fresh milk. A dose-reconstruction effort will be needed to estimate dose and the associated risk of thyroid cancer.

CANCER INCIDENCE

The petition to Congress to include residents of Guam in RECA claimed that "increased levels of radiation may have led to serious health and environmental problems for life." Figure C.5 shows the cancer incidence in Guam and the entire United States for various periods between 1990 and 1999. The incidence of cancer is not higher in Guam than in the entire United States.

SHIP DECONTAMINATION IN GUAM

Operation Crossroads consisted of the first nuclear explosions after the detonations in Hiroshima and Nagasaki near the end of World War II. Their purpose was to examine the effects of nuclear weapons on naval vessels, equipment, and animals. The tests were designed to have a fission yield equivalent to 21 kilotons (21 kT) of TNT, which was similar to the weapons used in Japan. At that time, Crossroads was the largest peacetime military operation ever conducted. It involved more than 45,000 persons, 220 ships, and 160 aircraft.

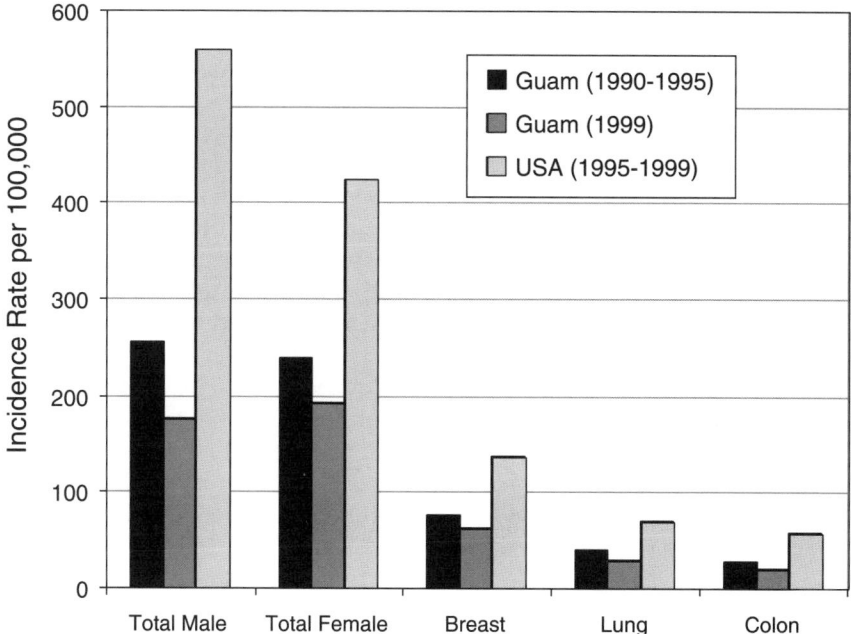

FIGURE C.5   Cancer incidence in Guam and entire United States.

A fleet of more than 90 surplus and captured ships anchored in the lagoon of Bikini Atoll in the Marshall Islands served as targets. The fleet included Allied and Axis vessels, such as the aircraft carrier USS Saratoga; the battleships USS Nevada, Pennsylvania, Arkansas, and New York; and the Japanese battleship Nagato. In addition to target vessels, there was a fleet of support ships, including large LCI and LCT infantry landing craft, and YMS mine sweepers; these ships were involved with inter-atoll dispatch, mail delivery and passenger transport.

On July 1, 1946 (local time), test Abel was dropped from a B-29 and detonated at an altitude of 160 meters. On July 23, 1946 (local time), test Baker was detonated. The bomb was encased in a watertight steel caisson and suspended 30 meters beneath a target ship. Shortly after the detonation, a huge water column containing bomb debris was formed. It expanded to 600 meters in diameter and reached an elevation of 2000 meters, and it held a million tons of water. At 10 seconds after detonation, water started to escape the stem of the water column and fall back toward the surface. A wave more than 30 meters high propagated from the stem.

The radioactive fission products created by Abel were atmospherically dispersed, so no extensive deposit of long-lived radioactivity was found on target vessels, and naval personnel could board the surviving targets within a day. After detonation of Baker, most of the radioactive inventory fell into the lagoon. After 2 days, officials recognized that many targets would remain highly radioactive for a long time. The nature and extent of the contamination were unexpected.

As nontarget support ships began to navigate the lagoon, they became contaminated with radioactivity below the water line. Conditions were ideal for ion exchange. Within 3 days, some vessels had a dose rate of 1 mGy per day in the hull of the ship near the water line (Operation Crossroads, 1946). In addition, saltwater lines and saltwater systems that continuously circulated lagoon water began to show increased exposure rates from penetrating gamma rays on external surfaces. It was recognized that algae, rust, sediment, and calcareous materials on or in the ship would absorb radioactivity from the contaminated seawater.

The US Navy monitored the ships closely. It became apparent that natural decay and normal steaming in uncontaminated water would not reduce the radioactivity to negligible amounts. All support ships that had been in Bikini lagoon from July 25 to August 10 were required to have extensive radiation monitoring before personnel could work on their hulls or interior saltwater systems. On September 9, 1946, commanding officers of all nontarget vessels were notified of precautions, monitoring, and clearance procedures. The commanders were frustrated by the disruption of naval operations caused by the almost universal contamination of nontarget vessels. On September 13, 1946, the chief of naval operations charged the Bureau of Ships with the task of developing methods and equipment for decontamination of radioactive ships. The procedures were to be developed with the assistance of scientists with the Manhattan Project.

Several radioactive ships were dispatched to the San Francisco Naval Shipyard. The Navy conducted experiments at the shipyard with the assistance of scientists from Stanford University and the University of California Radiation Laboratory. The tests resulted in adoption of procedures for decontaminating nontarget vessels.

A solution of hydrochloric acid, HCl, (1 normal) was used as a decontaminating agent for all saltwater systems in a ship. Each system was then drained, neutralized, and flushed thoroughly. The material containing the radioactivity was in solution or suspension and was removed from the ship when the acid solution was drained and flushed. Those liquids were released into the harbor, where the dilution factor and settling would reduce concentration of radioactivity to negligible levels.

The external hull of each ship was scraped in dry dock to remove all marine growth. The remaining paint and rust from the underwater hull was wet sandblasted with standard equipment. Sand and all material scraped from the ship's sides were collected and dumped into the sea.

Radiochemical analyses were performed at the University of California on numerous samples of solutions and sediments. They revealed that there was not sufficient long-lived alpha activity to pose a health problem from intake of radioactivity nor was there sufficient residual plutonium to be of concern for security purposes related to fissile materials.

Records indicate that about 18 vessels were dispersed to Guam for decontamination (four LCI[L], eight LCT and six YMS). No data indicated that radioactive materials affected sea life or entered the food chain of residents of Guam.

SUMMARY

An extensive radiologic survey of plants, animals, and soil in Micronesia was initiated after the termination of weapons testing in the Pacific. A report published in 1975 (Nelson, 1975) was reviewed at Lawrence Livermore Laboratory (Hamilton, 2001). The conclusion was that the estimated annual effective dose from residual fallout on Guam due to nuclear-weapons tests was only a small fraction of the dose that residents receive from natural sources of radiation, which is less than in many other locations around the world.

REFERENCES

Eisenbud, M. Radioactive Debris from Operation Ivy, US AEC, Health and Safety Laboratory, New York Operations Office, NYO-4522. 1953.

Hamilton, T., Radiation Fallout in Guam, Information Document Lawrence Livermore National Laboratory. 2001.

Harley, J. H., Hallden, N. A., Ong, L. D. Y. Summary of Gummed Film Results Through December 1959. HASL-93, Health and Safety Laboratory, U.S. AEC, New York Operations Office. 1960.

Nelson V. A. Radiological Survey of Plants, Animals and Soil in Micronesia, University of Seattle, Seattle WA, NVO-269-35, U.S. Department of Commerce. 1975.

Operation Crossroads: Radiological Decontamination of Target and Non-Target Vessels, Vol. 3, Technical Report XRD-185-87, NTIS Document No. AD 473 906. Radiological Safety Section to Technical Director, memorandum, September 25, 1946, LANL, App. 7, Sec. E; J. J. Fee. Washington, DC: U.S. Government Printing Office. 1946.

Appendix D

The Optimal Criterion for Positivity in Screening

Consider a population of individuals composed of two subpopulations: with disease, D, and with no disease, ND. Assume that the prevalence of disease is p_{dis} and that in a representative subset of the population we know, by some gold standard, which individuals have disease and which have no disease.

Consider that a diagnostic test T is applied to the representative subset and yields two distributions of results: one among patients with disease and the other among patients with no disease. Denote the two probability density distributions as $Dist_{dis}$ and $Dist_{nodis}$, respectively. $Dist_{dis}(x)$ denotes the probability of test result x in patients with disease; $Dist_{nodis}(x)$ denotes the probability of test result x in patients without disease. The distribution of test results in the population as a whole, $Dist_{pop}$, is the weighted average of $Dist_{dis}$ and $Dist_{nodis}$, with weights p_{dis} and $1 - p_{dis}$, respectively. The distributions can be seen in Figure 9.2.

The task in establishing a cutoff criterion (threshold T) for the test—that is, deciding how we classify patients—is in some sense to minimize the burden of misclassification. Among patients with disease, the probability of a positive result (i.e., a result that is > T) is the sensitivity or true-positive rate (TPR) and its complement is the false-negative rate (FNR). Among patients with no disease, the probability of a positive result (i.e., a result that is > T) is the false-positive rate (FPR), and its complement is the specificity.

Define the burden of a false positive as B_{fp} and the burden of a false negative as B_{fn}.

We achieve the goal of minimizing the overall burden by minimizing the expression

$$p_{dis} \times FNR \times B_{fn} + (1 - p_{dis}) \times FPR \times B_{fp}$$

or

$$p_{dis} \times (1-TPR) \times B_{fn} + (1-p_{dis}) \times FPR \times B_{fp}$$

Changing T will change both FNR and FPR in a fashion determined by the shape and the overlap of $Dist_{dis}$ and $Dist_{nodis}$. To minimize the overall burden of false-positive and false-negative results combined (with respect to changing T), one can differentiate the expression with respect to T and set the result to zero. That yields

$$(-p_{dis}) \times B_{fn} \times dTPR / dT + (1-p_{dis}) \times B_{fp} \times dFPR / dT = 0$$

Rearranging, we get:

$$(1-p_{dis}) \times B_{fp} \times dFPR / dT = p_{dis} \times B_{fn} \times dFPR / dT$$

or

$$(dTPR / dT) / (dFPR / dT) = ((1-p_{dis}) \times (B_{fp})) / (p_{dis} \times B_{fn})$$

This can be shown to equal:

$$dTPR / dFPR = ((1-p_{dis}) \times (B_{fp})) / (p_{dis} \times B_{fn}).$$

Another way of thinking about the cutoff criterion is to understand that it is the point t at which $Dist_{dis}(t) \times (p) \times B_{fn} = Dist_{nodis}(t) \times (1-p) \times B_{fp}$. That is an equivalent formulation of the same equation because dTPR/dT is simply probability density distribution $Dist_{dis}$, and dFPR/dT is simply probability density distribution $Dist_{nodis}$.

Now, if we plot TPR (vertical axis) against FPR (horizontal axis), we have the receiver operating characteristic (ROC) curve of the test. The slope of that curve at any point is simply dTPR/dFPR. Hence, the optimal operating point is the value of T where the slope of the curve (or its tangent) is numerically equivalent to

$$[(1-p_{dis}) \times (B_{fp})] / (p_{dis} \times B_{fn}).$$

Some authors use the term *cost of false positive* (C) in place of *burden of false positive* and the term *benefit of true positive* (B) in place of *burden of false negative*, all being greater than zero. In that case, the optimal operating point is the value of T where the slope of the ROC curve (or its tangent) is numerically equivalent to

$$[(1-p) \times C] / (p \times B).$$

The true and false-positive rates (from which one constructs an ROC curve) are the areas under the tails of the corresponding probability density distributions

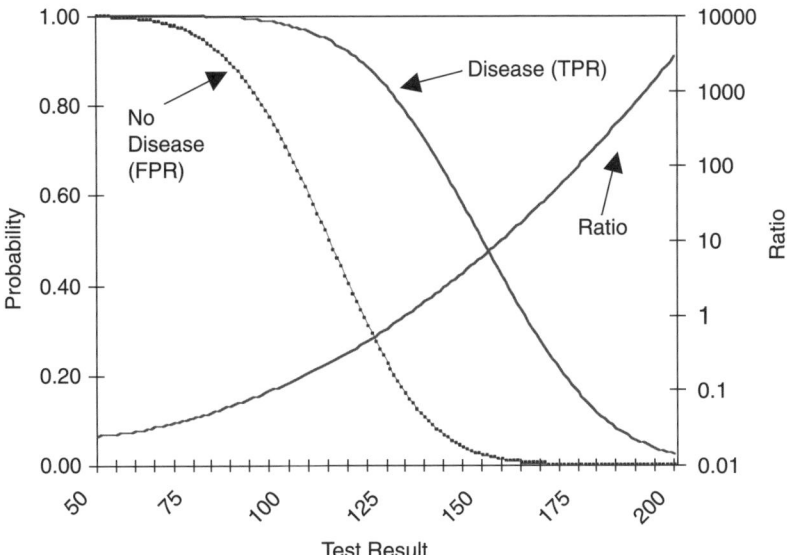

FIGURE D.1   True-positive and false-positive rates as a function of test result. Also shown is ratio of the probability of that test result in patients with disease to probability of that test result in patients without disease (plotted on logarithmic scale).

(or segments of the cumulative probability distributions). The slope of the ROC curve is the ratio of the height of the probability density distribution for patients with disease to the height of the probability density distribution for patients with no disease. If one plots that ratio on the vertical axis against the test result on the horizontal axis, one can determine the cutoff criterion that corresponds to any given slope; if one also plots the corresponding cumulative distributions against the test result, one can also find the corresponding optimal point on the ROC curve (Figure D.1).

AN EXAMPLE

Now consider an example of finding the best operating point (the best criterion of positivity) in a population to be screened. Assume that we are screening for a disease (perhaps a slow growing cancer) for which early detection provides a benefit of 0.5 years of survival; thus, B_{fn} is 0.5. Further assume that a false-positive result is associated with a risk of 0.05 years (perhaps because the population to be screened has a high prevalence of severe chronic pulmonary disease, which substantially increases the risk posed by surgery), making B_{fp} 0.05.

If one were considering screening a population in which the prevalence of disease is [subjunctive case is were] 10%, the best criterion for positivity would be the point on the ROC curve where its slope (or tangent) is $[(1 - 0.1) \times 0.05]/[0.1 \times 0.5]$ or 0.90, a point near the middle of most ROC curves, with modest true-positive and false-positive rates. If we use the distributions displayed in Figure 9.2 and the ROC curve displayed in Figure 9.3, the optimal criterion of positivity would correspond to a true-positive rate (sensitivity) of 80% and a false positive rate of 17% (a specificity of 83%). However, if one were considering screening a population in which the prevalence of disease is [subjunctive case is were] only 1%, then the best criterion for positivity would be the point on the ROC curve where its slope (or tangent) is $[(1 - 0.01) \times 0.05]/[0.01 \times 0.5]$ or 9.9, a point nearer to the origin for most ROC curves, with both true-positive and false-positive rates low. Again, if we use the distributions displayed in Figure 9.2 and the ROC curve displayed in Figure 9.3, the optimal criterion of positivity would correspond to a true-positive rate (sensitivity) of 42% and a false-positive rate of 1% (a specificity of 99%).

In the special case when both probability density distributions (patients with and without disease) are normal or Gaussian in shape, the slope of the corresponding ROC at any point (the ratio of the heights of the corresponding density distributions) can be solved algebraically, although the equation is fairly complex. Because the normal distribution is

$$\frac{1}{\sigma\sqrt{2\pi}} e^{-\frac{1}{2}\left(\frac{x-\bar{x}}{\sigma}\right)}$$

where x is the test result, $\bar{x}$ is the mean, and σ is the standard deviation, the optimal cutoff criterion will be the value of x where

$$\frac{1}{\sigma_{dis}\sqrt{2\pi}} e^{-\frac{1}{2}\left(\frac{x-\bar{x}_{dis}}{\sigma_{dis}}\right)^2} (p)B = \frac{1}{\sigma_{nodis}\sqrt{2\pi}} e^{-\frac{1}{2}\left(\frac{x-\bar{x}_{nodis}}{\sigma_{nodis}}\right)^2} (1-p)C$$

The value of x at which the equality holds can be found by successive approximations or using the "goal seek" function in a spreadsheet program.

Appendix E

Selected Cancer-Screening Recommendations

INTRODUCTION

This appendix presents summary information on recommendations for screening for various cancers that are relevant to the Radiation Exposure Compensation Act (RECA) and the Radiation Exposure Screening and Education Program (RESEP) of the Health Resources and Services Administration of the Department of Health and Human Services (DHHS). It supports and provides more information for Chapters 9 and 10 of the present report on issues in medical or compensational screening and RESEP screening activities, respectively; in particular, it documents how few cancers are amenable to medical screening in the conventional sense of the term.

We have elected to present screening guidelines and recommendations only from the US Preventive Services Task Force (USPSTF) and the Canadian Task Force on Preventive Health Care (Canadian TF) on screening for a variety of neoplastic diseases of interest to the RECA and RESEP efforts. Numerous guidelines on the relevant topics based on work done by various professional societies and associations and by voluntary disease and patient-advocacy groups are available; interested readers can go to the Web site of the National Guidelines Clearinghouse, supported by the DHHS Agency for Healthcare Research and Quality (AHRQ) at http://www.guideline.gov for more information. Generally, we believe that the USPSTF recommendations are the most thoroughly considered and backed by rigorous systematic reviews of published evidence, and they are clearly aimed at primary-care clinicians; those of the Canadian TF have also been developed through evidence-based methods.

In the material that follows, we list cancers in alphabetical order; information from the USPSTF appears before that from the Canadian TF. For the USPSTF, the entries below come from the Web site of AHRQ, which supports the current USPSTF; the specific URLs are not listed here, but the general one is http://www.ahrq.gov/clinic/cps3dix.htm#cancer (accessed May 4, 2004). Harris et al. (2001) present more information on USPSTF methods. The general URL for material from the Canadian TF is http://www.ctfphc.org/ (accessed May 4, 2004).

The USPSTF uses specific grades for quality of evidence and recommendations. The definitions for the grades are noted below (http://www.ahrq.gov/clinic/3rduspstf/ratings.htm#irec, accessed May 4, 2004), and the recommendations themselves follow. After them is the older grading system used by the Canadian TF.

USPSTF GRADES FOR QUALITY OF EVIDENCE

The USPSTF assigns one of three to the overall evidence for a service: good, fair, or poor:

"Good: Evidence includes consistent results from well-designed, well-conducted studies in representative populations that directly assess effects on health outcomes.

"Fair: Evidence is sufficient to determine effects on health outcomes, but the strength of the evidence is limited by the number, quality, or consistency of the individual studies, generalizability to routine practice, or indirect nature of the evidence on health outcomes.

"Poor: Evidence is insufficient to assess the effects on health outcomes because of the low number or power of studies, important flaws in their design or conduct, gaps in the chain of evidence, or lack of information on important health outcomes."

USPSTF GRADES FOR STRENGTH OF RECOMMENDATIONS

The USPSTF assigns one of five grades (A, B, C, D, and I) to its recommendations to reflect the strength of evidence and magnitude of net benefit (benefits minus harms):

"A. The USPSTF strongly recommends that clinicians provide [the service] to eligible patients. The USPSTF found good evidence that [the service] improves important health outcomes and concludes that benefits substantially outweigh harms.

"B. The USPSTF recommends that clinicians provide [this service] to eligible patients. The USPSTF found at least fair evidence that [the service] improves important health outcomes and concludes that benefits outweigh harms.

"C. The USPSTF makes no recommendation for or against routine provision of [the service]. The USPSTF found at least fair evidence that [the service] can improve health outcomes but concludes that the balance of benefits and harms is too close to justify a general recommendation.

"D. The USPSTF recommends against routinely providing [the service] to asymptomatic patients. The USPSTF found at least fair evidence that [the service] is ineffective or that harms outweigh benefits.

"E. The USPSTF concludes that the evidence is insufficient to recommend for or against routinely providing [the service]. Evidence that the [service] is effective is lacking, of poor quality, or conflicting and the balance of benefits and harms cannot be determined."

CANADIAN TF GRADES FOR QUALITY OF PUBLISHED EVIDENCE

"I. Evidence from at least 1 properly randomized controlled trial (RCT).

"II-1. Evidence from well-designed controlled trials without randomization.

"II.2. Evidence from well-designed cohort or case-control analytic studies, preferably from more than 1 center or research group.

"II-3. Evidence from comparisons between times or places with or without the intervention. Dramatic results in uncontrolled experiments could also be included here.

"III. Opinions of respected authorities, based on clinical experience, descriptive studies or reports of expert committees."

CANADIAN TF GRADES FOR RECOMMENDATIONS

"A. Good evidence to support the recommendation that the condition be specifically considered in a Periodic Health Examination (PHE).

"B. Fair evidence to support the recommendation that the condition be specifically considered in a PHE.

"C. Poor evidence regarding inclusion or exclusion of a condition in a PHE, but recommendations may be made on other grounds.

"D. Fair evidence to support the recommendation that the condition be specifically excluded from consideration in a PHE.

"E. Good evidence to support the recommendation that the condition be specifically excluded from consideration in a PHE."

We first present a summary of screening recommendations for cancers that may be relevant to RECA. Because the committee recommends, in Chapter 9, that persons administratively eligible for compensation under RECA should be offered the same screening as is recommended for the general population, we provide later in this appendix a table naming the screening protocols (i.e., conditions) that the USPSTF recommends for the general US population (or certain subgroups) with either an A or B rating (strongly recommends or recommends, respectively). The table reflects USPSTF published decisions as of March 2005. The full USPSTF statements of recommendations and rationales can be found on the AHRQ website at http://www.ahrq.gov/clinic/prevenix.htm.

APPENDIX E-1
SCREENING RECOMMENDATIONS FOR SPECIFIC CANCERS

Bladder Cancer
USPSTF
No information available.

Canadian Task Force
Fair evidence to exclude from Periodic Health Examination (PHE) for general population (*D*); poor evidence to include or exclude from the PHE for persons at high risk (*C*).

Breast Cancer
USPSTF
The USPSTF recommends screening mammography, with or without clinical breast examination (CBE), every 1-2 years for women aged 40 and older.

Rating: *B recommendation.*

Rationale: The USPSTF found fair evidence that mammography screening every 12-33 months significantly reduces mortality from breast cancer. Evidence is strongest for women aged 50-69, the age group generally included in screening trials. For women aged 40-49, the evidence that screening mammography reduces mortality from breast cancer is weaker, and the absolute benefit of mammography is smaller than it is for older women. Most, but not all, studies indicate a mortality benefit for women undergoing mammography at ages 40-49, but the delay in observed benefit in women younger than 50 makes it difficult to determine the incremental benefit of beginning screening at age 40 rather than at age 50.

The absolute benefit is smaller because the incidence of breast cancer is lower among women in their 40s than it is among older women. The USPSTF concluded that the evidence is also generalizable to women aged 70 and older (who face a higher absolute risk for breast cancer) if their life expectancy is not compromised by comorbid disease. The absolute probability of benefits of regular mammography increase along a continuum with age, whereas the likelihood of harms from screening (false-positive results and unnecessary anxiety, biopsies, and cost) diminish from ages 40-70. The balance of benefits and potential harms, therefore, grows more favorable as women age. The precise age at which the potential benefits of mammography justify the possible harms is a subjective choice. The USPSTF did not find sufficient evidence to specify the optimal screening interval for women aged 40-49 (see *Clinical Considerations*).

The USPSTF concludes that the evidence is insufficient to recommend for or against routine CBE alone to screen for breast cancer.

Rating: *I recommendation.*

Rationale: No screening trial has examined the benefits of CBE alone (without accompanying mammography) compared to no screening, and design characteristics limit the generalizability of studies that have examined CBE. The USPSTF could not determine the benefits of CBE alone or the incremental benefit of adding CBE to mammography. The USPSTF therefore could not determine whether potential benefits of routine CBE outweigh the potential harms.

The USPSTF concludes that the evidence is insufficient to recommend for or against teaching or performing routine breast self-examination (BSE).

Rating: *I recommendation.*

Rationale: The USPSTF found poor evidence to determine whether BSE reduces breast cancer mortality. The USPSTF found fair evidence that BSE is associated with an increased risk for false-positive results and biopsies. Due to design limitations of published and ongoing studies of BSE, the USPSTF could not determine the balance of benefits and potential harms of BSE.

Canadian Task Force
There is good evidence for screening women aged 50-69 years by clinical examination and mammography (*A*). The best available data support screening every 1-2 years.
Current evidence does not support the recommendation that screening mammography be included in or excluded from the periodic health examination of women aged 40-49 at average risk of breast cancer (*C*)

Cervical Cancer
USPSTF
The USPSTF strongly recommends screening for cervical cancer in women who have been sexually active and have a cervix.

Rating: *A recommendation.*

Rationale: The USPSTF found good evidence from multiple observational studies that screening with cervical cytology (Pap smears) reduces incidence of and mortality from cervical cancer. Direct evidence to determine the optimal starting and stopping age and interval for screening is limited. Indirect evidence suggests most of the benefit can be obtained by beginning screening within 3 years of onset of sexual activity or age 21 (whichever comes first) and screening at least every 3 years (*go to Clinical Considerations*). The USPSTF concludes that the benefits of screening substantially outweigh potential harms.

The USPSTF recommends against routinely screening women older than age 65 for cervical cancer if they have had adequate recent screening with normal Pap smears and are not otherwise at high risk for cervical cancer (go to *Clinical Considerations*).

Rating: *D recommendation*

Rationale: The USPSTF found limited evidence to determine the benefits of continued screening in women older than 65. The yield of screening is low in previously screened women older than 65 due to the declining incidence of high-grade cervical lesions after middle age. There is fair evidence that screening women older than 65 is associated with an increased risk for potential harms, including false-positive results and invasive procedures. The USPSTF concludes that the potential harms of screening are likely to exceed benefits among older women who have had normal results previously and who are not otherwise at high risk for cervical cancer.

The USPSTF recommends against routine Pap smear screening in women who have had a total hysterectomy for benign disease.

Rating: *D recommendation.*

Rationale: The USPSTF found fair evidence that the yield of cytologic screening is very low in women after hysterectomy and poor evidence that screening to detect vaginal cancer improves health outcomes. The USPSTF concludes that potential harms of continued screening after hysterectomy are likely to exceed benefits.

The USPSTF concludes that the evidence is insufficient to recommend for or against the routine use of new technologies to screen for cervical cancer.

Rating: *I recommendation*

Rationale: The USPSTF found poor evidence to determine whether new technologies, such as liquid-based cytology, computerized rescreening, and algorithm based screening, are more effective than conventional Pap smear screening in reducing incidence of or mortality from invasive cervical cancer. Evidence to determine both sensitivity and specificity of new screening technologies is limited. As a result, the USPSTF concludes that it cannot determine whether the potential benefits of new screening devices relative to conventional Pap tests are sufficient to justify a possible increase in potential harms or costs.

The USPSTF concludes that the evidence is insufficient to recommend for or against the routine use of human papillomavirus (HPV) testing as a primary screening test for cervical cancer.

Rating: *I recommendation.*

Rationale: The USPSTF found poor evidence to determine the benefits and potential harms of HPV screening as an adjunct or alternative to regular Pap smear screening. Trials are underway that should soon clarify the role of HPV testing in cervical cancer screening.

Canadian Task Force (dating from 1994)
Fair evidence to include in periodic health examination of sexually active women. (*B*)

Colorectal Cancer
USPSTF
The USPSTF strongly recommends that clinicians screen men and women 50 years of age or older for colorectal cancer.

Rating: *A recommendation.*

Rationale: The USPSTF found fair to good evidence that several screening methods are effective in reducing mortality from colorectal cancer. The USPSTF concluded that the benefits from screening substantially outweigh potential harms, but the quality of evidence, magnitude of benefit, and potential harms vary with each method.

The USPSTF found good evidence that periodic fecal occult blood testing (FOBT) reduces mortality from colorectal cancer and fair evidence that sigmoidoscopy alone or in combination with FOBT reduces mortality. The

USPSTF did not find direct evidence that screening colonoscopy is effective in reducing colorectal cancer mortality; efficacy of colonoscopy is supported by its integral role in trials of FOBT, extrapolation from sigmoidoscopy studies, limited case-control evidence, and the ability of colonoscopy to inspect the proximal colon. Double-contrast barium enema offers an alternative means of whole-bowel examination, but it is less sensitive than colonoscopy, and there is no direct evidence that it is effective in reducing mortality rates. The USPSTF found insufficient evidence that newer screening technologies (for example, computed tomographic colography) are effective in improving health outcomes.

There are insufficient data to determine which strategy is best in terms of the balance of benefits and potential harms or cost-effectiveness. Studies reviewed by the USPSTF indicate that colorectal cancer screening is likely to be cost-effective (less than $30,000 per additional year of life gained) regardless of the strategy chosen.

It is unclear whether the increased accuracy of colonoscopy compared with alternative screening methods (for example, the identification of lesions that FOBT and flexible sigmoidoscopy would not detect) offsets the procedure's additional complications, inconvenience, and costs.

Canadian TF
    Average Risk Individuals
    • Screening with the Hemoccult test: There is good evidence to include screening with Hemoccult test in the periodic health examination of asymptomatic patients over age 50 with no other risk factors [A, I]. However, there remain concerns about the high rate of false-positive results, feasibility and small clinical benefit of such screening (over 1000 individuals must be screened for 10 years to avert one death from colorectal cancer). For patients being screened with Hemoccult, it is recommended that they avoid red meat, cantaloupe and melons, raw turnip, radishes, broccoli and cauliflower, vitamin C supplements and aspirin and non-steroidal anti-inflammatory drugs for 3 days before fecal samples are collected. However, a recent meta-analysis of 4 RCTs found no improvement in positivity rates or change in compliance rates with moderate dietary restrictions.
    • Screening with sigmoidoscopy: There is evidence from case control studies, to recommend that flexible sigmoidoscopy be included in the periodic health examination of patients over age 50 [B, II-2, III]. There is insufficient evidence to make recommendations about whether only 1 or both of FOBT and sigmoidoscopy should be performed [C, I].
    • Screening with colonoscopy: There is insufficient evidence to include or exclude colonoscopy as an initial screen in the periodic health examination

[C, II-3]. Although colonoscopy is the best method for detecting adenomas and carcinomas, it may not be feasible to screen asymptomatic patients because of patient compliance and the expertise and equipment required and the potential costs. On the other hand, if colonoscopy were an effective screening strategy when performed at less frequent intervals, these issues might be of less concern.

Above Average Risk Individuals
• Individuals at Risk for Familial Adenomatous Polyposis (FAP): The Task Force recommends genetic testing of individuals at risk for FAP if the genetic mutation has been identified in the family and if genetic testing is available [B, II-3]. If the individual carries the mutation, then he or she should be screened with flexible sigmoidoscopy beginning at puberty [B, II-3]. Individuals from families where the gene mutation has been identified but are negative themselves, require screening similar to the average risk population. For at risk individuals where the mutation has not been identified in the family or where genetic testing is not available, screening with annual or biannual flexible sigmoidoscopy should be undertaken beginning at puberty. In all instances, genetic counseling should be performed prior to genetic testing.
• Individuals at Risk for Hereditary Non-Polyposis Colon Cancer (HNPCC): Patients in kindreds with the cancer family syndrome (HNPCC) have a high risk of colorectal cancer and a high incidence of right-sided colon cancer. Thus, colonoscopy rather than sigmoidoscopy is recommended for screening such patients. Based on Level III evidence, the Task Force recommends screening with colonoscopy in individuals from HNPCC kindreds [B, II-3]. Although higher levels of evidence are usually required to give a B recommendation, the Task Force realizes that it is unlikely that more rigorous studies could be performed in this cohort of patients given the high risk of cancer and relative infrequency of HNPCC. The ages when screening should begin and the frequency at which colonoscopy should be performed are unclear.
• Individuals with a Family History of Polyps or Colon Cancer: Patients who have only one or two first-degree relatives with colorectal cancer should be screened in the same way as average risk individuals. There is insufficient evidence to recommend colonoscopy for individuals who have a family history of colorectal polyps or cancer but do not fit the criteria for HNPCC [C, III]. While there is evidence that there is an increased prevalence of neoplasms in these individuals, there is insufficient information to recommend more intense screening than that of individuals at average risk. Further delineation of the risk for individuals with multiple affected family members and family members with early age of diagnosis of colorectal cancer is necessary.

• Because most screening options are multiphasic, it is preferable that there is adequate infrastructure to support the implementation, assure quality control and the timely follow-up of screened individuals.

Lung Cancer
USPSTF

The USPSTF concludes that the evidence is insufficient to recommend for or against screening asymptomatic persons for lung cancer with either low dose computerized tomography (LDCT), chest x-ray (CXR), sputum cytology, or a combination of these tests.

Rating: *I Recommendation.*

Rationale: The USPSTF found fair evidence that screening with LDCT, CXR, or sputum cytology can detect lung cancer at an earlier stage than lung cancer would be detected in an unscreened population; however, the USPSTF found poor evidence that any screening strategy for lung cancer decreases mortality. Because of the invasive nature of diagnostic testing and the possibility of a high number of false-positive tests in certain populations, there is potential for significant harms from screening. Therefore, the USPSTF could not determine the balance between the benefits and harms of screening for lung cancer.

Oral Cancers
USPSTF

The USPSTF concludes that the evidence is insufficient to recommend for or against routinely screening adults for oral cancer.

Rating: *I Recommendation.*

Rationale: The USPSTF found no new good-quality evidence that screening for oral cancer leads to improved health outcomes for either high-risk adults (i.e., those over the age of 50 who use tobacco) or for average-risk adults in the general population. It is unlikely that controlled trials of screening for oral cancer will ever be conducted in the general population because of the very low incidence of oral cancer in the United States. There is also no new evidence for the harms of screening. As a result, the USPSTF could not determine the balance between benefits and harms.

Canadian TF

With respect to screening by oral physical exam in 1994: Insufficient evidence to include or exclude from periodic health exam (*C*); annual examination by physician and/or dentist should be considered for men and women over age 60 years with risk factors for oral cancers and precancers; individual judgment should be exercised regarding the use of tolonium chloride for those identified as positive by oral physical exam.

The above recommendation has been updated in 1999
• Population screening: Fair evidence to exclude screening the general population for oral cancer by clinical examination (*D Recommendation*).
• Opportunistic screening: Insufficient evidence to recommend inclusion or exclusion of screening for oral cancer by clinical examination of asymptomatic patients (*C Recommendation*).
• For high risk patients, annual examination by physician or dentist should be considered. Major risk factors include a history of tobacco use and excessive alcohol consumption.

Ovarian Cancer
USPSTF
The USPSTF recommends against routine screening for ovarian cancer.

Rating: *D Recommendation.*

Rationale: The USPSTF found fair evidence that screening with serum CA-125 level or transvaginal ultrasound can detect ovarian cancer at an earlier stage than it can be detected in the absence of screening; however, the USPSTF found fair evidence that earlier detection would likely have a small effect, at best, on mortality from ovarian cancer. Because of the low prevalence of ovarian cancer and the invasive nature of diagnostic testing after a positive screening test, there is fair evidence that screening could likely lead to important harms. The USPSTF concluded that the potential harms outweigh the potential benefits.

Pancreatic Cancer
USPSTF
The USPSTF recommends against routine screening for pancreatic cancer in asymptomatic adults using abdominal palpation, ultrasonography, or serologic markers.

Rating: *D Recommendation.*

Rationale: The USPSTF found no evidence that screening for pancreatic cancer is effective in reducing mortality. There is a potential for significant harm due to the very low prevalence of pancreatic cancer, limited accuracy of available screening tests, the invasive nature of diagnostic tests, and the poor outcomes of treatment. As a result, the USPSTF concluded that the harms of screening for pancreatic cancer exceed any potential benefits.

Canadian TF
There is fair evidence that routine screening should be excluded from the periodic health examination (*D Recommendation*).

Prostate Cancer
USPSTF
The USPSTF concludes that the evidence is insufficient to recommend for or against routine screening for prostate cancer using prostate specific antigen (PSA) testing or digital rectal examination (DRE).

Rating: *I recommendation.*

Rationale: The USPSTF found good evidence that PSA screening can detect early-stage prostate cancer but mixed and inconclusive evidence that early detection improves health outcomes. Screening is associated with important harms, including frequent false-positive results and unnecessary anxiety, biopsies, and potential complications of treatment of some cancers that may never have affected a patient's health. The USPSTF concludes that evidence is insufficient to determine whether the benefits outweigh the harms for a screened population.

Canadian TF
There are two main philosophical views concerning early detection of cancer. One view holds that the major goal is to search aggressively for asymptomatic cancer and having found it, remove it. While the effectiveness of therapy may not be established, and its associated adverse affects may be recognized, the main mission is to detect cancer early. This view emphasizes the importance of developing tests which can detect cancer early, even if such tests may label many individuals falsely and subject them to subsequent unnecessary, invasive investigations.

The alternate view considers early detection and treatment as a single package and asks whether there is evidence that such combined efforts do more good than harm. This is the question of greatest importance, both from the individual patient's perspective as well as that of the population. Hence, while evaluating the performance of early detection tests is part of the picture, one must also evaluate the effectiveness of therapy and whether the use of available early detection tests ultimately provides overall net benefit to the patient. This is the view taken by the Canadian Task Force on the Periodic Health Examination.

Based on the absence of evidence for effectiveness of therapy and the substantial risk of adverse effects associated with such therapy; and the poor predictive value of screening tests, there is at present insufficient evidence to support wide-spread initiatives for the early detection of prostate cancer.

The Task Force does not recommend the routine use of PSA as part of a periodic health examination. While PSA can detect earlier cancer, it is associated with a substantial false positive rate. This, combined with poor evidence to support the effectiveness of subsequent therapy and clear evidence of substantial risk associated with such therapy, means that the wide-spread implementation of PSA would expose more men to uncertain benefit, but to definite risks. For these reasons the Task Force recommends that PSA be excluded from the periodic health examination (*D Recommendation*).

The Task Force debated recommending the exclusion of DRE from the periodic health examination because of its limited performance as an early detection test. However, DRE has been routine practice for many physicians for the early detection of prostate abnormalities and the available evidence was not considered sufficiently powerful to advise physicians who currently include DRE as part of a periodic health examination in men aged 50 to 70 to discontinue the practice. At the same time, the evidence is insufficient to advocate the inclusion of DRE for those physicians who do not currently include it as part of the periodic health examination for men aged 50 to 70. Hence, the decision to retain a *C Recommendation* for DRE—there is insufficient evidence to include DRE or exclude it from the periodic health exam.

- Based on the available evidence for TRUS, the Task Force recommends against the routine use of this procedure as part of a periodic health examination (*D Recommendation*).

These recommendations are made on the basis of the evaluation of the best available evidence using the Canadian Task Force guidelines, and the ethical imperative that early detection efforts must be proven to result in more good than harm before being incorporated into the periodic health examination.

Skin Cancers
USPSTF

The USPSTF concludes that the evidence is insufficient to recommend for or against routine screening for skin cancer using a total-body skin examination for the early detection of cutaneous melanoma, basal cell cancer, or squamous cell skin cancer.

Rating: *I recommendation.*

Rationale: Evidence is lacking that skin examination by clinicians is effective in reducing mortality or morbidity from skin cancer. The USPSTF could not determine the benefits and harms of periodic skin examination. (See *Clinical Considerations* for discussion of selected populations at high risk.)

Clinical Considerations
Benefits from screening are unproven, even in high-risk patients.

Clinicians should be aware that fair-skinned men and women aged >65, patients with atypical moles, and those with >50 moles constitute known groups at substantially increased risk for melanoma.

Clinicians should remain alert for skin lesions with malignant features noted in the context of physical examinations performed for other purposes.

Asymmetry, border irregularity, color variability, diameter >6 mm ("A," "B," "C," "D"), or rapidly changing lesions are features associated with an increased risk of malignancy. Suspicious lesions should be biopsied.

The USPSTF did not examine the outcomes related to surveillance of patients with familial syndromes, such as familial atypical mole and melanoma (FAM-M) syndrome.

Canadian Task Force (dating from 1994)
Routine screening for skin cancer by primary care providers using total-body skin examination is not recommended for the general population. Clinicians should remain alert for skin lesions with malignant features (i.e. asymmetry, border irregularity, color variability, diameter greater than 6 mm, or rapidly changing lesions) when examining patients for other reasons, particularly in those with established risk factors. Such risk factors include clinical evidence of melanocytic precursor or marker lesions (i.e. atypical moles, certain congenital moles), large numbers of common moles, immunosuppression, a family or personal history of skin cancer, substantial cumulative lifetime sun exposure, intermittent intense sun exposure or severe sunburns in childhood, or light skin, hair, and eye color, freckles, or poor tanning ability. Appropriate biopsy specimens should be taken of suspicious lesions (*C Recommendation*).

Currently, there is insufficient evidence to recommend for or against counseling patients to perform periodic self-examination of the skin. Clinicians may wish to educate patients with established risk factors for skin cancer (see above) concerning signs and symptoms suggesting cutaneous malignancy and the possible benefits of periodic self-examination (*C Recommendation*).

Persons with Family Melanoma Syndrome are at substantially increased risk for malignant melanoma. Clinicians examining these patients should be particularly alert to skin lesions with malignant features and should consider referral to skin cancer specialists for evaluation. For this very select subgroup there is fair evidence to offer total body skin examination (*B Recommendation*).

Testicular Cancer
USPSTF

The USPSTF recommends against routine screening for testicular cancer in asymptomatic adolescent and adult males.

Rating: *D Recommendation*

Rationale: The USPSTF found no new evidence that screening with clinical examination or testicular self-examination is effective in reducing mortality from testicular cancer. Even in the absence of screening, the current treatment interventions provide very favorable health outcomes. Given the low prevalence of testicular cancer, limited accuracy of screening tests, and no evidence for the incremental benefits of screening, the USPSTF concluded that the harms of screening exceed any potential benefits.

Canadian Task Force (dating from 1994)

Because no studies of screening for testicular cancer by physician or patient self-examination have been reported, there is insufficient evidence to include or exclude screening for this cancer in the periodic health examination of men (*C Recommendation*). Based on the low incidence of disease and the current high cure rate it is unlikely formal screening would improve the already excellent prognosis. Patients with a history of cryptorchidism, orchiopexy, or testicular atrophy should be informed of their increased risk for developing testicular cancer and counselled about screening options. The optimal frequency of such examinations has not been determined and is left to clinical discretion. Clinicians should advise adolescent and young adult males to seek prompt medical attention if a testicular mass is noted.

APPENDIX E-2: USPSTF GRADE A AND GRADE B SCREENING RECOMMENDATIONS FOR ADULTS

Disease	Screening Test	Comments
Abdominal aortic aneurysm	Abdominal ultrasound	Men who have ever smoked
Alcohol misuse	History	If present, behavioral counseling
Breast cancer	Mammogram	Age ≥ 40 years
Cervical cancer	Pap smear	Sexually active women with a cervix
Colorectal cancer	Colonoscopy, flexible sigmoidoscopy or FOBT	Age ≥ 50 years
Depression	Two-question quick screen	Mechanism for referral and followup required. Quality improvement program desirable
Dyslipidemia	Lipid profile	If present, also screen for diabetes
Hypertension	Blood pressure measurement	If present, also screen for diabetes
Obesity	Calculate body mass index (BMI)	Counsel if BMI ≥ 30
Osteoporosis	History, Risk assessment, DEXA	Age ≥ 65 years; ≥ 60 years if high risk
Syphilis infection	VDRL, RDR, or TP-PA	Persons at increased risk
Pregnant Women		
Asymptomatic bacteriuria	Urine culture	
Chlamydial infection	Culture or Antibody or DNA tests	
Hepatitis B infection	HBsAg test	
Rh incompatibility	Rh (D) blood typing	
Syphilis infection	VDRL, RDR, or TP-PA	

REFERENCES

Canadian TF (Canadian Task Force on Preventive Health Care). Evidence-Based Clinical Prevention. Available at: http://www.ctfphc.org/; accessed May 4, 2004.

Harris, R.P., Helfand, M., Woolf, S.H., Lohr, K.N., Mulrow, C.D., Teutsch, S.M., Atkins, D. Current Methods of the U.S. Preventive Services Task Force: A Review of the Process. Am. J. Prev. Med. 20(suppl 3):21-35. 2001.

USPSTF (U.S. Preventive Services Task Force). Guide to Clinical Preventive Services. Available at: http://www.ahrq.gov/clinic/cps3dix.htm#cancer; accessed May 4, 2004.

USPSTF (U.S. Preventive Services Task Force). U.S. Preventive Services Task Force Ratings: Strength of Recommendations and Quality of Evidence. Guide to Clinical Preventive Services, Third Edition: Periodic Updates, 2000-2003. Agency for Healthcare Research and Quality, Rockville, MD. Available at: http://www.ahrq.gov/clinic/3rduspstf/ratings.htm#irec; accessed May 4, 2004.

USPSTF (U.S. Preventive Services Task Force). Preventive Services. U.S. Preventive Services Task Force (USPSTF). Available at: http://www.ahrq.gov/clinic/prevenix.htm; accessed May 4, 2004.

Glossary

Absolute risk: The excess risk attributed to irradiation and usually expressed as the numeric difference between irradiated and nonirradiated populations (for example, 1 excess case of cancer/1 million people irradiated annually for each rad). Absolute risk may be given on an annual basis or lifetime (70-year) basis.

Absorbed dose: The energy imparted by ionizing radiation per unit mass of material irradiated. For purposes of radiation protection and assessing risks to human health, the quantity normally calculated is the average absorbed dose in an organ or tissue, equal to the total energy imparted to that organ or tissue divided by the total mass. The SI unit of absorbed dose is the joule per kilogram (J kg^{-1}), and its special name is the gray (Gy). In this report, absorbed dose is given in rads; 1 rad = 0.01 Gy.

Activation: The production of radionuclides by capture of radiation (for example, neutrons) in atomic nuclei.

Activity: The rate of transformation (or disintegration or decay) of radioactive material. The SI unit of activity is the reciprocal second (s^{-1}), and its special name is the Becquerel (Bq). In this report, activity is given in curies (Ci); 1 Ci = 3.7 × 10^{10} Bq.

Activity median aerodynamic diameter (AMAD):   The diameter in an aerodynamic particle size distribution for which the total activities on particles above and below this size are equal. A lognormal distribution of particles sizes usually is assumed.

Acute exposure: An exposure that took place over a short period of time— hours or days. Acute may also be used to refer to the short-term effects of exposure to radiation.

Aerosol: Extremely small liquid or solid particles suspended in air.

Alpha particle: An energetic nucleus of a helium atom, consisting of two protons and two neutrons, that is emitted spontaneously from nuclei in decay of some radionuclides; also called **alpha radiation** and sometimes shortened to **alpha** (for example, "alpha-emitting radionuclide"). Alpha particles are weakly penetrating and can be stopped by a sheet of paper or the outer dead layer of skin.

Atmospheric testing: Detonation of nuclear weapons or devices in the atmosphere or close to the earth's surface as part of the nuclear-weapons testing program.

Atom: The smallest particle of a chemical element that cannot be divided or broken up by chemical means. An atom consists of a central nucleus of protons and neutrons, and orbital electrons surrounding the nucleus.

Atomic bomb: A nuclear weapon that relies on fission only, in contrast with a thermonuclear ("hydrogen") bomb that uses fission and fusion.

Atomic Energy Commission (AEC): The agency of the US government that became the Department of Energy and the Nuclear Regulatory Commission.

Attributable risk percent: The percentage of disease that could be eliminated if a particular exposure were stopped.

Autoimmune disease: A disease caused by one's immune system's attacking the cells of one's own body rather than attacking foreign cells, such as germs.

Autoimmune hypothyroidism: An autoimmune disease that prevents the thyroid from producing enough thyroid hormone.

Autoimmune thyroiditis: Damage to the thyroid caused when the body's immune system attacks and destroys cells in the thyroid. It can be radiation-induced. If the damage is substantial enough, a person may develop signs and symptoms due to hypothyroidism. If there are no signs or symptoms of hypothyroidism, autoimmune thyroiditis is generally not a cause for concern about producing enough thyroid hormone.

Background radiation: Ionizing radiation that occurs naturally in the environment, including cosmic radiation; radiation emitted by naturally occurring radionuclides in air, water, soil, and rock; radiation emitted by naturally occurring radionuclides in tissues of humans and other organisms; and radiation emitted by human-made materials containing incidental amounts of naturally occurring radionuclides (such as building materials). Background radiation may also include radiation emitted by residual fallout from nuclear-weapons tests that has been dispersed throughout the world. The average annual effective dose due to natural background radiation in the United States is about 0.1 rem, excluding the dose due to indoor radon, and the average annual effective dose due to indoor radon is about 0.2 rem.

Badged dose: An estimate of a person's external radiation dose, specifically the deep equivalent dose from external exposure to photons, as derived from readings of exposure by one or more film badges assigned to the person.

Basal cells: A cell in the epidermis that give rise to more-specialized cells.

Basal-cell carcinoma: A malignant growth originating from basal cells that is most common in fair-skinned or sun-exposed areas; the most common form of skin cancer.

Becquerel (Bq):   The special name for the SI unit of activity; 1 Bq = 1 disintegration per second.

Benign tumor: A general category of tumors that do not invade surrounding tissue. Benign tumors are characterized by slow growth through expansion. They are not malignant or cancerous.

Beta particle: An energetic electron emitted spontaneously from the nucleus in decay of some radionuclides and produced by transmutation of a neutron into a proton; also called beta radiation and sometimes shortened to beta (for example, beta-emitting radionuclide). Beta particles are not highly penetrating, and the highest-energy beta radiation can be stopped by a few centimeters of plastic or aluminum.

Bias: The systematic tendency of a measurement or prediction of a quantity to overestimate or underestimate the actual value on the average.

Biological Effects of Ionizing Radiation (BEIR):   A series of National Research Council studies conducted by the committees on the Biological Effects of Ionizing Radiations (*BEIR VII* is the current study).

Biological response: A significant adverse effect in an organism resulting from exposure to a hazardous agent. The determination of whether an effect is significant or adverse sometimes involves subjective judgment. Often called a biologic endpoint or biologic effect in the literature.

Cancer: A malignant tumor of potentially unlimited growth that expands locally by invasion and systemically by **metastasis**.

Cancer risk: A theoretical risk of getting cancer if exposed to a substance every day for 70 years (a lifetime exposure). The true risk might be lower.

Carcinogen: An agent capable of inducing cancer.

Carcinoma: A malignant tumor that occurs in epithelial tissues, which cover the body or body parts and enclose and protect those parts, produce secretions and excretions, and function in absorption.

Case-control study: A study that compares exposures of people who have a disease or condition (cases) with people who do not have the disease or condition (controls). Exposures that are more common among the cases may be considered as possible risk factors for the disease.

Case study: A medical or epidemiologic evaluation of one person or a small group of people to gather information about specific health conditions and exposures.

Cataract: A clouding of the lens of the eye or its capsule that obstructs the passage of light.

Central estimate: A "best" estimate of the dose received, as distinct from an upper bound of the dose that accounts for uncertainty in that estimate.

Chance: A situation in which something happens unpredictably without discernable human intention or observable cause.

Chronic exposure: An exposure that occurred over a long period of time—weeks, months, or years.

Chronic lymphocytic leukemia: A slowly progressing form of leukemia characterized by an increased number of the white blood cells known as lymphocytes that studies have not shown to be caused by radiation in humans.

Code of Federal Regulations: Codification of general and permanent rules published in the Federal Register by executive departments and agencies of the federal government and published annually by the US Government Printing Office.

Cohort: A group of individuals having a common association or factor.

Cohort study: A study involving a group of people who either have or do not have a specified factor, such as exposure to a disease-causing agent. Such studies are usually used to compare disease rates.

Committed dose equivalent (CDE):   The dose equivalent to organs or tissues of reference that will be received from an intake of radioactive material by a person during the 50-year period after intake.

Committed effective dose equivalent (CEDE):   The sum of the products of the weighting factors applicable to each of the body organs or tissues that are irradiated and the committed dose equivalent (CDE) to the organs or tissues. It is a measure of the overall risk associated with internal deposition of radioactive material.

Computed axial tomography (CAT):   A scan that provides three-dimensional x-ray images of some part of the body. It is useful for diagnosing cancer and for planning radiation therapy treatments. Often called a *CT scan.*

Confidence interval: An estimate of the range within which the true value of an uncertain quantity is expected to occur in a specified percentage of measurements or predictions. For example, a 90% confidence interval of (x, y) means that, on the basis of available information, the probability is 0.9 that the true value lies between x and y.

Confidence limits: The highest and lowest boundaries in a confidence interval. As used here, a confidence interval accounts for the possibility that different groups of individuals might have different risk estimates even if they have the same range of dose estimates. Because there is uncertainty in risk estimates that are made for different radiation doses, scientists often include a confidence interval with a risk estimate.

Confounding factors: Any characteristic that makes it difficult to compare two or more distinct groups in an epidemiologic study. Confounding factors can mask a health effect so that the relationship of the effect and the exposure is not recognized. They can also make it appear as though there is an effect when, in fact, none exists.

Correlation: Most generally, the degree to which one phenomenon or variable is associated with or can be predicted from another. In statistics, usually refers to the degree to which a predictive relationship exists between variables. Correlation may be positive (both variables increase or decrease together) or negative or inverse (one variable increases when the other decreases).

Curie (Ci): The conventional unit of radioactivity, equal to 3.7×10^{10} Bq.

Deterministic effects: Effects that can be related directly to the radiation dose received. The severity increases as the dose increases. A deterministic effect typically has a threshold below which the effect will not occur.

Dose: A quantification of exposure to ionizing radiation, especially in humans. Units are rad, mrad, gray, and mgray.

Dose equivalent: The measure that indicates the degree of biologic damage caused by radiation. Dose equivalent is measured in rems, mrem, and Sv.

Dose rate: The quantity of absorbed dose delivered per unit time.

Dose and dose-rate effectiveness factor (DDREF): Dose and dose-rate reduction effectiveness factor, a measure of the extent to which radiation-related damage accruing at a high dose rate is ameliorated when the dose rate or dose is low. This value will presumably vary with the end point measured, but it is not known precisely for such end points as incidence of or death from cancer or any of the other effects also seen among the atomic-bomb survivors.

DS86: The currently used system of dosimetry to describe the exposure of the survivors of the atomic bombings of Hiroshima and Nagasaki; introduced in 1986.

Dose reconstruction: A scientific activity that estimates doses to people from releases of radioactivity or other pollutants. The reconstruction is usually done by determining how much material was released, how it was transported, and how people came into contact with it and the amount of radiation energy absorbed by their bodies.

Dose-response analysis: A statistical analysis to estimate values of parameters that describe the relationship between the dose of a hazardous agent (such as ionizing radiation) and an increase in a specified biologic response (such as a cancer or other health effect) above the normal (background) incidence. In assessing cancer risks in humans posed by exposure to ionizing radiation, for example, linear or linear-quadratic dose-response relationships are used most commonly.

Dosimeter: A portable instrument for measuring and registering total accumulated exposure to ionizing radiation.

Downwinder: A commonly used term for a person who may have been exposed to fallout due to the US nuclear-weapons testing program.

Effective dose: The sum over specified organs or tissues of the equivalent dose in each tissue modified by the tissue weighting factor, as defined in ICRP (1991). Supersedes **effective dose equivalent**.

Effective dose equivalent: The sum over specified organs or tissues of the average dose equivalent in each tissue modified by the tissue weighting factor. Now superseded by **effective dose**.

Element: A substance that cannot be separated by ordinary chemical methods. Elements are distinguished by the numbers of protons in the nuclei of their atoms.

Epidemiologic studies: Studies designed to examine associations—commonly, hypothesized causal relations. They are usually concerned with identifying or measuring the effects of risk factors or exposures. The common types of epidemiologic studies are case-control studies, cohort studies, and cross-sectional studies.

Epidemiology: The study of the incidence, distribution, and causes of health conditions and events in populations.

Equivalent dose: A quantity obtained by multiplying the absorbed dose by a radiation weighting factor to allow for the different effectiveness of the various types of ionizing radiation in causing late-effect harm in tissue. The equivalent dose is theoretical and has replaced the earlier **dose equivalent**. The equivalent dose is often expressed in sievert (Sv). It is also sometimes expressed in rem (an older unit). 100 rem = 1 Sv.

Estimate: A measure of or statement about the value of a quantity that is known, believed, or suspected to incorporate some degree of error.

Etiology: The causes of disease or the study of the causes.

Excess absolute risk: The increase in risk of disease posed by exposure to a specified dose, or the arithmetic difference in risk of disease between exposed and unexposed subjects. Usually expressed as increase in risk per unit dose. See **Excess relative risk**.

Excess relative risk: The increase in relative risk of disease posed by exposure to a specified dose. The mathematical distinction between excess absolute risk (calculated by simple subtraction) and excess relative risk is that the latter is calculated by dividing the risk of disease among exposed subjects by the risk among the unexposed and then subtracting 1.

Exposure: (A) A general term indicating human contact with ionizing radiation, radionuclides, or other hazardous agents. (B) For the purpose of measuring of ionizing photon radiation, the absolute value of the total charge of ions of one sign produced per unit mass of air when all electrons and positrons liberated or created by photons in air are completely stopped in air. Exposure is the quantity measured, for example, by a film badge. The SI unit of exposure is the coulomb per kilogram ($C\ kg^{-1}$). In conventional units used in this report, exposure is given in roentgens (R); $1\ R = 2.58 \times 10^{-4}\ C\ kg^{-1}$.

Exposure pathway: The physical course of a radionuclide or other hazardous agent from its source to an exposed person.

Exposure route: The means of intake of a radionuclide or other hazardous agent by a person (such as ingestion, inhalation, or absorption through the skin or an open wound).

External dose: The dose to organs or tissues of the body due to sources of ionizing radiation located outside the body, including sources deposited on the body surface.

External exposure: Radiation exposure from a source outside the body. The term refers to radiation, such as gamma rays and x rays, that can penetrate human skin and thus cause biological damage from outside the body.

Fallout: Deposition of radioactive particles produced by detonation of a nuclear weapon.

Fine-needle aspiration: A procedure in which a fine, hollow needle is inserted into tissue to extract a small amount of tissue for microscopic evaluation.

Fractionation of exposure: One of the terms used to describe how an exposure was delivered over time. Exposures can be either single (brief), repeated (fractionated), or continuous (chronic).

Gamma rays: Electromagnetic radiation emitted in de-excitation of atomic nuclei, frequently occurring as a result of decay of radionuclides; sometimes shortened to *gamma* (for example, gamma-emitting radionuclide). High-energy gamma radiation is highly penetrating and requires thick shielding, such as up to 1 m of concrete or a few tens of centimeters of steel.

Gastrointestinal tract: Organs of the digestive system, including the esophagus, stomach, small intestine, and upper and lower large intestine (colon).

Generalized anxiety disorder (GAD, DSM code 300.02): A relatively common anxiety problem, affecting 5% of the population, that turns daily life into a state of worry, anxiety, and fear. People with GAD experience pathologic anxiety, which is excessive and chronic and typically interferes with the ability to function in normal daily activities.

Genetic effect: The result of exposure to radioactivity or substances that cause damage to the genes of a reproductive cell (sperm or egg), which can then be passed from one generation to another.

Genetic injury or damage: Harm to a person's genes that can be passed on to later generations.

Germ cells: Reproductive cells—spermatozoa (sperm) in males and ova (eggs) in females.

Gonads: Reproductive organs—testes in males and ovaries in females.

Graves disease: A form of hyperthyroidism (an over-active thyroid).

Goiter: An enlargement of the thyroid gland.

Gray: The SI unit of absorbed dose named fo H. L. Gray an English radiation scientist; $1 \text{ Gy} = 1 \text{ J kg}^{-1} = 100 \text{ rad}$.

Half-life, biological: The time required for half the quantity of a material taken into the body to be eliminated from the body by biologic processes. For radionuclides, the biological half-time does not include elimination by radioactive decay.

Half-life, effective: The time required for the activity of a radioactive substance in the body to decrease to half its value because of the combined effects of biologic elimination and radioactive decay. The effective half-life facilitates evaluating radiation dose from inhaled and ingested radionuclides and applies when the biological and physical half-lives are constant. For an effective half-life of 1 hour, half of the radioactivity would be expected to be eliminated during the first hour. Of the radioactivity that remained, half would be expected to be eliminated during the second hour. That represents one-fourth of the radioactivity initially present. For each successive hour, the expected fractions of the initial radioactivity present that are eliminated would be 1/2, 1/4, 1/8, and so on. This type of decrease over time is called exponential.

Half-life, physical: The average time it takes for one-half of any given number of unstable atoms to decay. Half-lives of isotopes range from small fractions of a second to more than a billion years. For example, if on average 100 out of 200 radioactive atoms of a specified kind decay in 1 day (half-life = 1 day), then of the remaining 100 atoms, 50 would be expected to decay during the second day. Similarly, 25 of the remaining 50 atoms would be expected to decay during the third day. This type of decay is called exponential.

Half-life, radioactive: See **half-life, physical**.

Hashimoto's thyroiditis: An autoimmune disease of the thyroid. It is caused by lymphocytes entering the thyroid. The disease causes goiters, tissue damage, and hypothyroidism.

Health education: Programs designed with a community to help it know about health risks and how to reduce these risks.

Health effect: The result of exposure to substances (such as radiation) that cause any harm to a person's health. It includes diseases, cancers, birth defects, genetic effects, and death.

Health investigation: The collection and evaluation of information about the health of community residents. This information is used to describe or count the occurrence of a disease, symptom, or clinical measure and to evaluate the possible association between the occurrence and exposure to radiation or hazardous substances.

Health promotion: The process of enabling people to increase control over, and to improve, their health.

High-LET radiation: Radiation with linear energy transfer (LET) values above, say, 10 keV/μm. It produces much damage over a short distance in tissue or other material. In contrast, low-LET radiation produces only a small amount of damage when evaluated over the same amount of deposited energy. Alpha particles represent high-LET radiation. Gamma and x rays are low-LET radiation. It generally takes a larger absorbed dose of low-LET radiation than of high-LET radiation to produce a given amount of damage. Biologic damage produced by low-LET radiation is often more efficiently repaired than damage produced by high-LET radiation.

Hodgkin's disease: A type of lymphoma that appears to originate in a particular lymph node and to spread to the spleen, liver, and bone marrow and is characterized by progressive enlargement of the lymph nodes, spleen, and general lymph tissue.

Hyperparathyroidism: Disorder that is characterized by the excessive production of parathyroid hormones. This results in the body not being able to regulate the concentrations of calcium and phosphorus properly.

Hyperthyroidism: A disorder that is characterized by the excessive production of thyroid hormones. Symptoms include nervousness, constant hunger, weight loss, and tremors. Hyperthyroidism is not caused by radiation exposure.

Hypothyroidism: A condition caused by too little thyroid hormone in the body. Symptoms include fatigue, weight gain, intolerance to cold, decreased appetite, constipation, hoarseness, menstrual irregularities, dry skin, and hair changes.

Immune system disorders: Allergic reactions and disruption of the immune surveillance system whose prime function is to detect and eliminate diseased cells.

Incidence: The rate of occurrence of new cases of a specific disease in a specific period, calculated as the number of new cases during a specified period divided by the number of persons at risk of the disease during that period.

Ingestion dose pathway: The parts of the food chain or water system that might add to radiation exposure from eating food or drinking water.

Internal dose: The dose to organs or tissues due to sources of ionizing radiation in the body.

Internal radiation exposure: Exposure from taking a radioactive substance into the body by eating, drinking, or breathing.

International Atomic Energy Agency (IAEA): One of the specialized bodies of the United Nations charged with the responsibility of overseeing and setting standards and recommendations for the operation of nuclear activities and for radiation safety in the member states. It is headquartered in Vienna, Austria, and its members have played a major role in the accumulation and dissemination of the information derived from the Chornobyl accident and other accidents involving exposure to ionizing radiation.

International Commission on Radiological Protection (ICRP): A nongovernment agency headquartered in Sweden and the United Kingdom, and concerned with radiation protection in the workplace and of the general population. It was founded by the International Congress of Radiology in 1928. It is generally viewed as the world's leading source of authoritative statements on radiation protection.

International Commission on Radiation Units and Measurements (ICRU): A nongovernment agency headquartered in Bethesda, Maryland, and concerned with recommendations regarding harmonized measurement of radiation and responsible for recommending nomenclature for quantities, units and their special names, e.g., Bq, Gy, Sv.

International System of Units: A modern version of the meter-kilogram-second-ampere system of units that is published and controlled by the International Bureau of Weights and Measures; also referred to as SI units.

In utero: Means in the uterus or womb.

Iodine-131 (131I):   A radioactive isotope of iodine. Iodine is an element required in small amounts for healthy growth and development. It is mainly concentrated in the thyroid gland, where it is needed to synthesize thyroid hormones. ^{131}I is used as a radioactive tracer in nuclear medicine and is found in fallout from nuclear testing. ^{131}I has been demonstrated to cause thyroid cancer in children in moderate and high doses after the Chornobyl accident. Whether very low radiation doses cause thyroid cancer is uncertain. Iodine-131 has a relatively short physical half-life (8 days).

Ionizing radiation: Any radiation capable of displacing electrons from atoms or molecules, thereby producing ions. Examples are alpha particles, beta particles, gamma rays or x rays, and cosmic rays. The minimum energy of ionizing radiation is a few electron volts (eV); 1 eV = 1.6×10^{-19} joule (J).

Irradiate: To expose to radiation.

Isotope: A form of a particular chemical element determined by the number of neutrons in the atomic nucleus. An element may have many stable or unstable (radioactive) isotopes.

Latent period: The time after exposure that it takes for a radiation-induced cancer to be manifested. Latent periods, also called latency period, may vary widely between different types of cancer and within subgroups of one type of cancer.

Leukemia: The term used to describe a group of malignant, commonly fatal blood diseases characterized by an uncontrolled increase in the number of white cells (generally their immature forms) in the circulating blood.

Life Span Study (LSS):   Continuing followup of the population exposed to atomic-bomb detonations in Hiroshima and Nagasaki, Japan, and their progeny; conducted by the Radiation Effects Research Foundation (RERF).

Linear Energy Transfer (LET): The energy lost by a charged particle per unit distance traversed in a material. The SI unit of LET is joules per meter (J m^{-1}). For purposes of radiation protection, LET normally is specified in water and is given in units of keV μm^{-1}. For low-LET radiations—such as beta particles and the electrons associated with x-rays, gamma rays—little energy is lost in traversing a sheet of paper. For high-LET alpha particles emitted by plutonium isotopes, essentially all the particle's energy is lost in traversing a sheet of paper.

Linear model: The assumption that the effect of exposure to ionizing radiation is directly and simply proportional to dose.

Linear non-threshold model (LNT):   An empirical equation used to assign risk of cancer induction by a specified genotoxicant (including ionizing radiation). The equation has the form, risk = A + kD, where k is a risk coefficient, D is a measure of dose, and A represents the baseline risk, absent radiation. With this empirical model, any dose in excess of zero is presumed to be associated with an increased risk of cancer. Further, use of this model implies that doubling the dose will double the calculated excess risk. For low radiation doses and dose rates, there are large uncertainties about what the true risks to humans and how they relate to dose.

Linear-quadratic model: The assumption that the effect of exposure related not only to the dose received but also to the square of the dose.

Lymphocyte: A type of white blood cell that is found primarily in lymph nodes. Lymphocytes provide protection against some kinds of infections.

Lymphoma: Malignant tumors originating in cells of lymphatic tissues.

Malignancy/malignant neoplasm: A general category of neoplasm that invades surrounding tissue. A malignant tumor is generally characterized by invasive growth and is able to metastasize to distant tissue sites via the lymphatic and blood systems.

Malignant: Tending to infiltrate, metastasize, and terminate fatally.

Mean: The arithmetic average of a set of values, given by the sum of the values divided by the number of values. The mean of a distribution of values is the weighted average of possible values, each value weighted by its probability of occurrence in the distribution.

Median dose: The central estimate in a dose-estimate range. Half the possible doses are above the median; half are below it. A person's dose is more likely to be near the median than near the low or the high end of the range.

Medical monitoring: A program to screen a group of people who are at risk for specific diseases or conditions and to refer individuals for additional evaluation and treatment if needed. Monitoring does not include medical care.

Melanoma: A malignant, and often fatal, tumor in cells of the skin that synthesize dark pigments.

Metastasis: The spread of cancer through transfer of malignant cells from one organ or part to another part not directly connected with it.

Mill tailing: Naturally radioactive residue from the processing of uranium ore into yellowcake in a mill. Although the milling process recovers about 93% of the uranium, the residues, or tailings, contain several radioactive elements, including uranium, thorium, radium, polonium, and radon.

Minisatellite: Repeated segments of the same sequence of multiple triplet codons, each segment varying between 14 and 100 base pairs, useful as linkage markers because of their highly polymorphic nature and the fact that they are usually situated near genes. Minisatellites are inherently unstable and susceptible to mutation at a higher rate than other sequences of DNA.

Model: A construct (generally mathematical) that attempts to describe the events that underlie some biologic or physical phenomenon of interest, such as the occurrence of cancer after exposure to ionizing radiation.

Monte Carlo analysis: The computation of a probability distribution of an output of a model based on repeated calculations using random samplings of the model's input parameters (variables) from specified probability distributions.

Morbidity: A measure of a diseased condition or state; refers to illness, not death.

Morbidity rate: The rate at which people get a disease, usually expressed as the number of cases per 100,000 people per year.

Mortality: A measure of the number of people who die from a specific disease or condition.

Mortality rate: The rate at which people die from a disease, such as a specific type of cancer, usually expressed as the number of deaths from the disease per 100,000 deaths per year.

Multiple myeloma: The proliferation of plasma cells that often replace all other cells within bone marrow, leading to immune deficiency and, frequently, destruction of the outer layer of bone.

Mutation: A hereditary change in genetic material; it can be a change in a single gene (point mutation) or a change in chromosome characteristics.

National Academy of Sciences (NAS):   The National Academy of Sciences is a private, non-profit, self-perpetuating society of distinguished scholars engaged in scientific research.   Upon the authority of the charter granted by the Congress in 1863, the NAS has a mandate that requires it to advise the federal government on scientific and technical matters.

National Council on Radiation Protection and Measurements (NCRP):   A nongovernmental agency based in Bethesda, Maryland, with a charter similar to that of ICRP, but focusing in particular on issues related to radiation protection in the United States.

National Research Council (NRC):   The principal operating agency of the National Academy of Sciences and the National Academy of Engineering to serve the federal government and other organizations.

Neoplasm: Any new or abnormal growth, such as a tumor; *neoplastic disease* refers to any disease that forms tumors, whether malignant or benign.

Neural tube defects: A defect in the neural tube.   The neural tube develops into the spinal cord and brain.   Defects occur when the neural tube fails to close completely during the early stages of pregnancy.

Neutron: An elementary uncharged particle, of mass slightly greater than that of a proton, that is a constituent of atomic nuclei.

Nevada Test Site (NTS):   The region in Nevada set aside for the continental atmospheric nuclear weapons testing program.   Also referred to as the Nevada Proving Ground (NPG).

Noble gas: Any of a group of rare gases (helium, neon, argon, krypton, xenon, and radon) that exhibit great stability and very low chemical reaction rates.

Nuclear fallout: The descent of airborne particles of dust, debris, and radioactive substances from a nuclear bomb explosion.   Millions of curies of radioactivity in the form of dust and debris get carried into the upper atmosphere by the mushroom cloud.   Jet stream winds can carry fallout from bomb blasts around the world within a few months.

Nuclear weapon: A weapon that derives its explosive force from nuclear fusion or nuclear fission reactions.

Occupational radiation exposure: The radiation exposure of or dose to a person in the course of employment in which individual's assigned duties involve exposure to radiation and to radioactive materials. Occupational dose does not include dose received from background radiation, from being a patient in medical practice, from voluntary participation in medical research program, or from being a member of the general public.

Odds: A measure of the likelihood that an event will occur. For example, the odds of developing a particular disease.

Organ dose: The energy absorbed in a specific organ divided by its mass. This quantity is expressed in gray (Gy) or its submultiples.

Parathyroid: Any of four small glands next to or on the thyroid gland. The parathyroids secrete a hormone that helps to control the balance between calcium and phosphorus in the body.

Photon: A quantum of electromagnetic radiation, having no charge or mass, that exhibits both particle and wave behavior, especially a **gamma ray** or an **x ray**.

Posttraumatic stress disorder (PTSD, DSM code 309.81): A group of characteristic symptoms that follow exposure to an extreme traumatic stressor involving actual or threatened death or serious injury; another threat to one's physical integrity; witnessing of an event that involves the death of, injury of, or a threat to the physical integrity of another person; or learning about unexpected or violent death, serious harm, or threat of death or injury experienced by a family member or other close associate.

Power: The probability that a study can distinguish between a true exposure-to-disease relationship and a coincidence. The power of a study depends on the size of its population, the amount of radiation exposure, and the number of cases of the disease under investigation.

Prevalence: The number of cases of a specific disease existing in a particular population or area at a certain time.

Probability: The likelihood (chance) that a specified event will occur. Probability can range from 0, indicating that the event is certain not to occur, to 1, indicating that the event is certain to occur.

Probability of causation (PC): The probability that a specific disease in a person was caused by their exposure to a particular hazardous agent (such as ionizing radiation). PC is estimated as a quotient of two risks: $PC = R/(R + B)$, where R is the estimated risk of the disease in a person due to exposure to the particular hazardous agent and B is the estimated background (baseline) risk of the disease in that person from all other causes (that is, the risk in the absence of exposure to that agent). PC differs from risk in that it is conditional on the occurrence of a disease.

Progeny: The decay products resulting after a series of radioactive decay. Progeny can also be radioactive, and decay continues until a stable nuclide is formed.

Prospective study: A study in which two groups of people—one exposed and one nonexposed—are followed forward in time (prospective) to determine the possible linkage between exposure and health effects.

Public-health activities: Activities conducted to protect, promote, or restore public health. The activities can include such programs and campaigns as surveillance of disease, epidemiologic studies, disease registries, collection of vital statistics, disease-prevention programs, public and provider education, health inspections, and quality-assurance activities.

Quadratic model: The assumption that the effect of exposure to ionizing radiation is related to the square of the dose received.

Quality factor: A factor that depends on the linear energy transfer by which absorbed doses are multiplied to obtain a quantity that expresses the effectiveness for radiation-protection purposes of an absorbed dose on a common scale for all forms of radiation.

Rad: The special name for the conventional unit of absorbed dose; 1 rad = 100 ergs g^{-1} = 0.01 Gy.

Radiation: Energy emitted in the form of waves or particles. See also **ionizing radiation**.

Radiation Effects Research Foundation (RERF): The nonprofit research foundation sponsored by the governments of Japan and the United States that currently supervises the studies of the atomic-bomb survivors; the successor in 1975 of the Atomic Bomb Casualty Commission.

Radiation exposure: See exposure.

Radiation protection: The control of exposure to ionizing radiation by use of principles, standards, measurements, models, and such other means as restrictions on access to radiation areas or use of radioactive materials, restrictions on releases of radioactive effluents to the environment, and warning signs. Sometimes referred to as *radiologic protection.*

Radioactive: Exhibiting radioactivity.

Radioactive decay: The spontaneous transformation of the nucleus of an atom to a state of lower energy.

Radioactivity: The property or characteristic of an unstable atomic nucleus to spontaneously transform with the emission of energy in the form of radiation.

Radiobiology: A branch of biology that deals with the interaction of biologic systems and radiant energy or radioactive materials.

Radioepidemiological tables: A tabulation of estimated probabilities of causation of specific cancers in a person who receives various doses of ionizing radiation. See also **Probability of causation** and **risk**.

Radiogenic disease: A type of disease assumed on the basis of scientific studies to have an association with radiation exposure. A statement that a cancer is radiogenic does not imply that radiation is the only cause of the cancer but rather that radiation has been shown to be one of its causes. Exposure to other environmental substances could also cause the same type of cancer.

Radionuclide: A naturally occurring or artificially produced radioactive element or isotope.

Radiophobia: Abnormal fear of radiation.

Radiosensitivity: Susceptible to the injurious action of ionizing radiation.

Radon: A naturally occurring radioactive gas produced from uranium; decays to form **radon progeny**. Radon occurs naturally in many minerals and is a chief hazard of uranium mill tailings. Some radon can also be found in homes. Radon decays into other isotopes that emit alpha radiation.

Radon progeny: The radioactive products formed in the radioactive decay of radon; radionuclides which when inhaled can expose living cells to their emitted alpha particles.

Ratio: A measure of association calculated by dividing one amount by another.

Relative biological effectiveness (RBE): A factor used to compare the biological effectiveness of absorbed radiation doses due to different types of ionizing radiation for a defined biologic end point, such as cell killing; this factor is experimentally determined by using x or gamma rays as the standard of comparison. Thus, if 1 Gy of fast neutrons produced the same amount of cell killing as 5 Gy of gamma rays, the RBE of neutrons for cell killing would be 5. The RBE varies with the biologic end point used.

Relative risk: The ratio of the risk in one population to that in another. Relative risk indicates the increased or decreased degree of risk among exposed people compared with nonexposed people. A relative risk of 1 indicates no association between the exposure and the disease. A relative risk of 2 indicates that the exposed group is twice as likely as the nonexposed group to experience the health effect being studied.

Rem: The special name for the conventional unit of equivalent dose; 1 rem = 100 ergs g^{-1} = 0.01 Sv = 10 mSv. For gamma and beta radiation and x rays, 1 rem = 1 rad = 0.01 Gy = 10 mGy.

Risk: The probability of an adverse event. In regard to adverse effects of ionizing radiation on humans, the term usually refers to the probability that a given radiation dose to a person will produce a health effect (such as cancer) or the frequency of health effects produced by given radiation doses to a specified population within a specified period. The risk of cancer due to a given radiation dose generally depends on the cancer type, sex, age at exposure, and time since exposure (attained age), and it may depend on dose rate.

Risk factor: An aspect of personal behavior or lifestyle, an environmental exposure, or an inborn or inherited characteristic that is known from scientific evidence to be associated with a health effect.

Roentgen: The special name for the conventional unit of exposure; 1 R = 2.58 × 10⁻⁴ coulomb per kilogram (C kg⁻¹).

Sample size: The number of participants in a research study. The larger the sample size in a research study, the more power the study has to detect an association between exposure and a health effect.

Screening: The application of a test to detect a potential disease or condition in a person who has no known signs or symptoms of the disease or condition.

Sievert: The SI unit of equivalent dose named for R. M. Sievert a Swedish radiophysicists; 1 Sv = 1 J kg⁻¹ = 100 rem.

SI units: See **International System of Units.**

Somatic effects: The effects of radiation exposure that result from damage to nonreproductive cells. If the number of cells that suffer somatic effects is great enough, then the damage becomes clinically observable.

Squamous cell carcinoma: A malignant growth originating from plate-like cells found in the outer layer of the skin and usually occurring on the skin, lips, inside of the mouth, throat, or esophagus.

Standardized incidence ratio (SIR): The ratio of the number of observed cases divided by the number of expected cases. The word standardized means that there has been adjustment for one or more potential bias factors such as age and sex.

Standardized mortality ratio (SMR): The ratio of the number of deaths observed in a study population to the number of deaths expected if that population had death rates equivalent to those in some standard, general population (such as the US population). SMRs are typically calculated by using general population rates broken down by intervals of age and calendar time, and by age and race.

Statistical significance: The likelihood that an association between exposure and disease risk that a study finds did not occur by chance alone.

Statistics: A branch of mathematics dealing with the collection, analysis, interpretation, and presentation of numerical data.

Stochastic effect: An effect that occurs on a random basis independent of the size of dose. The effect typically has no threshold and is based on probabilities, with the chances of occurrence of the effect increasing with dose. Cancer is a stochastic effect.

Surveillance, Epidemiology, and End Results (SEER): Program developed as a result of the National Cancer Act of 1971, which mandated the collection, analysis, and dissemination of all data useful in the prevention, diagnosis, and treatment of cancer. It is a continuing project of the National Cancer Institute to collect cancer data routinely basis from designated population-based cancer registries in various areas of the country.

Thyroid-antibody test: A blood test that measures antibodies against the patient's thyroid tissue.

Thyroid burden: The total activity of a radionuclide in the thyroid.

Thyroid dose: The amount (or an estimate of the amount) of radiation, or energy, absorbed by the thyroid gland.

Thyroid gland: A two-lobed gland lying at the base of the throat that produces hormones essential for a variety of metabolic processes in the body. It secretes hormones that control body growth and metabolism. When iodine is ingested, much of it goes to the thyroid gland.

Thyroid nodules: Lumps in the thyroid gland that may be benign or cancerous. "Cold nodules" are non-functioning lumps in the thyroid gland. "Hot nodules" are overactive thyroid lumps. When a nodule is detected, it is important to diagnose the disease that has caused it. Benign thyroid tumors are often referred to as nodules.

Thyroid palpation: The procedure in which a physician characterizes the size, shape, and texture of the thyroid gland by manual examination of the neck.

Total effective dose equivalent (TEDE): The sum of the deep dose equivalent (DDE) for external exposures and the committed effective dose equivalent (CEDE) for internal exposures.

Tumor registry: A collection of records on the tumors that have been treated at a particular hospital or within a geographic area.

Uncertainty: The lack of sureness or confidence in results of measurements or predictions of quantities owing to stochastic variation or to a lack of knowledge founded on an incomplete characterization, understanding, or measurement of a system.

United Nations Scientific Committee on the Effects of Atomic Radiation (UNSCEAR): One of the specialized bodies of the United Nations charged with the responsibility of evaluating the effects of exposure to atomic (ionizing) radiation on behalf of the member nations.

Uranium (U): A radioactive element with atomic number 92 and, as found in natural ores, an average atomic weight of about 238. The two principal natural isotopes are ^{235}U (0.7% of natural uranium), which is fissionable, and ^{238}U (99.3% of natural uranium), which is fertile. Natural uranium also includes a minute amount of ^{234}U.

Variability: The variation of a property or quantity among members of a population. Variability is often assumed to be random and can be represented by a probability distribution.

Weighting factor (wT): For an organ or tissue (T), the proportion of the risk of stochastic effects resulting from irradiation of the organ or tissue to the total risk of stochastic effects when the whole body is irradiated uniformly.

Whole body: For purposes of estimating radiation dose, especially from external exposure, the head, trunk (including male gonads), arms above the elbow, and legs above the knee.

Working level (WL): Any combination of the short-lived progeny of radon in 1 liter of air, under ambient temperature and pressure, that results in the ultimate emission of 1.3×10^5 MeV of alpha-particle energy. It is approximately the total amount of energy released over a long period by the short-lived progeny in equilibrium with 100 pCi of radon. 1 WL = 2.08×10^{-5} J m^{-3}.

Working level month (WLM): A cumulative exposure equivalent to 1 working level for 1 working month (170 hours). 1 WLM = 2.08×10^{-5} J h m^{-3} $\times$ 170 h = 3.5×10^{-3} J h m^{-3}.

X radiation: (A) Electromagnetic radiation emitted in de-excitation of bound atomic electrons, frequently occurring in decay of radionuclides, referred to as characteristic x rays, or (B) electromagnetic radiation produced in deceleration of energetic charged particles (such as beta radiation) in passing through matter, referred to as continuous x rays or bremsstrahlung; also called x rays.

List of Abbreviations

AEC	Atomic Energy Commission
AR	Attributable Risk
AS	Assigned Share
BEIR	Biological Effects of Ionizing Radiation
BRER	Board on Radiation Effects Research
CDE	Committed Dose Equivalent
CEDE	Committed Effective Dose Equivalent
CFR	*Code of Federal Regulations*
CIRRPC	Committee on Interagency Radiation Research and Policy Coordination
CV	Coefficient of variation
DHHS	US Department of Health and Human Services
DOD	US Department of Defense
DOE	US Department of Energy
DOJ	US Department of Justice
DOL	US Department of Labor
DDREF	Dose and Dose-Rate Reduction Effectiveness Factor
DTRA	Defense Threat Reduction Agency
EEOICPA	Energy Employees Occupational Illness Compensation Program Act
ERR	Excess Relative Risk

GAO	General Accounting Office
HRSA	Health Resources and Services Administration
IARC	International Agency for Research on Cancer
IAEA	International Atomic Energy Agency
ICRP	International Commission on Radiological Protection
ICRU	International Commission on Radiation Units and Measurements
IOM	Institute of Medicine
IREP	Interactive Radio Epidemiological Program
LET	Linear Energy Transfer
LLE	Lose of Life Expectancy
MC	Monte Carlo
NAS	National Academy of Sciences
NCI	National Cancer Institute
NCRP	National Council on Radiation Protection and Measurements
NIH	National Institutes of Health
NIOSH	National Institute for Occupational Safety and Health
NRPB	National Radiological Protection Board
NRC	National Research Council
NTS	Nevada Test Site
PC	Probability of Causation
REF	Radiation Effectiveness Factor
RERF	Radiation Effects Research Foundation
RECA	Radiation Exposure Compensation Act
RECP	Radiation Exposure Compensation Program
RESEP	Radiation Exposure Screening and Education Program
REVCA	Radiation-Exposed Veterans Compensation Act
RR	Relative Risk
SD	Standard deviation
SMR	Standardized Mortality Ratio
SI	Système International (International System)
TEDE	Total Effective Dose Equivalent
UB	Upper bound

VA	US Department of Veterans Affairs
W_T	Weighting Factor
WHO	World Health Organization
WL	Working Level
WLM	Working Level Month

Committee and Staff Biographies

R. Julian Preston, PhD, *Chair,* has been director of the Environmental Carcinogenesis Division of the US Environmental Protection Agency since 1999. Before then, he served as the senior science advisor at the Chemical Industry Institute of Toxicology in Research Triangle Park, North Carolina, from 1991 to 1999. He was employed at the Biology Division of the Oak Ridge National Laboratory in Oak Ridge, Tennessee, from 1970 to 1991. He also served as associate director of the Oak Ridge–University of Tennessee Graduate School for Biomedical Sciences. He now holds adjunct professor appointments at Duke University (Integrated Toxicology Programs) and North Carolina State University (Department of Toxicology). Dr. Preston received his BA and MA from Peterhouse, Cambridge University, England, in genetics and his PhD from Reading University, England, in radiation genetics. Dr. Preston is a member of the Board of the National Council on Radiation Protection and Measurements, chairman of Committee 1 of the International Commission on Radiological Protection, and a member of the US Delegation to the United Nations Scientific Committee on the Effects of Atomic Radiation. Dr. Preston's research and current activities have focused on the mechanisms of radiation and chemical carcinogenesis and the approaches to incorporation of these types of data into cancer risk assessments.

Thomas B. Borak, PhD, is a professor in the Department of Environmental and Radiological Health Sciences at Colorado State University. He received a BS in physics from St. John's University (Minnesota) and a PhD in physics from Vanderbilt University. His research interests are in radiation physics and dosimetry. He has had scientific staff appointments at Fermilab, CERN, and Argonne

National Laboratory. He is a member of the American Physical Society, the Radiation Research Society, and the Health Physics Society, which he recently served on the Board of Directors. Dr. Borak is currently serving on the National Council on Radiation Protection and Measurements and is certified by the American Board of Health Physics. He has been a consultant to the governor of Colorado on low-level radioactive-waste management and nuclear criticality safety. Dr. Borak was also a member of the BRER Committee on Risk Assessment of Exposure to Radon in Drinking Water (1999).

Catherine Borbas, PhD, MPH, is executive director of the Healthcare Evaluation and Research Foundation in St. Paul, Minnesota. Prior committee memberships include the Committee on Methods for Setting Priorities for Guidelines for the Division of Health Care Services of the Institute of Medicine and the Committee to Review the NCI report on the Exposure of the American People to Iodine-131. Dr. Borbas has published in clinical-guidelines implementation and methods to influence clinical behavior. She earned her PhD in social work and a master's degree in public health from the University of Minnesota.

A. Bertrand Brill, MD, PhD, is a research professor in the Departments of Radiology and Physics at Vanderbilt University. Dr. Brill earned his MD at the University of Utah and his PhD in Biophysics at the University of California, Berkeley. He served in the US Public Health Service (PHS) in Japan at the Atomic Bomb Casualty Commission (ABCC) from 1957 to 1959 and as the PHS representative to ABCC until 1964. Dr. Brill's specialty is nuclear medicine, and his major research is in cancer imaging, radiation leukemogenesis, effects of radiation on thyroid function, and effects of diagnostic radioisotope studies, particularly iodine-131. He is a member of the National Cancer Institute-Columbia University Task Group doing a followup study of thyroid disease after the Chornobyl accident. He is a member of the Society of Nuclear Medicine Radiation Effects Committee, the Medical Internal Radiation Dose Committee (MIRD), and the American Thyroid Association. Dr. Brill served as a member of the National Research Council Committees on Atomic Casualties, the BEIR III committee, and the Committee on Assessment of Centers of Disease Control and Prevention Radiation Studies from DOE Contractor Sites.

Thomas Buhl, PhD, CHP, serves as chief scientist for the Health, Safety, and Radiation Protection Division of the Los Alamos National Laboratory (LANL). He has been a health physicist in the LANL radiation protection program since 1980, working in radiation-instrumentation development, in vivo bioassay measurements, environmental surveillance, dose assessment, and nuclear-accident dosimetry. He has been an adjunct professor in nuclear engineering at the University of New Mexico since 1994. He served as president of the American Academy of Health Physics (2004) and chair of the American Board of Health

Physics (1996). Dr. Buhl is a member of the Health Physics Society and the American Physical Society. Before joining LANL, he worked in the New Mexico Radiation Protection Program, designing and operating an environmental radiation monitoring program in the New Mexico uranium mining area from 1977 to 1980, and later serving as program director on a one-year leave of absence from LANL in 1983-1984. Dr. Buhl received his PhD in physics from the University of Wisconsin-Madison in 1971 and certification in health physics from the American Board of Health Physics in 1981.

Patricia A. Fleming, PhD, is senior associate dean of the College of Arts and Sciences and associate professor in the Department of Philosophy at Creighton University in Omaha, Nebraska. She received her bachelor's degree in sociology and philosophy from Marygrove Collge, Detroit, Michigan and her master's and doctorate from Washington University in St. Louis, Missouri. While there, she served as the assistant editor of the *Philosophy of Science Journal.* She has also served as an editor for the international journal *ESEP (Ethics in Science and Environmental Politics)* and as an external observer (thematic rapporteur) for the Organization for Economic Co-operation and Development (OECD)/Nuclear Energy Agency's (NEA) Forum on Stakeholder Conference in Ottawa, Canada. She is currently a Board Member of the Swedish-based international group VALDOC (Values on Decisions of Complexity). Her areas of specialization are philosophy of science, epistemology, and applied ethics. She has published and lectured internationally on the ethical and epistemological issues associated with the disposal of high-level nuclear waste, including the use of expert elicitation methodology in site characterization, waste management and indigenous populations, informed consent in stakeholder populations, and circularity in regulatory policy. She teaches courses in applied ethics—particularly ethics and public policy, medical ethics, environmental ethics, and the philosophy of science—at Creighton University.

Shirley Fry, MD, MPH, earned her medical degree from the University of Dublin, Ireland, and her MPH in epidemiology from the University of North Carolina, Chapel Hill. She was on the staff of the Center for Human Radiobiology at Argonne National Laboratory (1975-1978). She then joined the staff of the Medical and Health Sciences Division of Oak Ridge Associated Universities where she was a member of the Radiation Emergency Assistance Center/Training Site's medical response team and teaching faculty (1978-1995), Director of the Center for Epidemiologic Research (1984-1991), and Assistant Division Director (1991-1995). She later served as Scientific Director of the Washington-based International Consortium for Research on the Health Effects of Radiation, and currently is Clinical Professor, (honorary) in the Department of Radiation Oncology, Indiana University School of Medicine. Her experience and research interests are in the acute and long-term effects of ionizing radiation in human populations including the US radium dial workers, US nuclear industry workers, and survivors of radiation accidents including the Chornobyl reactor accident. She is the author or coauthor of a number of publications on subjects in these topic areas. She has served on national and international groups with interests in these areas, including the Institute of Medicine's Committee on Battlefield Exposure Criteria, the US/USSR Joint Commission on Chornobyl Nuclear Reactor Safety (JCCNRS)-Health Studies Group, and the International Agency for Research on Cancer's International Study of Cancer Risk Among Nuclear Workers. She currently is a member of the National Cancer Institute's Chornobyl Thyroid Advisory Group, and of the Radiation Advisory Committee of the US Environmental Protection Agency's Science Advisory Board. She is a member of the Health Physics Society, the Radiation Research Society, and the American College of Occupational and Environmental Medicine.

Richard Hornung, DrPH, received his doctorate in biostatistics from the University of North Carolina School of Public Health in 1985. His expertise includes survival analysis models, logistic regression, risk assessment, epidemiologic methods, and statistical methods in exposure assessment. He has over 25 years of experience in a wide variety of research, including radiation epidemiology, exposure prediction models, experimental design, environmental studies of lead and allergens, and occupational health. Dr. Hornung joined the Institute for Health Policy and Health Services (IHPHSR), University of Cincinnati, in 1997 after a 24-year career at the National Institute for Occupational Safety and Health (NIOSH), where he was chief of the Health-related Energy Research Branch in 1991-1996; the mission was to conduct epidemiologic studies of Department of Energy workers involved in the nuclear weapons program. He has also done extensive research involving the estimation of lung-cancer risk in uranium miners exposed to radon decay products. He is a member of the Environmental Protection Agency Science Advisory Board as a member of the Radiation Advisory Committee and has served as a member of the White House committee that helped to develop risk standards to be used for the Radiation Exposure Compensation Act (RECA), and as a consultant to BRER's BEIR IV committee and a reviewer for the BEIR VI report.

Kathleen N. Lohr, MPhil, PhD, is a Distinguished Fellow at RTI International and the founding codirector of the RTI-University of North Carolina Evidence-Based Practice Center. From 1996 to 2000 at RTI, she directed a program of research in health services and health policy involving more than 40 researchers in quality of care, evidence-based practice, Medicare and Medicaid evaluations, health communication, and similar fields; from 2000 to 2003, she was an RTI Chief Scientist. She also holds the rank of research professor in health policy and administration at the University of North Carolina (UNC) School of Public Health and is a senior research fellow at the UNC Cecil G. Sheps Center for Health

Services Research. At UNC, she is a coinvestigator in the Center for Education and Research in Therapeutics and the PROMIS (Patient-Reported Outcomes Measurement Information System) cooperative agreements. Before working at RTI, Dr. Lohr spent 9 years at the Institute of Medicine (IOM) where she was director of the Division of Health Care Services; she later served on the IOM committee to design a health-outcomes study for veterans of the Gulf War. From 1974 to 1987, she was an analyst with the RAND Corporation, chiefly on the RAND Health Insurance Experiment and in a variety of quality-of-care studies. She is a Fellow of Academy Health (formerly the Association for Health Services Research) and chairs its Distinguished Investigator Committee; she is also a member of advisory boards on quality-of-care measures and organ transplantation and other federally sponsored studies and a member of the advisory board for the North Carolina Partnership to Improve Math and Science education. Dr. Lohr serves as associate editor of *Quality of Life Research* and as a member of planning committees for the fourth (Sydney, Australia, 2001), fifth (Washington, DC, 2003), and sixth (Toronto, Canada, 2005) International Conferences on the Scientific Basis of Health Services. She has published in quality of care, clinical practice guidelines, evidence-based practice, and health status assessment. She earned a BA in sociology and an MA in education from Stanford University and an MPhil and PhD in public policy analysis from the Rand Graduate School. She was recently awarded the 2005 International Society of Pharmacoeconomics and Outcomes Research (ISPOR) Avedis Donabedian Outcomes Research Lifetime Achievement Award.

Stephen G. Pauker, MD, is vice chairman for clinical affairs and associate physician-in-chief at Tufts-New England Medical Center and professor of medicine and of psychiatry at Tufts University School of Medicine. He is an expert in clinical-decision making and evidence-based medicine. Dr. Pauker is a member of the Institute of Medicine. He has served on the Committee to Evaluate the Artificial Heart Program of the National Heart, Lung, and Blood Institute, the Workshop on the National Institutes of Health Consensus Development Process and the Use of Drugs in the Elderly, and the Committee on Thyroid Cancer Screening. His publications and research have addressed decisions about screening for cancer and other conditions. Dr. Pauker earned his MD at Harvard University in 1968 and trained in internal medicine and cardiology at Boston City and Massachusetts General Hospitals and the New England Medical Center, all in Boston. Dr. Pauker practices internal medicine and cardiology.

National Research Council Staff

Isaf Al-Nabulsi, PhD, is a senior program officer with the Board on Radiation Effects Research. She received her MS in radiation biology from Georgetown University and her PhD in biomedicinal chemistry from the School of Pharmacy of the University of Maryland at Baltimore. Her research interests include molecular mechanisms of DNA damage and repair, cytogenetic techniques, molecular mechanisms of tumor radioresponsiveness, the influence of hypoxic cells on the outcome of chemotherapy and radiotherapy, radiation dosimetry, and epidemiology. She joined the National Research Council staff in 2000 and has directed 12 studies that have produced five reports, six letter reports, and one interim report. She is a member of the Radiation Research Society and the Health Physics Society.